S0-AEF-770

TOXIC CHEMICALS

Risk Prevention through Use Reduction

Thomas E. Higgins
and
Jayanti A. Sachdev
Stephen A. Engleman

CRC Press is an imprint of the
Taylor & Francis Group, an **informa** business

CRC Press
Taylor & Francis Group
6000 Broken Sound Parkway NW, Suite 300
Boca Raton, FL 33487-2742

Printed in the United States of America on acid-free paper
10 9 8 7 6 5 4 3 2 1

International Standard Book Number: 978-1-4398-3915-7 (Hardback)

Visit the Taylor & Francis Web site at
http://www.taylorandfrancis.com

and the CRC Press Web site at
http://www.crcpress.com

Contents

Foreword

When I began my career in environmental engineering, the U.S. Environmental Protection Agency (USEPA) was just being formed, with Bill Ruckelshaus as its first administrator. At the time, the emphasis was on treating waste after it was produced. As a young professor, I continued in that vein. My research was focused on developing more efficient methods of removing toxic metals from wastewater and developing remediation technologies for Superfund sites.

When I returned to practicing engineering, I was involved with developing a waste minimization program for the Department of Defense, moving upstream from waste treatment to reducing or eliminating waste through changing manufacturing and maintenance processes and the toxic chemicals used. It was a unique concept at the time—reducing or even eliminating waste before it was produced. Based on that experience, I published and spoke widely about the importance of eliminating, rather than treating, waste. In a talk to the board of directors of a major manufacturer of tools, I followed a lawyer who gave them the status of various remediation sites and a rundown on the millions of dollars of liabilities associated with cleaning them. I began my talk by observing that I was following their lawyer responsible for their paternity suits, and that I was going to talk to them instead about the benefits of hazardous waste birth control. I subsequently compiled the lessons learned from my various projects in the *Waste Minimization Handbook.* I had first met Bill Ruckelshaus at a meeting at which he was discussing the need to regulate toxic chemicals. When I completed the *Waste Minimization Handbook*, he kindly agreed to write the foreword.

Subsequently, I lead a team of engineers and scientists in projects that expanded beyond hazardous waste minimization and treatment of toxic chemicals in air, water, and solid waste into pollution prevention. As our experience evolved, we wrote and published *The Pollution Prevention Handbook.* The book was a group effort, involving 24 other professionals who worked with me on projects in this evolving field.

During my career, I have had the opportunity of working in over 20 countries in six continents, assisting industries and governments in setting up pollution prevention programs. When the Soviet Union fell, I worked with newly privatized companies to improve manufacturing efficiency and compliance with Western environmental requirements. In Hungary, we consulted with six individual companies, which was helpful, but limited to these six companies. In moving on to Poland, we not only prepared industrial efficiency audits for individual companies but also provided training to Polish engineers and scientists, who performed industrial efficiency audits for many more companies under our supervision. By the time the program in Poland ended, we left a trained cadre of over 200 professionals who carried on the program, greatly expanding our individual effectiveness. From this experience, we moved on to the former Soviet Union. In Russia, we decided that our primary goal was training, and that all industrial efficiency audits would be performed by Russian trainees, with foreign nationals providing technical advice. We also provided seed money for our trainees to set up Russian companies in a way that would be sustainable. In addition to technical training, we provided training in business practices, something that was scarce in Russia at the time. In Uzbekistan, we assisted the state and federal governments and regulated industries in setting up regional pollution prevention programs.

The Soviet system had one aspect that was significantly different from past and current U.S. environmental regulation. In essence, they regulated the use of toxic chemicals based on a fee structure; that is, a company measured its annual releases of harmful constituents and paid a fee based on the quantity released. The cost per pound was based on the toxicity of chemicals released. The fee started at zero release to create an incentive to reduce chemical discharges to zero. The fee increased in steps as they approached what in the United States would be the permit level and increased again for even higher releases beyond the permit level. This approach has the potential to achieve greater environmental benefit at a lower expenditure of resources versus the "command-and-control" approach of allocating levels of pollution (permit levels) equally between all of the regulated community regardless of individual capability to achieve the permitted standard. It is ironic that the Communist bloc set up environmental regulation based on a market-based approach, while capitalist countries have based environmental protection on a system of command and control. The principal problem with the Soviet system was that enforcement was arbitrary.

While giving a pollution prevention talk in Brazil, I also gave a talk to the Companhia de Tecnologia de Saneamento Ambiental (CETESB), the environmental regulatory agency for the state of São Paulo. I left a copy of my *Pollution Prevention Handbook*, which they added to their growing library of materials for their Pollution Prevention Department, established in 1992. A few years later, they asked me to provide them with a few days of training on how to run a successful pollution prevention program. Three years ago, they decided that they wanted to expand on their program and wanted assistance in developing a program to reduce the impacts of toxic chemicals on public health and the environment. To develop this program, we first evaluated the effectiveness of similar programs in the United States and other countries.

The USEPA Toxics Release Inventory (TRI) was developed in response to the tragedy of Bhopal, where 3,787 people died and hundreds of thousands were sickened by the release of methyl isocyanate from a Union Carbide storage tank. The program required that U.S. companies report on their annual release of approximately 200 plus toxic chemicals. As a result of this program, companies were required to find out about the composition of products they used, calculate usage of the toxic chemicals, and report on releases of these chemicals to the environment. An unintended consequence of TRI reporting was that releases of these chemicals reduced over time as companies sought to reduce adverse publicity and reduce the cost of regulatory reporting. A subsequent U.S. voluntary program (33/50) resulted in over 50 percent reduction in releases of 17 highly toxic TRI chemicals over 4 years. Various state programs have expanded on the TRI requirements to include requirements that companies prepare pollution prevention plans and set toxic chemical reduction targets.

Although the TRI program has been successful as a first step in reducing some chemical releases to the environment, one enhancement to the program would involve measuring or reporting the release of toxic chemicals in products themselves. While the release of toxic chemicals to the environment around a plant has an adverse impact, putting a toxic chemical in a product for use in the home or office is more likely to have an impact on peoples' health. The chemical could then be released to the water, air, or landfill when the product is used or discarded. The root cause of toxic chemical impacts is use, not just release to a particular environmental medium. If industry does not use a toxic chemical, then it will not be released to the environment directly or indirectly through products. By using a chemical, a company increases the potential exposure to its workers through use, to its neighbors through releases, and to customers through its products.

Reducing or preferably eliminating the use of a single chemical takes a considerable effort. Effort is needed to find a suitable substitute, which may cause the cost of a product to increase. With thousands of toxic chemicals in use in hundreds of thousands of products, this approach looks infeasible on the surface. What makes this approach feasible, however, is that all chemicals do not have an equal toxic impact. Ingestion of 5 mg of mercury per year is expected to adversely affect development of a child, whereas it would take 17 pounds of tert-butyl alcohol. Moreover, some chemicals rapidly degrade after release to the environment, while others remain toxic (persistent) for millions of years. Some chemicals bioaccumulate or concentrate as they move up the food chain from plants to herbivores to carnivores. For instance, fish can contain 100,000 times the concentration of mercury when compared to the water in which they swim. These factors affect the impact that the release of a given amount of a chemical has on our health and well-being.

Another issue is that the TRI program requires reporting of individual chemicals, but the reporting is in pounds of chemical. Public reporting emphasizes the total pounds released, blurring the importance of the wide variation of toxicity of chemicals on the list. The state of Washington evaluates and prioritizes toxic chemicals released in the state and developed programs to eliminate usage of the most persistent, bioaccumulative, and toxic (PBT) chemicals, starting with mercury. From this experience, Washington State is also preparing a comprehensive chemicals policy that intends to revamp the context of the toxic substance laws and regulations such that all proposed products containing high-impact chemicals are discouraged. This comprehensive initiative is now in the development-and-implementation stage and includes emerging contaminants, such as pharmaceuticals and engineered nanoparticles.

In developing the toxic chemical use reduction program in São Paulo, we were charged with the goals of achieving the maximum reduction in toxic impact while minimizing the regulatory burden placed on the affected companies. We therefore developed an approach to measure the toxic impact of chemical use based on the annual use of a chemical multiplied by factors that accounted for persistence and toxicity. In our analysis of approximately 200 chemicals, we found that 1 chemical accounted for half, 5 chemicals accounted for 80 percent, and 10 chemicals accounted for 90 percent of the toxic burden. In preparing this book, our analysis of the most recent U.S. data found that one chemical (hexavalent chromium) accounted for over 99 percent of the potential risks associated with TRI releases in 2007.

We also recommended that the emphasis be placed on use of a toxic chemical rather than its release. It takes a fraction of the effort to inventory and report use than to report releases. To determine release, one has to track daily usage of each product in each manufacturing process and determine how much is released to each of the environmental media. Use accounts not only for potential release to the environment but also for potential worker exposure and the release into homes and offices in the products that are sold.

I once attended a meeting at which the USEPA Office of Water was looking for ideas on how to simplify and make regulations more effective. I suggested the Soviet system of charging for the toxic impact of chemical releases to the environment. The response was that it was an interesting idea but would not fit into our admittedly flawed regulatory structure, which was based on the separate regulation of each medium and firmly entrenched in the agency.

Why can other countries learn from our successes and failures, but we cannot? Reducing or eliminating the impacts of toxic chemical use requires a new approach.

I started writing this foreword while traveling from Brazil to my home near Washington, D.C., and was thinking about the lessons learned from reviewing various global existing programs, how we applied these lessons to propose a toxic chemical use reduction program for CETESB, and how these same concepts could be applied to the United States. This book lays out an analysis of the toxic impacts of chemical releases in the United States based on current TRI data and proposes an approach to ending our addiction to these chemicals. The approach is built on a review of previous and current global programs and takes the best components from each to develop a market-based approach to reducing our use. This market-based approach would move the downstream costs of toxic chemicals to the point of initial use. This program builds on the successful TRI program and focuses reduction efforts at the point of initial use and on the chemicals that have the greatest impact on toxicity.

The impetus and intention behind proposing the program in this book is to spur thinking among other professionals, especially federal-level policy makers, regarding the next level of toxic chemical use reduction, with the ultimate goal of instituting policy updates that build on and continue the success of the TRI and other associated programs. Regardless of how successful past programs have been, to continue progress in any field, especially those concerning public health, it is the responsibility of professionals like us to reflect on what has worked and what needs to be changed based on the world as it has evolved. Our proposal is certainly not set in stone; in fact, we welcome healthy discussion and suggestions on our specific proposed program elements and what modifications, if any, would maximize the effectiveness of the next level U.S.-based toxic chemical usage reduction program. Due to the fact that this is an initial proposal to generate a new way of thinking and move toward certain toxic chemical use reduction policy adjustments, we also recognize that if select or even all program elements proposed in this book are deemed appropriate, additional refinement of details will be required as part of actual program implementation to convert these ideas on paper into actual policy. Some thoughts regarding the additional conceptual-level steps that could be taken toward this objective to further refine our initial thoughts over time as part of a formal program, as warranted, are also provided in this book.

In addition, the analysis of the specific toxicity of any given chemical typically requires a detailed analysis of exposure pathways and individuals affected. For the purposes of a toxic chemical usage reduction program, we have simplified the analysis to come up with a single, relative, chemical-specific toxicity factor to rank chemicals by their overall effect. The important point to keep in mind is that the objective of this approach was not to come up with an absolute number that we claim to represent the actual and fixed toxic impact of a chemical. Rather, the proposed methodology was developed solely for the purposes of coming up with a logical means of applying published and routinely used toxicological data to then develop chemical-specific "toxicity" factors. These factors can then be used to conduct an apples-to-apples comparison across chemicals to *relatively* rank the potential effect among various toxic chemicals, such that reduction in the use of toxic chemicals can be targeted and managed accordingly. We do not claim to be perfect in this analysis. We welcome suggestions on how to improve our analysis or how to use the same or additional toxicity data to relatively rank chemicals differently. The absence of ranking chemicals by some toxicity measure, and therefore not prioritizing which ones to focus on eliminating, is not an option. Without some type of toxicity ranking, all chemicals will be considered to be equally "bad." The undertaking to reduce and even eliminate use of all chemicals would therefore be overwhelming, and ultimately nothing will be done to reach the next step in reducing overall toxicity.

Thomas E. Higgins

Acknowledgments

This book reflects the work, views, and opinions of the authors. This book does not represent the work, views, or opinions of any of the authors' current or past employers, their clients, or their staff.

We wish to acknowledge the assistance of Lucila Telles, Keisha Voigt, Amy Halloran, and Wendy Longley-Cook, who helped prepare background material for the book and provided advice and encouragement.

We are grateful to Flávio de Miranda Ribeiro, Meron Petro Zajac, and Marcelo de Souza Minelli of Companhia de Tecnologia de Saneamento Ambiental (CETESB) for the opportunity to evaluate options for reducing the impacts of toxic chemicals for the state of São Paulo. While work for CETESB formed some of the background for this book, the program suggested in this book is not the same program developed for or being considered by CETESB for implementation in São Paulo, Brazil.

We are grateful to Maria Victoria Peeler, Senior Policy Specialist, Hazardous Waste and Toxics Reduction, Washington State Department of Ecology, who helped us to understand the programs she worked on for reducing toxic chemicals, particularly mercury, in the state of Washington.

We are indebted to Michael DiGiore, chief, Office of Pollution Prevention and Right to Know, N.J. Department of Environmental Protection, for his help in collecting information on the program in New Jersey.

We appreciate the help that Nate and Stacy Higgins provided in assisting with research and editing of the manuscript.

Finally, and most important, we appreciate the overall and continuous support, encouragement, and understanding of our families. You endured our lengthy physical and mental absences as we conducted research for and wrote this book.

About the Authors

Thomas E. Higgins (author), Jayanti A. Sachdev (coauthor), and Stephen A. Engleman (coauthor) are environmental engineering professionals with a broad range of private and public sector experience. This book reflects the work, views, and opinions of the authors. This book does not represent the work, views, or opinions of any of the authors' current or past employers, their clients, or their staff.

Thomas E. Higgins, PhD, PE received three degrees from the University of Notre Dame: a BS in civil engineering, an MS in environmental engineering, and a PhD in environmental engineering with minors in law and chemistry. After graduating, Higgins designed wastewater treatment plants for nuclear power plants. He, then, was a professor of civil and environmental engineering at Arizona State University, where he taught courses in environmental engineering and performed research in treatment of industrial and hazardous waste.

During the past 25 years, Higgins has worked in the private sector, leading research efforts to develop processes for treating or eliminating waste, designing treatment plants, and helping companies and governments set up programs to eliminate waste before it is produced. One project was a study for the Department of Defense that evaluated technologies and recommended policy changes that resulted in substantial reduction in hazardous waste, air emissions, and wastewater; resulting in net savings of over $100 million per year in avoided costs for waste disposal and operations.

Based on this and subsequent experience with other clients, Higgins developed waste minimization tools and led teams performing waste minimization assessments, implementing projects, and providing training to clients in these methods. These experiences and lectures led first to the publication of over 80 articles and two books, *The Hazardous Waste Minimization Handbook* and *The Pollution Prevention Handbook.*

He has assisted companies and governmental agencies in setting up pollution prevention programs in the United States, Germany, Great Britain, Italy, Canada, Hungary, Poland, Russia, Uzbekistan, Taiwan, mainland China, Australia, and Brazil.

Higgins recently led an effort to set up a program for the state of Sâo Paulo, Brazil to reduce the impacts of toxic chemicals. The project developed a program of toxic chemical use reduction, adapting methods used in the United States and other countries that had resulted in reduction of exposure by workers and the public to hazardous chemicals. This project resulted in a program tailored to the needs of Brazil. As a result of this project, Higgins applied these methods to develop a program for the United States. This book is the result of that evaluation.

Jayanti A. Sachdev, PE received two degrees from Virginia Polytechnic Institute and State University: a BS in biology (completed in 3 years) and an MS in environmental sciences and engineering. During her education, Sachdev held an internship at the U.S. Geological Survey (USGS) in Reston, Virginia. Her contributions at USGS included surface water and associated biological tissue analysis in support of studying the effect of pesticides on endocrine systems, research on the impacts of various groundwater sampling and analysis technologies to determine the effects on detected metal concentrations, and U.S. Environmental Protection Agency STORET (Storage and Retrieval) water quality database work. After graduating, Sachdev worked in the private sector and was responsible for quantification of environmental contamination at sites contaminated by past releases of various chemicals and associated risk evaluation and analysis of remediation technology applications as well as environmental database management programming.

Subsequent contributions over the past 12 years have included project and program management as well as technical work in the areas of site investigations and remediation, air quality compliance, greenhouse gas and climate change, industrial wastewater engineering, and pollution prevention.

Specifically, Sachdev managed and conducted technical work under various federal- and state-level programs to evaluate and remediate sites contaminated by past releases of various chemicals with the ultimate goal of the protection of human health and the environment under various potential current or future land uses. The driver behind the environmental contamination assessments as well as the ultimate associated remediation of various environmental media (groundwater, surface water/sediment, soil) was based on both human health and ecological risk evaluations. Close regulatory (both federal and state level) and local community-level collaboration was required on these projects.

Sachdev managed and conducted technical work associated with air emissions inventorying, air permitting, and greenhouse gas and climate change services for various clients in support of the ultimate goal of reducing emissions into the environment and promoting overall sustainable practices. An example of one of Sachdev's projects was to conduct an evaluation and recommend options for reducing emissions in company operations, including hazardous waste minimization opportunities, material substitution options, process control options, and add-on control options.

She was also responsible for environmental data collection and associated analysis to assist power plant utilities in evaluating potential toxics issues in plant-generated process wastewater.

More recently, Sachdev worked with Higgins to set up a program for the state of Sao Paulo, Brazil, to reduce the impacts of toxic chemicals.

Stephen A. Engleman, PE earned two degrees from the University of California at Berkeley: a BS in mechanical engineering (graduating with honors) and an MS in mechanical engineering. While earning his master's degree, Engleman conducted research on innovative technologies using high-temperature incineration in the destruction of wastes that are difficult to eliminate. After graduating, he performed dispersion modeling of toxic clouds from potential catastrophic industrial accidents. He also performed emissions testing of various industrial sources, including hazardous waste incinerators and cement kilns that burned tires.

Engleman assisted industrial companies to understand and comply with the complex air quality regulations in the Los Angeles area. His work included assisting facilities in identifying the most effective air pollution control technology available and for 4 years provided on-site air quality support for a major Southern California petroleum refinery.

Engleman also spent 3 years in Paris, France, managing international due diligence reviews and conducting environmental compliance and liability assessments of complex industrial sites. The assessments involved sites in Europe, North America, South America, Asia, Africa, and Australia and covered air, waste, water, soil, and groundwater and special material (PCB [polychlorinated biphenyl], asbestos, CFC [chlorofluorocarbon], radioactive materials) issues. Representative industrial audits include chemical manufacturing and distribution, paper and pulp, pharmaceutical, printing, metalwork, and plastic film manufacturing.

Engleman currently assists companies with air quality compliance services, greenhouse gas and climate change services, and environmental auditing. He recently worked as a senior technical resource with Higgins to set up a program for the state of Sâo Paulo, Brazil to reduce the impacts of toxic chemicals.

Acronyms and Abbreviations

AB	Assembly Bill
ADEQ	Arizona Department of Environmental Quality
AIM	analog identification methodology
ARB	Air Resources Board
AROW	Automotive Recyclers of Washington
ASTM	American Society for Testing and Materials
atm	Atmosphere
avg	average
BAF	bioconcentration adjustment factor
BCF	bioconcentration factor
BMP	best management practice
BTU	British thermal unit
CAES	Center for Advanced Energy Systems
CAS	Chemical Abstracts Service
CEC	Commission for Environmental Cooperation
CEPA	Canadian Environmental Protection Act
CETESB	Companhia de Tecnologia de Saneamento Ambiental (Environmental Sanitation Technology Company, the São Paulo State Environment Agency)
CFO	chief financial officer
CIC	Carcinogen Identification Committee
CRTK	Community Right-to-Know Act
CSP	certified safety professional
DART	Development and Reproductive Toxicant (Identification Committee)
DDT	dichlorodiphenyltrichloroethane
DEP	Department of Environmental Protection
DNR	Department of Natural Resources
DOE	Department of Ecology
E-PRTR	European Pollutant Release and Transfer Registry
ELVS	End-of-Life Vehicle Solutions Corporation
EMS	environmental management system
EPA	Environmental Protection Agency
EPCRA	Emergency Planning and Community Right-to-Know Act
EPER	European Pollutant Emission Register
ESHB	Engrossed Substitute House Bill
ETF	effective toxicity factor
FTE	full-time employee
FSU	former Soviet Union
GDP	gross domestic product
HAP	hazardous air pollutant
HL	half-life
HRA	Hampshire Research Associates
ILO	International Labor Organization
IPCC	Intergovernmental Panel on Climate Change

IUR	inhalation unit risk
LLC	Limited Liability Company
LP	Limited Partnership
Ltd.	Limited
MassDEP	Massachusetts Department of Environmental Protection
MEK	methyl ethyl ketone
MEP	Manufacturing Extension Partnership
MERA	Mercury Education Reduction Act
MF	mobility factor
MININW	Marine Industries Northwest Incorporated
MOU	Memorandum of Understanding
MSDS	Material Safety Data Sheet
NA	not available
NEC	Northern Engraving Corporation
NIOSH	National Institute for Occupational Safety and Health
NIST	National Institute of Standards and Technology
NJ DEP	New Jersey Department of Environmental Protection
NJME	New Jersey Program for Manufacturing Excellence
NJTAP	New Jersey Technical Assistance Program
NOEC	No observed effect concentration
NOX	nitrogen oxide
NPI	National Pollutant Inventory
NPO	nonproduct output
NPRI	National Pollutant Release Inventory
OECD	Organization for Economic Cooperation and Development
OEHHA	Office of Environmental Health Hazard Assessment
OIA	Office of Innovation and Assistance
OPP	Office of Pollution Prevention
OPPTD	Office of Pollution Prevention and Technology and Development
OSHA	Occupational Safety and Health Administration
OTA	Office of Technical Assistance
P2	pollution prevention
PAH	polycyclic aromatic hydrocarbon
PBDE	polybrominated diphenyl ether
PBT	persistent, bioaccumulative, and toxic
PCB	polychlorinated biphenyl
PCT	polychlorinated terphenyl
PE	professional engineer
PF	persistence factor
PFOS	perfluorooctane sulfonate
PhD	doctor of philosophy
POTW	publicly owned treatment works
PPA	Pollution Prevention Act
PPI	Progressive Policy Institute
PPIS	Pollution Prevention Incentives for States
ppm	part per million
Prop 65	California Proposition 65

PRTR	Pollutant Release and Transfer Register
QSAR	Quantitative structure-activity relationships
RCRA	Resource Conservation and Recovery Act
RCW	Revised Code of Washington
REACH	Registration, Evaluation, Authorization, and Restriction of Chemicals
RECLAIM	Regional Clean Air Incentives Market
RETAP	Retired Engineer Technical Assistance Program
RfCi	inhalation reference concentration
RfDo	oral reference dose
RTK	Right-to-Know Act
SARA	Superfund Amendments and Reauthorization Act
SB	Senate Bill
SCAQMD	South Coast Air Quality Management District
sed	sediment
SFO	oral slope factor
SIC	Standard Industrial Classification
SOX	sulfur oxide
TF	toxicity factor
TPPA	Toxic Pollution Prevention Act (of Minnesota)
TREE	Technical Resources for Engineering Efficiency
TRI	Toxics Release Inventory
TU	toxicity unit
TUR	Toxics Use Reduction
TURA	Toxics Use Reduction Act
TURI	Toxics Use Reduction Institute
UN-ECE	United Nations Economic Commission for Europe
USEPA	United States Environmental Protection Agency
US FDA	U.S. Food and Drug Administration
VOC	volatile organic compounds
VP	vapor pressure
VPBT	very persistent, bioaccumulative, and toxic
VSM	value stream mapping
WA	Washington
WAC	Washington Administrative Code
WCRTK	Worker and Community Right-to-Know (Act)
WMS	Washington Manufacturing Services
WSDA	Washington State Dental Association
WSHA	Washington State Hospital Association

1 Introduction

There is a growing concern regarding the impact of exposure to toxic chemicals. While the immediate effects of breathing cyanide fumes or drinking concentrated arsenic are quickly apparent, the long-term exposure to trace quantities of a carcinogen is just as deadly and, to the typical individual, just as feared.

Some have attempted to reduce their individual exposure to toxic chemicals by buying "green products" or consuming "organic food." It is our opinion that this by itself is neither effective nor efficient. It requires that each individual have access to the specific chemical composition of each product with which one comes in contact as well as information on the toxicity of these specific chemicals. Existing programs that have resulted in requirements for reporting product composition are presented in Chapter 2.

One important path by which we are exposed to toxic chemicals is through the products that we purchase and consume. If there is an appropriate understanding of the effects of that exposure, then the risk is one that an individual can choose to take, assuming the benefit of product use is greater than the risk. When a toxic chemical is released to the air, water, or land in a community, individuals are involuntarily exposed to risks. While these releases may have the economic benefits of reduced product cost or local employment, they can pass on to the general public the additional medical costs and suffering associated with the resulting exposure. It took the tragic release of a toxic gas from a pesticide plant in Bhopal, India, with the deaths of thousands, to result in legislation to require that companies (and eventually government agencies) report on the release of toxic chemicals to the surroundings. This legislation (the Community Right-to-Know Act of 1986) resulted in requirements that companies report to their communities on the storage of toxic chemicals as well as all release, not just the major ones, such as the one that occurred in Bhopal. Chemical releases are compiled in a Toxics Release Inventory (TRI), which provides the public with access to data on annual releases of individual chemicals for each facility included in the program. Before this legislation took effect, we, and in many cases the companies themselves, did not know the quantities of toxic chemicals that they used or released to the environment because, at the time, composition information was not available for the products that the companies used to produce other products. Chapter 3 lays out the history of toxic chemical release reporting, presenting national data on U.S. toxic chemical release for 2007, the most recent year that data were available at the time this book was written.

Since the first TRI program was established in the United States, similar pollutant release and transfer register (PRTR) programs have been established in 29 other countries. Chapter 4 presents information on these PRTR programs.

The requirement that companies and governmental agencies report toxic chemical release has had the unintended beneficial effect of reducing the use of the individual chemicals and their release. One reason is that elimination or reduction of use of a chemical below the level of required reporting eliminates a company from having to go through the process of calculating release and reporting. This saves on the cost of compliance as well as adverse publicity and public response to a reported release. The effects of the TRI program itself on reducing release are presented in Chapter 5.

Although the TRI program has resulted in some degree of reduction of the use of individual chemicals and their release, the emphasis of the program has been on total pounds chemical of released, not on the toxicity of those releases. Reports generally stress total pounds of release or rank the chemicals released by total pounds. This, in essence, assumes that chemicals are either toxic (and included in the inventory) or not. In reality, not all chemicals included on the TRI list are equally toxic. It is intuitively obvious that a pound of arsenic has more potential impact than a pound of saccharin or nitrate, yet all three are included in the TRI and generally counted equally. Companies find themselves in the news as having the most pounds of toxic release to the community, with little analysis done on the actual risks associated with the release.

To determine the toxic effect of a given release, a toxicologist looks at the pathway by which an individual or group is exposed to a toxin (i.e., oral ingestion for a solid or liquid and inhalation for a gas or aerosol). Based on this pathway, a model of dose is compared to cancer or noncancer toxicity dose-response factors, the weight of the receptor, age, and other susceptibility factors to estimate risk to the individual or group of individuals. There is a need for a simplified index for quantifying the toxic impact of a chemical, one that can be used to evaluate the combined effects of release of a TRI chemical, even if it is not used to determine the effect of an individual release. We need such an index so that individuals can determine their relative and potential risk when evaluating chemical composition of products. We also need this index to better evaluate the relative risks associated with TRI chemical release and subsequently to prioritize chemicals for reduction based on their cumulative or countrywide impact on the health of a nation. Chapter 6 utilizes available oral and inhalation cancer and noncancer measures of toxicity to develop a method for the relative ranking of toxicity (toxicity factor) for the TRI chemicals.

The impact of a chemical release is dependent not only on the toxicity of the compound but also on the likelihood that it will be ingested or inhaled. The likelihood of a particular chemical being ingested or inhaled increases proportionally to the mobility of the chemical, that is, its tendency to dissolve and enter the water we drink or evaporate and enter the air that we breathe. The tendency to enter the water is measured by its solubility. The tendency to evaporate is measured by its vapor pressure. Chapter 7 presents the use of solubility and vapor pressure data to develop a mobility factor, to be integrated into the toxicity factor for each TRI chemical.

A chemical that is persistent, that is, does not degrade or decompose in the environment, will accumulate in the environment over time, increasing the potential for exposure compared with a chemical that rapidly degrades. A measure of how rapidly a chemical decomposes is the half-life or time needed for half of a given amount of the chemical to degrade. Chapter 8 presents data on the half-lives of the TRI chemicals as well as a method for converting these to a persistence factor, which is then integrated into the toxicity factor for each TRI chemical.

Compounds that concentrate in the food chain can provide a much higher dose of toxic chemicals than may be apparent from the initial release to the environment. Bioconcentration is a process by which an organism living in water develops a chemical concentration higher than that of the water. This is the result of the intake or absorption of the chemical from the water being higher than the rate of excretion and metabolism of the chemical. This occurs when a toxin remains unchanged as it moves up the food chain. For example, although mercury is only present in small amounts in seawater, it is absorbed by algae (generally as methyl mercury). Mercury is efficiently absorbed but only very slowly excreted by organisms (Croteau, Luoma, and Stewart 2005). Mercury builds up in the adipose (fatty) tissue of successive levels in the food chain. At each level, mercury in the tissue of organisms that are eaten accumulates in the tissue of the animals until they in turn are eaten by organisms at the next level, who then add to their own mercury contamination. The higher the level in the food chain, the higher the concentration of mercury is in the fish. This process explains why predatory fish such as swordfish and sharks or birds like osprey and eagles have higher concentrations of mercury in their tissue than could be accounted for solely by direct exposure. For example, herring contain mercury levels at approximately 0.01 part per million (ppm), whereas sharks contain mercury levels at greater than 1 ppm. In Chapter 9, bioconcentration data are used to develop a bioconcentration adjustment factor, to be integrated into the toxicity factor for each TRI chemical.

A nervous radio interviewer once asked John Dillinger, "Why do you rob banks?" His well-known answer was, "Because that is where the money is." Similarly, if our goal is to keep track of and reduce use or exposure to toxic chemicals, we should measure and report not just volumes released, but rather quantify the chemical release in units that account for the mass released, the toxicity (toxicity factor), mobility (mobility factor), persistence (persistence factor), and ability to bioconcentrate (bioconcentration adjustment factor). Chapter 10 presents one method for integrating these factors to estimate the relative impact of TRI chemical release (effective toxicity factor/toxicity units), with the goal of developing a rational method of prioritizing toxic chemicals for reduction. As this is an initial proposed approach, we welcome suggestions on how to improve our analysis or use the same or additional toxicity data to rank chemicals differently. The important point to keep in mind is that the objective of this approach was not to come up with an absolute number that we claim to represent the actual and fixed toxic impact of a chemical. Rather, the proposed methodology was developed solely for the purposes of coming up with a logical means of applying published and routinely used toxicological data to then develop chemical-specific "toxicity" factors. In this way, we can conduct an apples-to-apples comparison across chemicals and rank the potential effect among various toxic chemicals such that reduction in the use of toxic chemicals can be targeted and managed accordingly.

The problem with basing programs on total poundage of chemicals is that reduction efforts may end up being concentrated on high-volume, low-toxicity release. One could conceivably reduce total volume of a chemical by replacing a low-toxicity compound with a smaller volume of a much more toxic compound. Chapter 11 reviews programs that concentrate on reducing the release of chemicals with the greatest adverse toxic impact.

Although the TRI has been successful in initially reducing the use of toxic chemicals, to continue and to build on this success, release of toxic chemicals contained in the products sold to customers (vs. just release to environmental media), actual use of toxic chemicals by manufacturers (vs. just release), and worker exposure now need to be addressed. It is important to note that it is expensive to track release by each source in the factory. As part of the TRI process, the company must determine use of the chemical to determine if the release of that chemical needs to be quantified and reported; however, the data determined on use itself are not reported. Chapter 12 presents a comparison between use and release reporting and programs aimed at reducing use. If we limit or eliminate use, this would contribute to reduced worker exposure and release, both directly to the air, water, and land and indirectly through products.

One method of reducing use or release of toxic chemicals is through requirements that companies prepare pollution prevention plans and set goals for reducing use or release. Such existing programs are reviewed in Chapter 13.

Providing technical assistance is an important component of pollution prevention programs. They provide assistance to small businesses as they search for information on alternative chemicals to use and proven process changes needed to eliminate the use of toxic chemicals. These technical assistance programs provide a central repository of industry-specific best management practices that can help diverse and otherwise competing companies. Existing technical assistance programs are reviewed in Chapter 14.

In recent years, alternative market-based approaches (vs. the traditional "command-and-control" strategy) for achieving environmental protection have evolved. Chapter 15 reviews these approaches and how they might apply to a program aimed at reducing or eliminating the use of toxic chemicals.

In Chapter 16, we lay out a proposed program for reducing the use of toxic chemicals. This program is an initial proposal that builds on the existing TRI program and uses lessons learned from this program, similar programs overseas, and other U.S. state-based toxic chemical use reduction programs aimed at minimizing the impacts of toxic chemicals. The impetus and intention behind proposing the program in this book is to spur thinking among other professionals, particularly federal-level policy makers, about the "next level" of toxic chemical use reduction with the ultimate goal of instituting policy updates that build on and continue the success of the TRI and other associated programs. Regardless of how successful past programs have been, to continue progress in any field (especially those in which public health is concerned), it is the responsibility of professionals like us to reflect on what has worked and what needs to be changed based on the world as it has evolved. Our proposal is certainly not set in stone; in fact, we welcome healthy discussion and suggestions regarding our proposed program elements and what modifications, if any, would maximize the effectiveness of the next-level U.S.-based toxic chemical use reduction program. Because this is an initial proposal to generate a new way of thinking and move toward certain toxic chemical use reduction policy adjustments, we also recognize that if select or even all proposed program elements proposed in this book are deemed appropriate, additional refinement of details will be required as part of actual program implementation to convert these ideas into actual policy.

Some thoughts regarding the additional conceptual-level steps that could be taken toward this objective are also provided as part of Chapter 16.

Any program that will have an impact on our use of toxic chemicals is going to have a cost. Changing processes can be expensive, and there is competition for resources with projects that bring new products to market. For instance, converting chloralkali plants from mercury cells to membrane systems can cost over a billion U.S. dollars. It is hard for a toxic chemical use reduction program to compete with projects that bring new products to the market with this level of cost. On the other side of the balance sheet, however, a toxic chemical use reduction program will also have numerous other returns on investment as well as numerous and important health benefits (e.g., avoided medical costs for exposures that do not occur, for cancers that are avoided, for productive lives that are extended). The costs and potential benefits of a toxic chemical use reduction program are laid out in Chapter 17.

BIBLIOGRAPHY

Croteau, M., S. N. Luoma, and A. R. Stewart. 2005. Trophic transfer of metals along freshwater food webs: Evidence of cadmium biomagnification in nature. *Limnology and Oceanography* 50 (5): 1511–1519.

U.S. Environmental Protection Agency. 1997. *Mercury Study Report to Congress. Vol. IV: An Assessment of Exposure to Mercury in the United States*. EPA-452/R-97-006. U.S. Environmental Protection Agency, Office of Air Quality Planning and Standards and Office of Research and Development. http://nepis.epa.gov/Exe/ZyNET.exe/2000EIMV.TXT?ZyActionD=ZyDocument&Client=EPA&Index=1995+Thru+1999&Docs=&Query=&Time=&EndTime=&SearchMethod=3&TocRestrict=n&Toc=&TocEntry=&QField=pubnumber^%22452R97006%22&QFieldYear=&QFieldMonth=&QFieldDay=&UseQField=pubnumber&IntQFieldOp=1&ExtQFieldOp=1&XmlQuery=&File=D%3A\zyfiles\Index%20Data\95thru99\Txt\00000009\2000EIMV.txt&User=ANONYMOUS&Password=anonymous&SortMethod=h|-&MaximumDocuments=10&FuzzyDegree=0&ImageQuality=r75g8/r75g8/x150y150g16/i425&Display=p|f&DefSeekPage=x&SearchBack=ZyActionL&Back=ZyActionS&BackDesc=Results%20page&MaximumPages=1&ZyEntry=1&SeekPage=x (accessed January 17, 2009).

2 Toxic Chemical Composition Reporting

INTRODUCTION

Concerned individuals have attempted to reduce their individual exposure to toxic chemicals by buying significantly more expensive "green" products or consuming "organic" food. Although the development of these products is an initial step forward in the right direction, it is our opinion that this puts an impossible burden on the individual consumer. First, without conducting any research, it requires that individual consumers fully understand and trust the numerous types of green or organic labels in the market today. The reality is that some types of labels are regulated, requiring adherence to a set of standards to use the label, and "trustworthy," whereas others are not. Taking it one step further requires that the individual first determine the complete chemical composition using a list of compounds coupled with the amount of each compound in the product for each of the products purchased, information that is not currently on product labels. Then, it requires that the individual determine all of the relevant toxicity information on each of these compounds and assess the relevant exposure pathways to make a decision on the potential hazards of using the product, also not currently on product labels. For instance, hexavalent chromium is a carcinogen if inhaled but is relatively harmless if swallowed because it is converted in the stomach to the relatively benign trivalent form. Moreover, the benign trivalent form is an essential nutrient and frequently purchased as a dietary supplement.

There are literally hundreds of thousands of chemicals that we encounter in our daily lives. In addition to chemicals contained in products themselves, when a toxic chemical is released to the air, water, or land in a community, individuals are involuntarily exposed to the associated risks. Avoiding all chemicals is not an effective or even possible method of shielding oneself from toxic exposure. Alleviating the burden of reducing or eliminating toxic chemical exposure on the individual consumer level to the extent that is feasible requires the modification of product labels to include chemical composition and toxicity data as well as the concurrent development of a federal program that is aimed at reducing the overall use of the most highly toxic (and used) chemicals in the manufacturing process in the first place. If the use is limited, worker exposure and releases will be limited, to the air, water, and land and through products. A first step in label modification and development of such a reduction program is to know exactly what is in the manufactured products that one consumes and in what amounts. The remainder of this chapter discusses three existing programs that require composition reporting on products.

Subsequent to this chapter, this book details approaches for the next steps of appropriately prioritizing the toxicity of various chemicals based on relevant toxicity data and for chemical composition reporting to assist individual consumers in making more efficient and informed decisions about the products that they are buying. Also, details are provided on a federal-level program that could be implemented to reduce the use of at least the most highly toxic chemicals used in the manufacturing process to reduce exposure in the first place.

CALIFORNIA: PROPOSITION 65

In November 1986, California voters approved the Safe Drinking Water and Toxic Enforcement Act of 1986, also known as Proposition 65 (Prop 65). This act requires the state to publish a list of chemicals known to cause cancer or birth defects or other reproductive harm and to update the list annually.

Any company that is located in or does business in California is prohibited from knowingly discharging listed chemicals into sources of drinking water. They must also provide "clear-and-reasonable" warning before exposing anyone to a listed chemical. Businesses are responsible for developing their own warnings.

Basis for Program

The Prop 65 chemical list contains a wide range of naturally occurring and synthetic chemicals that are known to cause cancer or birth defects or other reproductive harm. The chemicals include those present in household products as well as those used in manufacturing and construction or chemicals that are by-products of chemical processes such as motor vehicle exhaust.

To be added to the list, a chemical first has to meet guidelines identifying it as a carcinogen or reproductive toxicant. At that point, there are three main ways that chemicals are added to the list.

1. There are two independent committees of scientists and health professionals that can assign a chemical to the list: the Carcinogen Identification Committee (CIC) and the Development and Reproductive Toxicant (DART) Identification Committee. Committee members are appointed by the governor.
2. If either CIC or DART identifies an organization as an "authoritative body," that organization can classify a chemical as a carcinogen or as a chemical with the potential to cause reproductive harm. The following organizations have been designated as authoritative bodies: the U.S. Environmental Protection Agency (USEPA), U.S. Food and Drug Administration (U.S. FDA), National Institute for Occupational Safety and Health (NIOSH), National Toxicology Program, and International Agency for Research on Cancer.
3. An agency of the state or federal government identifies a chemical as a possible carcinogen or reproductive toxicant. This generally applies to prescription drugs identified by the U.S. FDA.

Program Requirements

Once a chemical is on the list, any company with more than 10 employees that is located in California or does business in California is prohibited from knowingly discharging listed chemicals into sources of drinking water. These companies must also provide a clear-and-reasonable warning before knowingly and intentionally exposing anyone to a listed chemical. Once a chemical is listed, businesses have 12 months to comply with warning requirements and 20 months to comply with the discharge prohibition. Businesses are responsible for developing their own warnings and do not have to report any Prop 65 chemicals to the California Office of Environmental Health Hazard Assessment (OEHHA). Governmental agencies and public water systems are exempt from this act.

A company does not have to follow the requirements of Prop 65 if it is determined that a chemical presents "no significant risk." For a carcinogen, this is a level of the chemical that is calculated to result in one excess case of cancer in an exposed population of 100,000, assuming lifetime exposure at the level in question. For a reproductive toxicant, it is defined as a level of an exposure to the chemical that could be increased by 1,000 and still not produce birth defects or reproductive harm. These threshold levels are called *safe harbor numbers*. Businesses may or may not determine the level of a chemical present in their product. If they provide a warning that the listed chemical is present, it can either mean that levels of the chemical have been evaluated and is present above the safe harbor number or mean that the chemical is present at some level but the company did not find it worthwhile to evaluate.

Role of Stakeholders

Prop 65 was introduced via the California State initiative process. An initiative measure is proposed by California citizens, who must present a signed petition to the California secretary of state outlining the text of the proposed statute or amendment to the state constitution. The secretary of state then submits the measure at the next general election.

The California attorney general's office enforces Proposition 65. Any district attorney or city attorney (for cities with a population of at least 750,000) may also enforce Proposition 65. Any individual acting in the public interest may also enforce Proposition 65 by filing a lawsuit against a business alleged to be in violation of this law. Lawsuits have been filed by the attorney general's office, district attorneys, consumer advocacy groups, and private citizens and law firms. Penalties for violating Proposition 65 by failing to provide notices can be as high as $2,500 per day for each violation in addition to any other penalty established by law.

Financial Impacts

The Safe Drinking Water and Toxic Enforcement Fund is established in the California State Treasury. The director of the lead agency designated by the governor may expend the funds in the Safe Drinking Water and Toxic Enforcement Fund, on appropriation by the legislature, to implement and administer this chapter.

Any company that falls under the requirements of Prop 65 is financially responsible for all chemical analysis of their products and waste as well as for providing a clear-and-reasonable warning before knowingly and intentionally exposing anyone to a listed chemical. The warning can include labeling of a consumer product, posting signs at the workplace, distributing notices at a rental housing complex, and publishing notices in a newspaper.

All civil and criminal penalties collected pursuant to this chapter shall be apportioned in the following manner:

- Fifty percent of all civil and criminal penalties collected under Prop 65 will be deposited in the Hazardous Substance Account in the General Fund.
- Twenty-five percent will be paid to the office of the city attorney, city prosecutor, district attorney, or attorney general, whichever office brought the action.
- Twenty-five percent will be paid to OEHHA and used to fund the activity of the California state environmental agency, the local health officer, or other local public officer or agency that investigated the matter that led to the bringing of the action.

Because the penalties for violating Proposition 65 can be costly, most businesses settle before a case goes to trial. Businesses usually pay the plaintiff's attorney fees and cost. Plaintiffs might also ask for restitution—money that goes to a public interest group, usually in lieu of penalties.

The settlement usually will also include some provision for the company to reformulate the product, give some type of warning, or remove the product from the market.

Effectiveness of Program

Prop 65 has provided residents of California with information that allows them to limit their exposures to listed chemicals beyond restrictions established by other state and federal laws. It has also encouraged manufacturers to remove listed chemicals from their products. This law benefits consumers; however, it adds an additional cost to companies that do business in the state of California. The list of chemicals, combined with the burden of proof on individual companies, can easily lead to lawsuits. Companies must test their products, come up with potential replacements for the toxic chemicals, reduce discharges, notify the public when necessary, and pay for any enforcement actions against their company.

Prop 65 has reduced the amount of certain chemicals in commonly used products. According to OEHHA, air emissions of certain chemicals, including ethylene oxide, hexavalent chromium, and chloroform, from facilities in California have also been significantly reduced as a result of Proposition 65. Warnings regarding the danger of alcohol to fetuses are one of the most widespread results of this program.

EUROPEAN UNION'S REGISTRATION, EVALUATION, AUTHORIZATION, AND RESTRICTION OF CHEMICALS PROGRAM

Chemicals pose a quandary for the European Union (EU). Production of chemicals is Europe's third largest industry, employing 1.7 million people directly. There are over 100,000 different chemicals used in the European market, and a number of these chemicals have been linked to certain health conditions, such as asbestos connected to lung cancer and benzene to leukemia. Of the 100,000 chemicals used in the market, only a small portion has adequate information on carcinogenicity or other toxicity. In addition, new substances are introduced to the market annually. The EU program for Registration, Evaluation, Authorization, and Restriction of Chemicals (REACH) has four basic components for dealing with toxic chemicals, as listed in its title.

The first part of the program, registration, is described further here as it focuses on composition reporting. The remaining three parts of the program are based on focusing on impact chemicals; therefore, these are described further in Chapter 11.

Basis for Program

REACH is based on Council Directives 67/548/EEC of June 27, 1967, related to regulating classification, packaging, and labeling of dangerous substances; 76/769/EEC of July 27, 1976, associated with restricting marketing and use of dangerous substances; 1999/45/EC of May 31, 1999, related to the classification, packaging, and labeling of dangerous preparations; and 793/93 of March 23, 1993, related to the evaluation and control of the risks of existing substances. These directives identified problems of disparities in the laws of individual countries and the need to do more to protect health and the environment from chemicals.

Program Requirements

Registration

REACH proposes to require companies that wish to sell their products to EU countries to provide composition information if

1. The substance is present in those products in quantities totaling over 1 metric ton per producer or importer per year; and
2. The substance is present in those products above a concentration of 0.1 percent by weight.

This regulation goes beyond the usual requirements of companies providing Material Safety Data Sheets (MSDSs), which do not require composition information, but rather general impacts of the chemicals such as flammability, toxicity, exposure risks, and ranges of composition of certain toxic compounds.

Registration requires a manufacturer or importer to notify an authority in advance of the intention to produce or import a substance and to submit required information in advance of importing a product. The appropriate EU authority will put this information into an electronic database, assign a registration number and then screen the registered substances for properties raising particular concern.

Registration information is to include the following:

1. Data/information on the identity and properties of the substance, including data on toxicological and ecotoxicological properties
2. Intended uses, estimated human and environmental exposure
3. Production quantity envisaged
4. Proposal for the classification and labeling of the substance
5. Safety Data Sheet
6. Preliminary risk assessment covering the intended uses
7. Proposed risk management measures

TOXICS RELEASE INVENTORY

The Toxics Release Inventory (TRI) is a U.S. program (discussed further in Chapter 3) that requires regulated entities to report on their environmental releases of approximately 650 chemicals. The TRI includes those chemicals that are considered to be the most highly toxic; however, it also includes chemicals that are not necessarily toxic.

This regulation does not directly require that companies report the composition of their products to the public, but because they are required to keep track of each of these chemicals in the materials that they purchase, the program has ended up placing a requirement for suppliers of materials to report the concentrations of these chemicals (i.e., composition reporting) in their products to those individual companies required to report on their releases.

Therefore, although in essence the TRI has become an "internal company" requirement for tracking of the chemical composition of chemicals used in their products, including those chemicals that are highly toxic, the actual emphasis of the TRI program to date, as far as public reporting goes, has only been on the total pounds of releases of these chemicals and not on specific product composition reporting and not based on the relative toxic ranking of the chemical releases reported.

BIBLIOGRAPHY

Commission of the European Communities. 2001. *White Paper: Strategy for a Future Chemicals Policy*. Brussels, BE.

Commission of the European Communities. 2003. *Commission Staff Working Paper: Regulation of the European Parliament and of the Council Concerning REACH—Extended Impact Assessment*. Brussels, BE.

Council of the European Union. 2006. *EU Common Ground on REACH.*

Council of the European Union. 2006. *The REACH Proposal Process Description.*

Council of the European Union. 2006. *Questions and Answers on REACH.*

Prop 65 Clearinghouse. 2007. The Reports. http://www.prop65clearinghouse.com/ (accessed November 14, 2006).

Prop 65 News—Online Edition. 2005. Proposition 65 Made Simple. http://www.prop65news.net/Pubs/brochure/Proposition%2065%20Made%20Simple.htm (accessed November 7, 2006).

State of California Environmental Protection Agency, Office of Environmental Health Hazard Assessment. 1986. Safe Drinking Water and Toxic Enforcement Act of 1986, Chapter 6.6, Added by Proposition 65, 1986 General Election. Sacramento, CA.

State of California Environmental Protection Agency, Office of Environmental Health Hazard Assessment. 2006. *California Code of Regulations*, Title 22, Division 2, Part 2, Subdivision 1, Chapter 3. Sacramento, CA.

State of California Environmental Protection Agency, Office of Environmental Health Hazard Assessment. 2006. Safe Drinking Water and Toxic Enforcement Act of 1986, List of Chemicals Known to the State to Cause Cancer or Reproductive Toxicity.

State of California Environmental Protection Agency, Office of Environmental Health Hazard Assessment. 2007. Frequently Asked Questions about Proposition 65. http://www.oehha.ca.gov/prop65/p65faq.html (accessed November 6, 2006).

State of California Environmental Protection Agency, Office of Environmental Health Hazard Assessment. 2007. Proposition 65. http://www.oehha.ca.gov/prop65.html (accessed November 6, 2006).

U.S. Environmental Protection Agency. 2010. Toxics Release Inventory Program. http://www.epa.gov/TRI/ (accessed January 20, 2010).

3 Toxic Chemical Release Reporting

INTRODUCTION

At midnight on December 2, 1984, citizens of Bhopal, India, awoke with a burning sensation in their lungs. The local Union Carbide pesticide plant had an accident; an estimated 42 tons of methyl isocyanate gas was released from a holding tank. The resulting plume exposed more than 500,000 people to varying concentrations of the toxic chemical, with the result that at least 3,787 died within 72 hours and additional thousands subsequently died from gas-related diseases. Shortly thereafter, there was a chemical release at a sister plant in West Virginia.

These events resulted in demands by industrial workers and communities for information on hazardous materials stored in and released from industrial plants in their communities. Public interest and environmental organizations around the country accelerated demands for information on toxic chemicals being released outside industrial facilities.

The response in the United States became known as the Emergency Planning and Community Right-to-Know Act (EPCRA) of 1986. EPCRA was enacted to facilitate emergency planning by local agencies, to minimize the effects of potential toxic chemical accidents, and to provide the public with information on releases of toxic chemicals in their communities.

A significant requirement of EPCRA was that users of certain toxic chemicals annually report their releases of these chemicals to the land, air, and water. This information is compiled into the Toxics Release Inventory (TRI). At first, the requirement was applied to chemical industries, but this has been expanded to other industry groups that use these toxic chemicals. TRI is a database available to the public; it contains detailed information on toxic chemical releases by individual industrial facilities. The TRI program has been amended by the Pollution Prevention Act of 1990 and is administered by the Environmental Protection Agency (EPA). Originally, federal facilities were exempt from the requirement, but President Clinton, on August 3, 1993, signed Executive Order 12856, Federal Compliance with Right-to-Know Laws and Pollution Prevention Requirements, which removed the exemption and required all federal facilities to comply, regardless of industrial classification.

BASIS FOR PROGRAM

Facilities that manufacture, process, or use toxic chemicals above specified amounts must report annually on disposal or other releases and other waste management activities related to these chemicals. This report is filed with a specified format. Reports must be filed with both the EPA and the applicable state agency. Most states allow reporters to submit both reports at once through programs provided on the EPA Web site.

A facility must report to TRI if it

- Operates within any of the following industry sectors:
  - Manufacturing (Standard Industrial Classification [SIC] codes 20–39)
  - Metal mining (SIC code 10, except 1011, 1081, and 1094)
  - Coal mining (SIC code 12, except 1241)
  - Electrical utilities that combust coal or oil for the purpose of generating power for distribution in commerce (SIC codes 4911, 4931, and 4939)
  - Resource Conservation and Recovery Act (RCRA) Subtitle C hazardous waste treatment and disposal facilities (in SIC code 4953)
  - Chemical wholesalers (SIC code 5169)
  - Petroleum terminals and bulk stations (SIC code 5171)
  - Solvent recovery services (SIC code 7389)
  - A federal facility in any SIC code
- Employs 10 or more full-time-equivalent employees and
- Manufactures or processes more than 25,000 lb or otherwise uses more than 10,000 lb of any listed chemical during the calendar year, except for chemicals that are persistent, bioaccumulative, and toxic (PBT) for which the thresholds are 0.1 g for dioxin and dioxin-like compounds and 10 or 100 lb for other PBT chemicals

The Pollution Prevention Act (PPA) of 1990 mandated collection of data on toxic chemicals that are treated on site, recycled, and combusted for energy recovery. Together, these laws require facilities in certain industries, which manufacture, process, or use toxic chemicals above specified amounts, to report annually on disposal or other releases and other waste management activities related to these chemicals. Each year, the EPA makes TRI data available to the public on two Internet sites: TRI Explorer (http://www.epa.gov/triexplore) and Envirofacts (http://www.epa.gov/enviro).

For the year 2007, some 21,996 facilities reported releasing some 4.1 billion pounds of chemicals under the TRI program. Figure 3.1 shows how these releases were reported by industrial category. The largest releases were by industrial categories that were not included in the original TRI program.

The list of toxic chemicals subject to reporting under section 313 of EPCRA is not static. The original list consisted of 306 chemicals. Over the years, some chemicals, such as sodium sulfate and methyl ethyl ketone (MEK), have been removed, but others have been added. The current TRI toxic chemical list contains 581 individually listed chemicals and 30 chemical categories, including 3 delimited categories containing 58 chemicals. The total number of chemicals and chemical categories is 666.

PROGRAM REQUIREMENTS

Information reported annually by facilities includes

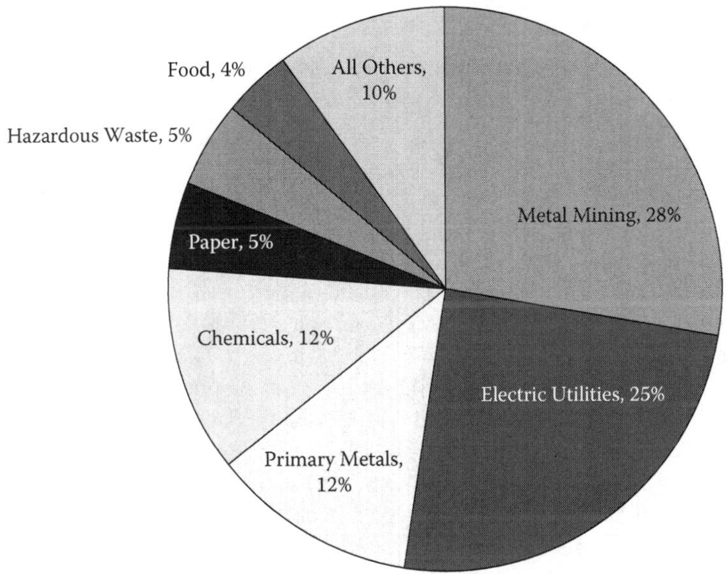

FIGURE 3.1 U.S. TRI releases for 2007 by industry category.

- Basic information identifying the facility, including name, location, type of business, and name of parent company
- Name and telephone number of a contact person
- Environmental permits held
- Amounts of each listed chemical disposed of or released to the environment at the facility
- Amounts of each chemical sent from the facility to other locations for recycling, energy recovery, treatment, or disposal or other release
- Amounts of each chemical recycled, burned for energy recovery, or treated at the facility
- Maximum amount of chemical present on site at the facility during the year
- Types of activities conducted at the facility involving the toxic chemical
- Source reduction activities
- General information about the manufacture, process, and otherwise use of the listed chemical at the facility
- Information about methods used to treat waste streams containing the toxic chemicals at the site and the efficiencies of those treatment methods
- Information regarding the amount of toxic chemicals sent off site for further waste management, facilities, and the destination of these transfers

There are two different reporting forms, Form R and Form A. The simplified Form A can be used if the chemicals in question are not PBTs and when the amount manufactured, processed, or otherwise used is less than or equal to 1 million pounds and the reportable amount is less than or equal to 500 lb/yr.

REFINEMENTS TO TRI

Since the program was established, the EPA has made many refinements to the program. The following timeline highlights many of the key changes to the program over the years.

October 1986: The Superfund Amendments and Reauthorization Act (SARA) Title III, commonly known as the EPCRA, was signed into law. Section 313 of the act established the TRI program.

February 1988: The EPA published a rule in the *Federal Register* (53 *Federal Register* 4500) that, under EPCRA Section 313, certain industrial facilities are required to report releases of listed toxic chemicals to the EPA annually.

November 1990: Congress passed the PPA, which among other requirements, expanded the TRI program to require providing additional information on toxic chemicals in waste and reporting on source reduction methods.

Beginning in 1991: Covered facilities were required to report quantities of TRI chemicals recycled, combusted for energy recovery, and treated on and off site. At this time, TRI reporting was only required of the manufacturing sector (SIC codes 20–39).

August 1993: By Executive Order 12856, federal facilities were required to report under TRI starting in 1994 regardless of the industrial classification (SIC code) of the facility.

Reporting Year 1993: TRI reporting was required for 316 chemicals and 20 chemical categories. Certain RCRA chemicals and certain hydrochlorofluorocarbons (HCFCs) were added to the TRI chemicals list.

November 1994: The EPA promulgated the Chemical Expansion Final Rule (59 FR61431) Phase I, expanding TRI by 286 new chemicals and categories. This expansion of the chemical list raised the number of chemicals and chemical categories reported under TRI to over 600.

July 1996: The EPA deleted di-(2-ethylhexyl) adipate (DEHA) (CAS No. 103-23-1), also known as *bis*-(2-ethylhexyl) adipate; diethyl phthalate (DEP); and nonaerosol forms of hydrochloric acid (HCl) from the TRI list because it was determined that there was not sufficient evidence that they met the required toxic criteria.

May 1997: The EPA implemented Phase II (expansion of the facility list) of an expansion of the TRI program by promulgating the Industry Expansion Final Rule (62 *Federal Register* 23834), which added metal mining; coal mining; electric utilities; commercial hazardous waste treatment; chemicals and allied products, wholesale; petroleum bulk terminals and plants, wholesale; and solvent recovery services to the list of facilities that must report under TRI. The EPA estimated that about 6,600 additional facilities would submit more than 37,000 additional Form R reports because of the addition of these industry groups.

April 1998: The EPA consented under court order to delete dimethyldichlorosilane, methyltrichlorosilane, and trimethylchlorosilane (chlorosilanes) from the TRI list of chemicals.

October 1999: Several actions were taken to expand the TRI program:

- The reporting thresholds were lowered for certain PBT compound chemicals.
- A category of dioxin and dioxin-like compounds was added to the chemical list, and a 0.1-g reporting threshold was established for the category.
- Certain other PBT chemicals were added to the reporting list, and lower reporting thresholds were established.
- Reporting of vanadium was expanded to all forms of the metal except when contained in alloys. Vanadium was formerly required to be reported only if it were in the form of fume or dust.

June 2000: Phosphoric acid was deleted from the list of chemicals subject to TRI reporting due to a court order.

January 2001: The reporting thresholds for lead and lead compounds were lowered to 100 pounds.

May 2001: Chromite ore mined in the Transvaal Region of South Africa and the unreacted ore component of the chromite ore processing residue (COPR) were deleted from TRI reporting requirements because the EPA determined that there was not sufficient evidence that they met the required toxic criteria.

June 2005: MEK was deleted from the list of chemicals subject to TRI reporting in response to a court order.

December 2006: The Toxics Release Inventory Burden Reduction Final Rule was passed, expanding the eligibility of facilities to report under the simplified Form A. It expanded the reporting threshold for non-PBT chemicals and, to a lesser extent, PBT chemicals except dioxin and dioxin-like compounds.

May 2007: The requirement for reporting of dioxin and dioxin-like compounds was expanded to require reporting on the mass of each individual member of the dioxin category released in addition to the total mass of the entire category released. This expansion was implemented to allow the EPA to perform and publish toxic equivalency (TEQ) computations.

April 2009: The Toxics Release Inventory Burden Reduction Final Rule was reversed as required by the Omnibus Appropriations Act of 2009 enacted on March 11, 2009. This action eliminated the expanded use of Form A that was implemented in December 2006.

ROLE OF STAKEHOLDERS

Facilities are responsible for completing the reports and filing them with the EPA and the applicable state agency within the published deadlines. This can be a time-consuming process, even with the free software that the EPA provides for this purpose. Facilities are allowed to estimate releases, but this still involves considerable recordkeeping and calculation time. Once the report is filed, facilities must maintain copies of all documents for at least 3 years.

After the deadline, the EPA and state personnel must review the reports to ensure accuracy.

Once all the reporting is complete, the EPA posts the results online, within easy access of the public. The data can be searched by facility, area, chemical type, amount, and other variables.

FINANCIAL IMPACTS

Violators of the requirements of this regulation are liable for a civil penalty of up to $25,000 each day for each violation. Facilities are financially responsible for their own reporting requirements.

The cost to the EPA to process the Form A Certification Statements can be broken down into fixed costs and variable costs. Fixed costs include full-time employees (FTEs) and other recurring EPA costs. The variable component depends on the number of forms. It reflects total data-processing costs divided by the total number of paper reports processed in the 2002 reporting year. The EPA expends $26 in variable costs for each form processed. A total of 1,800 Form A Certification Statements were filed in the 2002 reporting year. Thus, the total annual burden to the EPA is estimated to be $46,800 in variable costs for the Form A Certification Statement. The cost to facilities is outlined in Table 3.1.

To estimate the EPA burden and cost to process the Form R Certification Statements, costs are separated into a fixed component and a variable component. Activities and expenses that are not greatly affected by marginal changes in numbers of reports are treated as fixed. These include rent for the EPCRA reporting center, development costs for data access tools, compliance assistance measures, and other activities and expenses. The variable component is the amount that varies depending on the number of forms. The variable component reflects total extramural data-processing costs divided by the total number of reports processed in the 2003 reporting year. For each form processed, $0.8 million in fixed costs and 26.3 FTEs are required to conduct the EPA activities described plus an additional $35 in variable costs for each. The cost to facilities, as estimated by the EPA, in both time and money is outlined in Tables 3.2 and 3.3.

LIMITATIONS OF TRI PROGRAM

TRI chemicals vary widely in toxicity or their potential to produce toxic effects. Publication of the data tends to center on total pounds released rather than toxic impact. As a result, some high-volume releases of less-toxic chemicals may appear to be more serious than lower-volume releases of highly toxic chemicals, when just the opposite may be true. This is confounded because of the large number of chemicals and complexity in taking toxicity into account.

The potential for exposure may be greater the longer the chemical remains unchanged in the environment. Sunlight, heat, or microorganisms may or may not decompose the chemical. Smaller releases of a persistent, highly toxic chemical may create a more serious problem than larger releases of a chemical that is rapidly converted to a less-toxic form.

TABLE 3.1
Form A Annual Burden and Cost per Facility (Assuming 1, 2, or 3 Chemicals)

Activity	Number of Chemicals Reported on Each Form A					
	1 Chemical		2 Chemicals		3 Chemicals	
	Hours	Cost	Hours	Cost	Hours	Cost
First-Year Filers						
Rule familiarization: first-time filers	34.5	$1,644	34.5	$1,644	34.5	$1,644
Calculations/certification: first-time filers	44.5	$2,081	89	$4,162	133.5	$6,243
Form A completion: first-time filers	1.6	$70	1.6	$70	1.6	$70
Recordkeeping/submission	3	$122	6	$244	9	$366
Total per facility	83.6	$3,917	131.1	$6,120	178.6	$8,323
Average per chemical	83.6	$3,917	65.5	$3,060	59.5	$2,774
Subsequent Year Filers						
Calculations/certification: subsequent year filers	16.2	$759	32.4	$1,518	48.6	$2,277
Form A completion: subsequent year filers	1.3	$58	1.3	$58	1.3	$58
Recordkeeping/submission	3	$122	6	$244	9	$366
Total per facility	20.5	$939	39.7	$1,820	58.9	$2,701
Average per chemical	20.5	$939	19.8	$910	19.7	$900

Source: EPA 2005b.

TABLE 3.2
Form R Average Annual Burden Hour Estimate by Activity

Category	Activity	Management	Technical	Clerical	Total Hours
Facility level	Compliance determination: all facilities	1	3	0	4
	Rule familiarization: first-time filers	12	22.5	0	34.5
	Supplier notification	0	7	17	24
Per Form R	Calculations and report completion: first-time filers, PBTs	20.3	43.9	2.7	66.8
	Calculations and report completion: first-time filers, non-PBTs	20.5	44.4	2.8	67.6
	Calculations and report completion: subsequent year filers, PBTs	14.1	30.4	1.9	46.3
	Calculations and report completion: subsequent year filers, non-PBTs	7.5	16.1	1.0	24.6
	Recordkeeping/submission: all filers	0	4	1	5

Source: EPA 2005b.

TABLE 3.3
Form R Average Annual Cost Estimate by Activity

Category	Activity	Management	Technical	Clerical	Total Cost
Facility level	Compliance determination: all facilities	$53	$134	$0	$187
	Rule familiarization: first-time filers	$637	$1,008	$0	$1,644
	Supplier notification	$0	$314	$402	$715
Per Form R	Calculations and report completion: first-time filers, PBTs	$1,076	$1,965	$63	$3,104
	Calculations and report completion: first-time filers, non-PBTs	$1,088	$1,986	$66	$3,140
	Calculations and report completion: subsequent year filers, PBTs	$748	$1,360	$44	$2,152
	Calculations and report completion: subsequent year filers, non-PBTs	$400	$720	$24	$1,144
	Recordkeeping/submission: all filers	$0	$179	$24	$203

Source: EPA 2005b.

Finally, the largest potential for exposure to the toxic chemicals, exposure to products when used by the consumer or end user, is exempt. Companies do not need to report releases in product. In fact, ironically, the TRI process can result in increased product exposure as one method of reducing releases is to ensure that more of the toxic chemical ends up in the product. Table 3.4 shows the releases of TRI chemicals for 2007.

TABLE 3.4
TRI Releases in 2007 (United States) (Pounds)

Chemical	Land	Water	Air	Totals
Zinc compounds	449,798,408	275,418,082	6,087,716	731,304,206
Hydrochloric acid	175,725	1,153,947	501,202,584	502,532,256
Lead compounds	311,541,457	162,591,053	802,709	474,935,219
Nitrate compounds	19,441,928	250,759,751	488,230	270,689,909
Barium compounds	160,130,216	74,427,623	2,035,867	236,593,707
Manganese compounds	158,365,170	59,674,346	1,790,399	219,829,915
Copper compounds	61,429,475	99,542,178	702,823	161,674,476
Ammonia	4,254,594	32,338,312	116,890,143	153,483,048
Methanol	2,313,184	19,476,031	129,347,741	151,136,956
Sulfuric acid	473,324	351,360	137,258,870	138,083,554
Arsenic compounds	7,707,484	86,564,733	106,957	94,379,174
Hydrogen fluoride	396,375	4,902,314	67,608,746	72,907,435
Zinc (fume or dust)	68,384,722	92,739	536,755	69,014,216
Chromium compounds	32,800,182	15,373,456	459,442	48,633,079

TABLE 3.4 (continued)
TRI Releases in 2007 (United States) (Pounds)

Chemical	Land	Water	Air	Totals
Vanadium compounds	24,798,974	18,877,669	600,897	44,277,540
Toluene	2,184,184	752,997	38,779,350	41,716,531
Styrene	2,847,810	224,410	37,676,447	40,748,666
Aluminum (fume or dust)	38,287,204	3,570	1,611,090	39,901,864
N-Hexane	544,148	36,307	34,397,735	34,978,189
Nickel compounds	17,121,445	11,927,666	639,983	29,689,094
Xylene (mixed isomers)	2,303,230	400,098	22,892,342	25,595,670
Manganese	23,460,768	1,667,961	394,704	25,523,433
Formaldehyde	462,233	12,224,204	9,247,247	21,933,684
Lead	18,745,918	1,968,780	225,646	20,940,344
Carbonyl sulfide	2,600	—	19,899,493	19,902,093
Ethylene	92	1,172	18,576,660	18,577,924
Certain glycol ethers	1,846,895	213,492	16,416,033	18,476,420
Nitric acid	3,796,000	12,761,347	1,437,032	17,994,379
Acetonitrile	274,750	17,207,770	459,450	17,941,970
Copper	13,413,973	962,700	542,441	14,919,114
Formic acid	2,113,897	11,117,309	703,160	13,934,367
N-Butyl alcohol	152,149	820,000	12,924,925	13,897,074
Chromium	8,841,743	2,771,841	155,826	11,769,410
Acetaldehyde	12,864	668,924	10,627,638	11,309,426
Propylene	18,541	1,939	11,235,456	11,255,935
Antimony compounds	4,426,132	6,550,800	36,433	11,013,365
Asbestos (friable)	10,430,282	—	99	10,430,381
Carbon disulfide	1,687	7,846	8,926,379	8,935,912
Benzene	484,614	2,468,305	5,512,448	8,465,367
Barium	7,929,271	52,194	261,902	8,243,367
Nickel	6,918,643	1,050,827	243,513	8,212,983
1,2,4-Trimethylbenzene	422,885	71,288	6,748,251	7,242,424
Acrylonitrile	10,436	6,603,291	446,109	7,059,836
Cyanide compounds	756,100	5,688,660	426,779	6,871,538
Mercury compounds	5,371,841	1,341,220	120,578	6,833,638
Chlorodifluoromethane	107,102	2,798	6,572,634	6,682,534
Phenol	1,619,180	1,144,353	3,914,234	6,677,767
Ethylene glycol	2,974,388	1,216,778	2,423,600	6,614,766
Acrylamide	11,551	6,137,905	11,791	6,161,247
Cobalt compounds	3,892,790	2,146,821	55,390	6,095,001
1-Chloro-1,1-difluoroethane	4,534	1,372	6,039,136	6,045,042
Dichloromethane	576,532	78,616	5,248,093	5,903,242
Chlorine	247,549	297,999	5,097,675	5,643,223
Cyclohexane	447,355	196,636	4,639,355	5,283,345
Ethylbenzene	277,547	872,841	3,692,714	4,843,102
Methyl isobutyl ketone	136,705	35,656	4,645,854	4,818,216

continued

TABLE 3.4 (continued)
TRI Releases in 2007 (United States) (Pounds)

Chemical	Land	Water	Air	Totals
Biphenyl	4,414,473	23,944	362,286	4,800,703
N-Methyl-2-pyrrolidone	756,438	2,226,079	1,730,576	4,713,092
Aluminum oxide (fibrous forms)	4,624,919	9,817	51,624	4,686,360
Acrylic acid	91,498	4,249,341	284,170	4,625,008
Trichloroethylene	71,948	54,944	4,358,309	4,485,202
Sodium nitrite	771,787	3,452,157	96,297	4,320,241
Arsenic	3,093,087	107,946	953	3,201,986
Cadmium compounds	2,330,300	632,113	9,467	2,971,879
Naphthalene	1,081,717	175,223	1,593,938	2,850,878
Selenium compounds	1,556,457	719,586	567,189	2,843,232
Methyl methacrylate	103,072	251,992	2,378,307	2,733,371
Cyclohexanol	258	2,469,045	147,231	2,616,534
N,N-Dimethylformamide	1,758,916	371,583	301,408	2,431,907
Hydrogen cyanide	814	1,269,170	1,039,877	2,309,862
Tetrachloroethylene	346,298	121,577	1,769,989	2,237,864
Vanadium	1,987,403	48,723	81,043	2,117,170
Vinyl acetate	27,410	67,019	2,013,725	2,108,154
Polychlorinated biphenyls	2,090,003	178	189	2,090,371
Chloromethane	449	238,268	1,776,100	2,014,817
Creosote	1,604,695	2,152	343,795	1,950,641
Thallium compounds	862,748	1,031,134	5,112	1,898,994
1,3-Butadiene	4,511	1,061	1,782,512	1,788,084
Triethylamine	37,415	1,278,738	428,432	1,744,585
Acrolein	—	1,533,883	162,993	1,696,876
Diethanolamine	883,679	495,808	184,103	1,563,591
Diisocyanates	1,283,713	40	188,700	1,472,453
Selenium	1,413,581	1,746	28,385	1,443,711
Molybdenum trioxide	1,017,199	289,174	111,336	1,417,709
Polycyclic aromatic compounds	907,994	16,028	441,362	1,365,384
Antimony	1,298,716	6,704	7,269	1,312,688
P-Xylene	6,182	41	1,305,316	1,311,540
Cresol (mixed isomers)	13,743	474,403	752,142	1,240,288
4,4′-Isopropylidenediphenol	1,001,447	7,546	122,920	1,131,912
Di(2-Ethylhexyl) phthalate	965,435	9,198	151,937	1,126,569
Cumene	44,069	6,137	1,056,549	1,106,755
Acetamide	259	1,095,619	1,616	1,097,494
Cadmium	815,821	118,488	1,202	935,512
Aniline	5,053	778,970	136,583	920,606
Cobalt	812,185	22,056	41,051	875,292
Beryllium compounds	557,285	290,581	5,027	852,894
Nicotine and salts	519,725	552	323,792	844,069
tert-Butyl alcohol	43,907	380,375	364,918	789,200
sec-Butyl alcohol	6,953	70,270	700,628	777,850

TABLE 3.4 (continued)
TRI Releases in 2007 (United States) (Pounds)

Chemical	Land	Water	Air	Totals
Chloroprene	343	20,082	721,428	741,853
Acetophenone	48,530	570,396	108,008	726,934
Chloroethane	351	12,946	702,506	715,803
Maleic anhydride	294,774	39,508	372,782	707,065
Chloroform	83,836	23,440	599,279	706,555
Ozone	—	—	704,712	704,712
Pyridine	3,111	666,750	32,506	702,367
Decabromodiphenyl oxide	650,475	1,690	43,696	695,861
Nitrobenzene	5,099	571,709	24,312	601,120
Methyl tert-butyl ether	59,349	4,281	521,835	585,465
Silver compounds	458,020	103,193	10,722	571,936
Freon 113	263	1,229	563,990	565,482
2-Chloro-1,1,1,2-tetrafluoroethane	—	8,606	542,752	551,358
Chlorobenzene	29,551	60,994	455,163	545,708
Chlorine dioxide	27,907	—	517,384	545,291
Allyl alcohol	516	496,150	29,550	526,216
Atrazine	501,658	50	13,364	515,072
Butyraldehyde	84	115,477	395,006	510,567
1,2-Dichloroethane	4,549	111,282	334,022	449,853
Hydroquinone	14,710	399,307	16,971	430,989
Bromine	9,003	404	401,075	410,482
Mixture	119,124	167	264,014	383,305
Phenanthrene	319,337	1,579	57,640	378,556
Vinyl chloride	29,824	81	342,729	372,635
Bromomethane	1,266	8,161	354,144	363,571
Dimethylamine	1,646	246,156	114,722	362,524
Tetrabromobisphenol A	288,078	8	55,187	343,273
Silver	336,893	783	2,184	339,861
Propylene oxide	2,301	9,599	326,698	338,598
Dibutyl phthalate	20,132	262,516	52,839	335,487
Propionaldehyde	13,225	97,531	223,682	334,438
1,1-Dichloro-1-fluoroethane	11	92	327,218	327,321
O-Xylene	2,628	13,310	309,558	325,496
Trichlorofluoromethane	26,398	53	291,600	318,052
M-Xylene	3,419	32	314,289	317,740
Dichlorotetrafluoroethane (CFC-114)	91	—	312,780	312,871
Carbon tetrachloride	94,023	48,868	165,466	308,357
Ethylene oxide	3,448	17,277	285,235	305,961
Benzoyl peroxide	295,426	250	2,733	298,409
Dimethyl phthalate	32,348	2,010	252,958	287,316
Phosphorus (yellow or white)	232,577	5,952	38,465	276,993
Phthalic anhydride	32,436	864	235,697	268,997

continued

TABLE 3.4 (continued)
TRI Releases in 2007 (United States) (Pounds)

Chemical	Land	Water	Air	Totals
Dichlorodifluoromethane	—	5	268,440	268,445
3-Iodo-2-propynyl butylcarbamate	259,710	—	1,474	261,184
Thallium	226,748	—	27	226,775
M-Cresol	513	163,793	28,467	192,773
Methyl acrylate	500	7,742	178,404	186,646
1,4-Dioxane	2,796	56,996	125,341	185,132
Dicyclopentadiene	1,039	52,123	121,656	174,818
Anthracene	158,670	833	12,394	171,896
Sodium dimethyldithiocarbamate	136,408	35,000	63	171,471
Fluorine	1,440	97,777	69,627	168,844
Thiram	100,828	385	65,606	166,819
2-Methoxyethanol	97,911	36,153	22,769	156,833
Epichlorohydrin	3,617	20,917	131,279	155,813
Chlorothalonil	147,115	2,192	4,881	154,189
2,2-Dichloro-1,1,1-trifluoroethane	200	18,945	127,240	146,385
Isobutyraldehyde	79	19	141,834	141,932
Lithium carbonate	129,627	2,119	5,832	137,578
1,2-Dichloro-1,1,2-trifluoroethane	—	—	137,476	137,476
Butyl acrylate	23,920	21	113,227	137,168
Nitroglycerin	12,030	3	118,405	130,438
2-Methyllactonitrile	—	95,332	30,837	126,169
Titanium tetrachloride	75,692	—	47,854	123,546
2-Mercaptobenzothiazole	98,393	21,018	2,208	121,620
1,2-Dichloropropane	257	4,871	110,583	115,710
1,3-Phenylenediamine	112,291	0	2,870	115,161
Sulfuryl fluoride	0	0	112,245	112,245
P-Cresol	600	82,023	22,234	104,857
Mercury	88,356	3,124	10,504	101,984
Benzo(G,H,I)perylene	75,411	2,268	20,607	98,286
Urethane	90,045	—	4,765	94,810
Polychlorinated alkanes	92,385	9	2,024	94,418
Methacrylonitrile	—	92,257	10	92,267
1,2-Dichlorobenzene	20,258	31,957	37,904	90,119
Cumene hydroperoxide	14,960	42,145	30,588	87,693
Pendimethalin	84,638	27	1,671	86,336
2-Ethoxyethanol	12,185	10,275	59,959	82,419
1,1,1-Trichloroethane	24,112	69	57,298	81,479
1,4-Dichlorobenzene	736	291	78,239	79,266
Ethyl acrylate	7,379	265	70,492	78,135
Fomesafen	69,111	—	4	69,115
4,4′-Methylenedianiline	44	59,161	8,218	67,423
Nitrilotriacetic acid	61,470	2,480	21	63,971
Propiconazole	59,896	—	3	59,899

TABLE 3.4 (continued)
TRI Releases in 2007 (United States) (Pounds)

Chemical	Land	Water	Air	Totals
Vinylidene chloride	1,003	847	55,163	57,013
Dichlorotrifluoroethane	53,391	—	70	53,461
Catechol	2,032	47,547	2,425	52,005
Diphenylamine	18,163	1,368	32,145	51,677
Toluene diisocyanate (mixed isomers)	23,306	—	26,720	50,026
Chlorophenols	1,579	47,129	488	49,196
2-Chloro-1,1,1-trifluoroethane	136	22	48,523	48,681
Ethoprop	48,151	—	160	48,311
Picric acid	102	44,902	11	45,015
Hexachlorobenzene	24,959	17,724	335	43,018
P-Phenylenediamine	2,587	32,873	5,840	41,300
Dicamba	159	39,005	112	39,276
2-Methylpyridine	15	27,560	11,563	39,138
2,4-Dimethylphenol	567	22,154	15,711	38,433
Allyl chloride	8,920	60	26,208	35,188
Nabam	—	35,000	54	35,054
Metham sodium	32,491	—	2,551	35,042
2-Nitrophenol	378	32,834	20	33,232
Ethylenebisdithiocarbamic acid, salts, and esters	500	—	30,777	31,277
Iron pentacarbonyl	—	—	30,204	30,204
Thiabendazole	29,524	—	502	30,026
2-Nitropropane	2,721	—	25,850	28,571
Pentachloroethane	501	—	27,012	27,513
Methyl iodide	139	881	25,725	26,745
1,2,4-Trichlorobenzene	758	3,700	21,881	26,339
Propargyl alcohol	12	25,438	486	25,936
Diaminotoluene (mixed isomers)	13,132	8,054	4,368	25,554
Peracetic acid	14,742	—	8,283	23,025
1,1,2-Trichloroethane	704	8	21,655	22,367
Monochloropentafluoroethane	142	—	21,287	21,429
2,4-D	18,827	308	1,871	21,006
O-Cresol	707	8,409	11,002	20,118
Toluene-2,4-diisocyanate	9,145	331	9,479	18,955
2,4-Diaminotoluene	17,000	5	1,215	18,220
Dibenzofuran	10,107	4	7,329	17,440
Diuron	17,037	5	327	17,369
Hydrazine	550	14,851	1,358	16,759
Metribuzin	16,585	88	45	16,718
3,3-Dichloro-1,1,1,2,2-pentafluoropropane	1,800	—	14,697	16,497
O-Toluidine	126	785	15,437	16,348
Sodium azide	16,089	—	11	16,100

continued

TABLE 3.4 (continued)
TRI Releases in 2007 (United States) (Pounds)

Chemical	Land	Water	Air	Totals
Quinoline	1,489	12,972	1,364	15,825
Aluminum phosphide	15,235	8	225	15,468
Phosgene	—	—	15,290	15,290
M-Dinitrobenzene	7,310	—	7,350	14,660
1,2-Dichloro-1,1-difluoroethane	30	8	14,594	14,632
Tetramethrin	—	—	14,616	14,616
Beryllium	13,323	665	197	14,185
2,4-Dinitrotoluene	11,121	150	2,323	13,594
2,4-Dichlorophenol	10	9,031	4,500	13,541
Boron trifluoride	8,753	—	4,638	13,391
Benzyl chloride	162	447	12,714	13,323
1,2-Phenylenediamine	10,330	—	2,519	12,849
Bromotrifluoromethane	—	—	11,682	11,682
Dimethoate	1,098	5	10,419	11,522
Dichlorobenzene (mixed isomers)	1,084	—	8,871	9,955
Sodium dicamba	500	5	9,036	9,541
N-Methylolacrylamide	2,324	510	6,442	9,276
Potassium dimethyldithiocarbamate	—	—	8,956	8,956
Trade secret chemical	7,525	1,000	339	8,864
1-Chloro-1,1,2,2-tetrafluoroethane	—	—	8,626	8,626
Malathion	1,984	5	6,568	8,557
Chloroacetic acid	1,786	—	6,572	8,358
Dichloropentafluoropropane	180	—	8,000	8,180
Dinitrobutyl phenol	127	7,354	651	8,132
Trifluralin	6,628	—	667	7,295
Ethylidene dichloride	502	11	6,740	7,253
3-Chloro-2-methyl-1-propene	—	—	6,536	6,536
Chlordane	6,270	22	61	6,353
4,4′-Methylenebis(2-chloroaniline)	2,325	—	3,908	6,233
Allylamine	250	1	5,843	6,094
Hexachlorocyclopentadiene	5,043	24	923	5,990
1,4-Dichloro-2-butene	10	4,721	996	5,727
2-Phenylphenol	5,715	—	—	5,715
1,3-Dichloropropylene	271	1	5,423	5,695
Diethyl sulfate	—	—	5,346	5,346
Dichlorofluoromethane	7	—	5,035	5,042
Methylene bromide	500	1,147	3,351	4,998
1,2-dichloroethylene	357	1	4,041	4,399
Benzoyl chloride	948	—	3,369	4,317
1,2-Dibromoethane	256	51	3,929	4,236
Dinitrotoluene (mixed isomers)	3,805	—	298	4,103
Crotonaldehyde	255	2,664	1,089	4,008
2,3-Dichloropropene	—	—	3,812	3,812

TABLE 3.4 (continued)
TRI Releases in 2007 (United States) (Pounds)

Chemical	Land	Water	Air	Totals
Toluene-2,6-diisocyanate	1,614	83	2,098	3,795
Oxydiazon	3,424	—	255	3,679
Chloromethyl methyl ether	—	—	3,600	3,600
Bromochlorodifluoromethane	—	—	3,491	3,491
Quintozene	2,883	7	224	3,115
Chloropicrin	—	—	3,081	3,081
Folpet	3,013	—	34	3,047
Methyl parathion	896	250	1,619	2,765
Pentachlorophenol	2,229	426	85	2,740
4,4′-Diaminodiphenyl ether	2,149	214	345	2,708
Dimethyl sulfate	500	—	2,126	2,626
Boron trichloride	2,422	—	70	2,492
1,3-Dichloro-1,1,2,2,3-pentafluoropropane	1,800	—	500	2,300
1,1,1,2-Tetrachloroethane	448	—	1,801	2,249
Norflurazon	—	2,206	—	2,206
Diazinon	1,727	5	462	2,194
Bis(tributyltin) oxide	2,000	1	—	2,001
Ethylene thiourea	1,853	—	92	1,945
1-(3-Chloroallyl)-3,5,7-triaza-1-azoniaadamantane chloride	1,662	—	270	1,932
1,1,2,2-Tetrachloroethane	463	—	1,398	1,861
1,2-Butylene oxide	—	—	1,828	1,828
1,3-Dichlorobenzene	255	698	874	1,827
Dimethylamine dicamba	—	1,100	683	1,783
Hexachloroethane	1,051	105	595	1,751
Acephate	750	—	986	1,736
Captan	1,122	10	590	1,722
Dichlorvos	1,326	—	389	1,715
C.I. direct blue 218	1,659	—	—	1,659
3,3′-Dichlorobenzidine dihydrochloride	1,563	—	2	1,565
Lindane	1,373	—	182	1,555
N,N-Dimethylaniline	90	941	515	1,546
Propyleneimine	—	—	1,482	1,482
1,2,3-Trichloropropane	57	291	1,126	1,474
Pentachlorobenzene	1,247	2	215	1,464
2,6-Dinitrotoluene	1,277	59	7	1,343
Thiourea	320	500	513	1,333
O-Dinitrobenzene	1,234	12	26	1,272
P-Dinitrobenzene	1,234	12	26	1,272
Propanil	750	—	511	1,261
Methyl isocyanate	250	—	1,009	1,259
Toxaphene	1,167	20	25	1,212

continued

TABLE 3.4 (continued)
TRI Releases in 2007 (United States) (Pounds)

Chemical	Land	Water	Air	Totals
Dazomet	949	179	70	1,198
2,4-d 2-ethylhexyl ester	74	2	1,083	1,158
Heptachlor	1,105	20	8	1,133
Aldrin	1,092	—	37	1,128
Methoxychlor	1,024	—	26	1,050
Safrole	500	—	500	1,000
Octachlorostyrene	815	174	—	989
Hexachloro-1,3-butadiene	512	5	417	934
Thiodicarb	—	5	885	890
Carbaryl	275	23	549	847
Bis(2-Chloro-1-methylethyl) ether	70	—	718	788
P-Chloroaniline	500	—	261	761
2,4-Dinitrophenol	424	317	16	757
N-Nitrosodi-N-propylamine	500	—	251	751
2-Acetylaminofluorene	500	—	250	750
Benzal chloride	500	—	208	708
Warfarin and salts	602	—	100	702
Hexachlorophene	609	—	81	690
Hexazinone	—	396	256	652
Benzoic trichloride	10	—	636	646
Permethrin	303	—	341	644
O-Anisidine	—	250	388	638
Tetracycline hydrochloride	631	—	—	631
Tribenuron methyl	58	—	568	626
Saccharin	550	—	51	601
Pronamide	587	—	11	598
Thiophanate-methyl	515	—	13	528
2,4,6-Trichlorophenol	255	26	232	513
4-Nitrophenol	135	368	9	512
Dihydrosafrole	10	—	500	510
Bis(2-Chloroethoxy)methane	500	—	—	500
2,4,5-Trichlorophenol	500	—	—	500
N-Nitrosopiperidine	500	—	—	500
Dipotassium endothall	—	—	500	500
Fluorouracil	500	—	—	500
N-Nitrosodi-N-butylamine	500	—	—	500
N-Nitrosodiethylamine	500	—	—	500
Simazine	393	12	86	491
Styrene oxide	380	1	85	466
Chlorotrifluoromethane	—	—	415	415
Carbofuran	221	—	170	391
trans-1,3-Dichloropropene	—	—	389	389
Picloram	39	4	336	379

TABLE 3.4 (continued)
TRI Releases in 2007 (United States) (Pounds)

Chemical	Land	Water	Air	Totals
Alachlor	—	4	369	373
Aldicarb	239	—	133	372
Potassium N-methyldithiocarbamate	88	—	270	358
Tetrachlorvinphos	—	—	355	355
Bis(2-Chloroethyl) ether	255	2	90	347
4-Aminoazobenzene	—	335	—	335
2,4-D Butoxyethyl ester	—	—	327	327
Ethyl chloroformate	—	—	325	325
Paraldehyde	215	—	108	323
P-Nitroaniline	255	—	66	321
Dioxin and dioxin-like compounds	303	13	3	319
Ametryn	—	11	302	313
Methyl chlorocarbonate	—	—	301	301
Dichlorobromomethane	—	—	296	296
Hydramethylnon	251	—	23	274
Mecoprop	106	—	161	267
Dimethylcarbamyl chloride	255	—	5	260
Tris(2,3-Dibromopropyl) phosphate	10	—	250	260
P-Cresidine	—	250	10	260
Propane sultone	10	—	250	260
Tebuthiuron	250	—	10	260
4-Dimethylaminoazobenzene	6	—	250	256
1,2-Dibromo-3-chloropropane	255	—	—	255
Isosafrole	255	—	—	255
Diallate	255	—	—	255
5-Nitro-O-toluidine	255	—	—	255
3,3′-Dimethoxybenzidine dihydrochloride	—	—	255	255
Prometryn	—	—	255	255
Potassium bromate	—	—	250	250
Trichlorfon	—	—	250	250
Oxyfluorfen	224	—	—	224
Myclobutanil	194	—	18	212
Paraquat dichloride	136	—	66	202
Bromoform	—	—	191	191
Cyanazine	173	—	16	189
Benfluralin	—	—	175	175
Linuron	128	—	14	142
Dimethipin	—	—	139	139
Propoxur	114	—	15	129
Calcium cyanamide	88	—	39	127
Amitrole	113	—	10	123

continued

TABLE 3.4 (continued)
TRI Releases in 2007 (United States) (Pounds)

Chemical	Land	Water	Air	Totals
Methoxone	109	—	13	122
Dimethyl chlorothiophosphate	—	—	110	110
4,6-Dinitro-O-cresol	103	—	1	104
Ferbam	92	—	9	101
Chlorendic acid	88	—	8	96
Parathion	81	—	8	89
Methazole	—	—	60	60
Desmedipham	—	—	51	51
Ethyl dipropylthiocarbamate	—	—	50	50
Propargite	—	—	48	48
Chlorsulfuron	—	—	42	42
Phenytoin	—	—	40	40
Triallate	—	—	38	38
Cyfluthrin	—	—	34	34
Chlorobenzilate	10	—	21	32
Bromoxynil octanoate	—	—	27	27
Isodrin	20	—	2	22
Phenothrin	—	—	20	20
Chlorimuron ethyl	—	—	19	19
Benzidine	10	—	6	16
Chlorotetrafluoroethane	—	—	15	15
1,1-Dimethyl hydrazine	10	—	5	15
Carboxin	—	—	14	14
Bromoxynil	—	—	13	13
Chlorpyrifos methyl	—	—	12	12
4-Aminobiphenyl	—	10	1	11
Abamectin	—	7	3	10
O-Toluidine hydrochloride	10	—	—	10
3,3′-Dimethylbenzidine	10	—	—	10
1,2-Diphenylhydrazine	10	—	—	10
3,3′-Dimethoxybenzidine	10	—	—	10
Malononitrile	10	—	—	10
N-Nitroso-N-ethylurea	10	—	—	10
N-Nitroso-N-methylurea	10	—	—	10
N-Nitrosodiphenylamine	—	—	10	10
Naled	—	—	10	10
Quinone	10	—	—	10
Thioacetamide	10	—	—	10
Triclopyr triethylammonium salt	—	—	10	10
Trypan blue	10	—	—	10
2,4-D Sodium salt	—	—	9	9
Bromacil	—	—	8	8
Trichloroacetyl chloride	—	—	6	6

TABLE 3.4 (continued)
TRI Releases in 2007 (United States) (Pounds)

Chemical	Land	Water	Air	Totals
1,1,1,2-Tetrachloro-2-fluoroethane	—	—	6	6
Phosphine	—	—	5	5
1,1,2,2-Tetrachloro-1-fluoroethane	—	5	—	5
Lactofen	—	—	5	5
S,S,S-Tributyltrithiophosphate	—	2	3	5
Bifenthrin	—	—	4	4
Ethyleneimine	—	—	4	4
Pentobarbital sodium	—	—	4	4
Merphos	—	—	3	3
Nitrapyrin	—	—	2	2
2,6-Xylidine	—	—	1	1
Diglycidyl resorcinol ether	—	—	1	1
Fenoxycarb	—	—	1	1
Fluometuron	—	—	1	1
Piperonyl butoxide	—	—	1	1
Propachlor	—	—	1	1
Strychnine and salts	—	—	1	1
Trans-1,4-Dichloro-2-butene	—	—	1	1
1-Bromo-1-(bromomethyl)-1,3-propanedicarbonitrile	—	—	—	—
2,4-D Isopropyl ester	—	—	—	—
2,4-Db	—	—	—	—
Acifluorfen, sodium salt	—	—	—	—
Alpha-Naphthylamine	—	—	—	—
Amitraz	—	—	—	—
Bis(Chloromethyl) ether	—	—	—	—
C.I. basic green 4	—	—	—	—
C.I. solvent orange 7	—	—	—	—
Cupferron	—	—	—	—
Cyhalothrin	—	—	—	—
D-trans-Allethrin	—	—	—	—
Dazomet, sodium salt	—	—	—	—
Dibromotetrafluoroethane	—	—	—	—
Dichloran	—	—	—	—
Diflubenzuron	—	—	—	—
Disodium cyanodithioimidocarbonate	—	—	—	—
Fenarimol	—	—	—	—
Fenbutatin oxide	—	—	—	—
Fenpropathrin	—	—	—	—
Fluazifop butyl	—	—	—	—
Maneb	—	—	—	—
Methiocarb	—	—	—	—

continued

TABLE 3.4 (continued)
TRI Releases in 2007 (United States) (Pounds)

Chemical	Land	Water	Air	Totals
Methoxone sodium salt	—	—	—	—
Methyl hydrazine	—	—	—	—
Methyl isothiocyanate	—	—	—	—
Totals	1,531,614,787	1,242,969,132	1,311,649,055	4,086,232,974

Source: EPA 2005.

BIBLIOGRAPHY

Eckerman, I. 2004. *The Bhopal Saga—Causes and Consequences of the World's Largest Industrial Disaster*. Hyderabad, IN: India: Universities Press.

U.S. Environmental Protection Agency. 2004. Public Data Release Brochure. http://www.epa.gov/tri/tridata/tri04/brochure/brochure.htm (accessed November 9, 2006).

U.S. Environmental Protection Agency. 2005a. *Toxic Chemical Release Inventory: Alternate Threshold for Low Annual Reportable Amounts; Toxic Chemical Release Reporting Information Collection Request Supporting Statement.* OMB Control No. 2070–0143, Washington, DC: EPA ICR No. 1704.08.

U.S. Environmental Protection Agency. 2005b. *Toxic Chemical Release Inventory: Toxic Chemical Release Reporting Information Collection Request Supporting Statement.* OMB Control No. 2070–0093. Washington, DC: EPA ICR No. 1363.14.

U.S. Environmental Protection Agency. 2005c. Toxic Chemical Release Reporting: Community Right to Know. Code of Federal Regulations, Title 40, Chapter 1, Part 372. http://www.access.gpo.gov/nara/cfr/waisidx_06/40cfr372_06.html (accessed November 9, 2006).

U.S. Environmental Protection Agency. 2006. *Toxic Chemical Release Inventory Reporting Forms and Instructions. Section 313 of the Emergency Planning and Community Right-to-Know Act (Title III of the Superfund Amendments and Reauthorization Act of 1986).* Washington, DC: EPA 260-B-06-001.

U.S. Environmental Protection Agency. 2009a. Making Year-to-Year Comparisons of TRI Data. http://www.epa.gov/triexplorer/yearsum.htm (accessed November 13, 2009).

U.S. Environmental Protection Agency. 2009b. TRI Historical Archive: 1980s. http://www.epa.gov/tri/archive/1980s.html (accessed November 13, 2009).

U.S. Environmental Protection Agency. 2009c. TRI Historical Archive: 1990s. http://www.epa.gov/tri/archive/1990s.html (accessed November 13, 2009).

U.S. Environmental Protection Agency. 2009d. TRI Historical Archive: 2000s. http://www.epa.gov/tri/archive/2000s.html (accessed November 13, 2009).

U.S. Environmental Protection Agency. 2009e. TRI Total Release Trend 1988–2004. http://www.epa.gov/triexplorer/ (accessed January 20, 2010).

U.S. Environmental Protection Agency. *The Toxics Release Inventory (TRI) and Factors to Consider When Using TRI Data.* May 3, 2006. Washington, DC.

U.S. Environmental Protection Agency. Toxic Release Inventory (TRI) Program Homepage. http://www.epa.gov/tri/index.htm (accessed November 9, 2006).

U.S. Environmental Protection Agency. TRI Explorer (version 4.5). http://www.epa.gov/triexplorer/ (accessed November 9, 2006).

4 TRI Programs in Other Countries

INTRODUCTION

Since the first Toxics Release Inventory (TRI) program was established in the United States, pollutant release and transfer register (PRTR) programs have been established in approximately 30 countries (Figure 4.1). A PRTR is an inventory of potentially toxic or hazardous chemicals that are released to the air, water, or land or transferred off site for treatment or disposal. Normally, industrial facilities quantify their releases to the environment as part of the program. In some cases, fugitive or diffuse sources such as agriculture or motor vehicles are also included in the inventory.

In 1993, the Organization for Economic Cooperation and Development (OECD), an intergovernmental organization, began work to encourage the development of PRTRs. OECD works with governments, industry, and nongovernmental organizations to develop practical tools that facilitate efforts by member countries, provide outreach to nonmember countries, and coordinate international activities.

OECD produces documents describing the experiences of countries that have developed PRTRs, current and emerging uses of PRTR data, how PRTRs differ, and the identification, selection, and adaptation of release estimation techniques that industry can use to calculate pollutant releases and transfers. The OECD coordinates PRTR activities between the industrialized nations of Europe, North America, and Asia-Pacific through its PRTR Task Force. The goal of the task force is to enable the OECD member countries to provide and improve information about the implementation of PRTRs.

FIGURE 4.1 Countries with active PRTR programs. (From Organization for Economic Co-operation and Development, Task Force on Pollutant Release and Transfer Registers, 2008. http://www.prtr.net/links_e.cfm.)

According to the OECD Council Recommendation [C(96)41/FINAL], as amended by [C(2003)87], the core elements of a PRTR system are

1. A listing of chemicals, groups of chemicals, and if appropriate, other relevant categories, all of which are pollutants when released or transferred
2. Integrated multimedia reporting of releases and transfers (air, water, and land)
3. Reporting of data by source if the reporting sources are defined
4. Reporting on a periodic basis, preferably annually
5. Making data available to the public

The following principles are considered when establishing a PRTR program:

1. PRTR systems should provide data to support the identification and assessment of possible risks to humans and the environment.
2. The PRTR data should be used to promote prevention of pollution at the source and help national governments evaluate the progress of environmental policies and goals.
3. Government and interested parties should cooperate to develop a set of goals and objectives for the system and estimate potential benefits and costs.
4. PRTR systems should include coverage of an appropriate number of substances.
5. Public and private sectors should be included.
6. PRTR systems should be integrated to the degree practicable with existing information sources, such as licenses or operating permits.
7. Both voluntary and mandatory reporting mechanisms should be considered to meet the goals and objectives of the system.
8. The comprehensiveness of a PRTR in helping to meet environmental policy goals should be taken into account (e.g., fugitive/diffuse sources).
9. The results should be made accessible to all affected and interested parties on a timely and regular basis.
10. The program should allow for midcourse evaluation and flexibility to alter the program in response to changing needs.
11. The system should allow for verification of inputs and outputs and be capable of identifying the geographical distribution of releases and transfers.
12. The program should allow, as far as possible, comparison and cooperation with other national PRTR systems and possible harmonization with similar international databases.
13. A compliance mechanism should be agreed upon by affected and interested parties to best meet the needs of the goals and objectives.
14. The process of establishing the PRTR system and its implementation and operation should be transparent and objective.

The Protocol on Pollutant Release and Transfer Registers has been signed by at least 38 member states and the European Community.

Countries having PRTR programs include Australia, Austria, Belgium, Canada, Chile, Cyprus, Czech Republic, Denmark, Estonia, Finland, France, Germany,

Greece, Hungary, Italy, Ireland, Japan, Latvia, Lithuania, Luxembourg, Malta, Mexico, Netherlands, Norway, Poland, Portugal, Slovakia, Slovenia, Spain, Sweden, Switzerland, the United Kingdom, and the United States (U.S. Environmental Protection Agency 2009; OECD 2009).

What follows is a description of several existing PRTR programs.

CANADIAN NATIONAL POLLUTANT RELEASE INVENTORY PROGRAM

The National Pollutant Release Inventory (NPRI) is Canada's legislated, publicly accessible inventory of pollutant releases and transfers. It comprises information reported by facilities to Environment Canada under the Canadian Environmental Protection Act, 1999 (CEPA 1999), together with air pollutant emission estimates compiled for facilities not required to report and nonindustrial sources such as motor vehicles, residential heating, forest fires, and agriculture.

The NPRI is Canada's principal effort for tracking and public reporting of releases of toxic substances and other substances of concern. It is used to identify and monitor sources of pollution in Canada as well as to develop indicators for the quality of air, water, and land. Information collected through the NPRI is used by Environment Canada in its chemicals management programs, and it is made publicly available to Canadians each year. Public access to the NPRI motivates industry to prevent and reduce pollutant releases. NPRI data help the government of Canada to track progress in pollution prevention, evaluate releases and transfers of substances of concern, identify and take action on environmental priorities, conduct air quality modeling, and implement policy initiatives and risk management measures.

The first report by Canada's NPRI was released in 1995 and contained pollutant release and transfer information reported for 1993. In 2007, this list consisted of 347 uniquely listed substances or substance groups, compared with 178 when the NPRI program was established.

The NPRI collects information only from industrial, commercial, institutional, and other facilities that meet reporting requirements. These reporting requirements are based on the number of employees at the facility; the quantity of the substances manufactured, processed, used, or released; and the type of activities performed at the facility. For the 2007 reporting year, over 8,500 industrial, commercial, and other facilities reported to the NPRI on their releases, disposals, and transfers for recycling of approximately 10 billion pounds of toxic substances and other substances of concern (Environment Canada 2009b). On-site releases for 2008 are shown in Table 4.1.

AUSTRALIAN NATIONAL POLLUTANT INVENTORY

The Australian National Pollutant Inventory (NPI) is a database of emissions managed by the Australian government. The stated purpose of the NPI is to maintain and improve air and water quality, minimize environmental impacts associated with hazardous waste, and improve the sustainable use of resources.

TABLE 4.1
Canadian On-Site Releases for 2008

Substance Name	2008 On-Site Releases (Tonnes Unless Noted)
1,1,1,2-Tetrachloroethane	0.001
1,1,2,2-Tetrachloroethane	0.001
1,1,2-Trichloroethane	0.125
1,1-Methylenebis(4-isocyanatocyclohexane)	0.001
1,2,4-Trichlorobenzene	0.714
1,2,4-Trimethylbenzene	1,093
1,2,4-Trimethylbenzene	1,094
1,2-Dichloroethane	0.67
1,2-Dichloroethane	5.1
1,2-Dichloropropane	0.001
1,3-Butadiene	40
1,3-Butadiene	41
1,4-Dioxane	27
1-Nitropyrene	0.289
2,6-Di-t-butyl-4-methylphenol	0.031
2-Butoxyethanol	504
2-Butoxyethanol	596
2-Ethoxyethanol	0.04
2-Ethoxyethyl acetate	1.3
2-Methoxyethanol	0.061
3-Methylcholanthrene	1.7
5-Methylchrysene	1.6
7,12-Dimethylbenz(a)anthracene	0.783
7H-Dibenzo(c,g)carbazole	831
Acenaphthene	9,160
Acenaphthylene	27,593
Acetaldehyde	1,044
Acetonitrile	63
Acetophenone	2.5
Acetylene	148
Acrolein	108
Acrylamide	0.024
Acrylic acid (and its salts)	0.098
Acrylonitrile	10
Adipic acid	26
Alkanes, C10-13, chloro	18
Allyl alcohol	2.6
alpha-Pinene	1,910
Aluminum (fume or dust)	671
Aluminum oxide (fibrous forms)	2.8
Ammonia (total)	69,709
Aniline (and its salts)	0.065

TABLE 4.1 (continued)
Canadian On-Site Releases for 2008

Substance Name	2008 On-Site Releases (Tonnes Unless Noted)
Aniline (and its salts)	0.073
Anthracene	13
Antimony (and its compounds)	9.2
Asbestos (friable form)	715
Benzene	754
Benzene	773
Benzo(a)anthracene	12,971
Benzo(a)phenanthrene	27,767
Benzo(a)pyrene	10,214
Benzo(b)fluoranthene	19,633
Benzo(e)pyrene	15,267
Benzo(g,h,i)perylene	8,910
Benzo(j)fluoranthene	3,788
Benzo(k)fluoranthene	6,365
beta-Phellandrene	1,278
beta-Pinene	1,186
Biphenyl	3.8
Bis(2-ethylhexyl)adipate	1.4
Bis(2-ethylhexyl)phthalate	6.6
Bromine	10
Bromomethane	0.001
Butane (all isomers)	16,898
Butene (all isomers)	1,291
Butyl acrylate	1.7
Butyl benzyl phthalate	5.9
Butyraldehyde	10
C.I. basic red 1	0.001
C.I. food red 15	0.001
Calcium fluoride	76
Carbon disulfide	2,827
Carbon tetrachloride	0.002
Carbonyl sulfide	5064
Catechol	0.028
CFC-12	0.61
Chlorine	571
Chlorine dioxide	447
Chloroacetic acid (and its salts)	0.1
Chlorobenzene	0.087
Chlorobenzene	2
Chloroethane	0.073
Chloroform	161

continued

TABLE 4.1 (continued)
Canadian On-Site Releases for 2008

Substance Name	2008 On-Site Releases (Tonnes Unless Noted)
Chloromethane	732
Chromium (and its compounds)	104
Cobalt (and its compounds)	10
Copper (and its compounds)	858
Creosote	0.061
Cresol (mixed isomers and their salts)	38
Cumene	89
Cyanides (ionic)	2.8
Cycloheptane (all isomers)	727
Cyclohexane	990
Cyclohexanol	0.527
Cyclohexene (all isomers)	350
Cyclooctane (all isomers)	383
Decabromodiphenyl oxide	0.005
Decane (all isomers)	360
Dibenz(a,h)acridine	0.734
Dibenz(a,j)acridine	9.3
Dibenzo(a,e)fluoranthene	0.248
Dibenzo(a,e)pyrene	0.906
Dibenzo(a,h)anthracene	1,899
Dibenzo(a,h)pyrene	0.747
Dibenzo(a,i)pyrene	157
Dibenzo(a,l)pyrene	0.406
Dibutyl phthalate	1.2
Dichloromethane	135
Dicyclopentadiene	3.3
Diethanolamine (and its salts)	16
Diethyl phthalate	0.016
Diethylene glycol butyl ether	82
Diethylene glycol ethyl ether acetate	19
Dihydronapthalene (all isomers)	1.6
Dimethyl phthalate	0.047
Dimethylamine	0.499
Dimethylether	114
Di-n-octyl phthalate	0.069
Dioxins and furans: total (g I-TEQ)*	32
Diphenylamine	0.279
D-Limonene	696
Dodecane (all isomers)	2
Ethyl acetate	3,193
Ethyl acrylate	0.133
Ethyl alcohol	22,366

TABLE 4.1 (continued)
Canadian On-Site Releases for 2008

Substance Name	2008 On-Site Releases (Tonnes Unless Noted)
Ethylbenzene	949
Ethylene	1,242
Ethylene	1,985
Ethylene glycol	5,563
Ethylene glycol butyl ether acetate	57
Ethylene glycol hexyl ether	32
Ethylene oxide	1.3
Fluoranthene	70,436
Fluorene	18,506
Fluorine	67
Formaldehyde	1,278
Formaldehyde	2,487
Formic acid	32
Furfuryl alcohol	28
HCFC-123 and all isomers	8
HCFC-124 and all isomers	1.3
HCFC-141b	5.2
HCFC-142b	721
HCFC-22	181
Heavy alkylate naphtha	90
Heavy aromatic solvent naphtha	458
Heptane (all isomers)	4,048
Hexachlorobenzene (grams)	9,029
Hexane (all isomers excluding n-hexane)	3,928
Hexene (all isomers)	562
Hydrazine (and its salts)	6.8
Hydrochloric acid	8,243
Hydrogen cyanide	14
Hydrogen fluoride	3,332
Hydrogen sulphide	3,464
Hydrotreated heavy naphtha	478
Hydrotreated light distillate	1,120
i-Butyl alcohol	203
Indeno(1,2,3-c,d)pyrene	5,828
Iron pentacarbonyl	0.202
Isophorone diisocyanate	0.007
Isoprene	18
Isopropyl alcohol	1,365
Isopropyl alcohol	1,681
Light aromatic solvent naphtha	1,254
Lithium carbonate	0.878

continued

TABLE 4.1 (continued)
Canadian On-Site Releases for 2008

Substance Name	2008 On-Site Releases (Tonnes Unless Noted)
Maleic anhydride	0.413
Manganese (and its compounds)	1,765
Methanol	11,812
Methanol	15,703
Methyl acrylate	0.105
Methyl ethyl ketone	1,671
Methyl ethyl ketone	1,902
Methyl isobutyl ketone	265
Methyl isobutyl ketone	402
Methyl methacrylate	66
Methyl tert-butyl ether	34
Methylenebis(phenylisocyanate)	15
Mineral spirits	217
Molybdenum trioxide	4
Myrcene	99
N,N-Dimethylformamide	10
Naphtha	344
Naphthalene	121
n-Butyl acetate	1,514
n-Butyl alcohol	557
n-Hexane	5,559
n-Hexane	5,691
Nickel (and its compounds)	272
Nitrate ion in solution at pH ≥ 6.0	62,791
Nitric acid	27
Nitrilotriacetic acid (and its salts)	0.069
Nitroglycerin	1.7
N-Methyl-2-pyrrolidone	104
Nonane (all isomers)	715
Nonylphenol and its ethoxylates	58
Octane (all isomers)	2,406
Octylphenol and its ethoxylates	0.482
o-Dichlorobenzene	0.746
p,p′-Isopropylidenediphenol	11
p,p′-Methylenebis(2-chloroaniline)	0.006
PAHs, total Schedule 1, Part 2	4,006
p-Dichlorobenzene	7.5
p-Dichlorobenzene	7.6
Pentane (all isomers)	13,679
Pentene (all isomers)	1,032
Perylene	645
Phenanthrene	130,911

TABLE 4.1 (continued)
Canadian On-Site Releases for 2008

Substance Name	2008 On-Site Releases (Tonnes Unless Noted)
Phenol (and its salts)	873
Phosphorus (total)	7,082
Phosphorus (yellow or white)	7.1
Phthalic anhydride	0.271
Polymeric diphenylmethane diisocyanate	8.2
Propane	9,148
Propylene	696
Propylene	1,719
Propylene glycol butyl ether	55
Propylene glycol methyl ether acetate	365
Propylene oxide	0.116
Pyrene	61144
Pyridine (and its salts)	0.048
Quinoline (and its salts)	0.445
sec-Butyl alcohol	24
Selenium (and its compounds)	12
Silver (and its compounds)	0.93
Sodium fluoride	12
Sodium nitrite	1.4
Solvent naphtha light aliphatic	1,454
Solvent naphtha medium aliphatic	407
Stoddard solvent	865
Styrene	1,771
Styrene	1,924
Sulfur hexafluoride	2.5
Sulfuric acid	6,185
Terpene (all isomers)	24
tert-Butyl alcohol	26
Tetrachloroethylene	48
Tetrahydrofuran	130
Thorium dioxide	0.04
Titanium tetrachloride	0.04
Toluene	4,385
Toluene	4,517
Toluene-2,4-diisocyanate	0.002
Toluenediisocyanate (mixed isomers)	1.9
Total reduced sulfur (TRS)	11,847
Trichloroethylene	181
Triethylamine	4.2
Trimethylbenzene (all isomers excluding 1,2,4-trimethylbenzene)	1,155

continued

TABLE 4.1 (continued)
Canadian On-Site Releases for 2008

Substance Name	2008 On-Site Releases (Tonnes Unless Noted)
Trimethylfluorosilane	4.8
Vanadium (except when in an alloy) and its compounds	131
Vinyl acetate	95
Vinyl acetate	96
Vinyl chloride	1.7
VM & P naphtha	157
White mineral oil	113
Xylene (all isomers)	5,679
Xylene (all isomers)	5,931
Zinc (and its compounds)	1,141

Source: Environment Canada 2009a.

* The total values for dioxins and furans (in g I-TEQ) above represent the weighted sum of 17 individual dioxin and furan congeners listed on the NPRI.

The NPI contains data on 93 substances that have been identified as important due to their possible effect on human health and the environment. Emissions from industrial sources are reported annually by each facility that exceeds certain fuel, electricity, and NPI substance use thresholds. Releases from residential and transportation-related sources are estimated by government agencies.

The NPI is used to enhance environmental quality; increase public and industry understanding of the types and quantities of toxic substances emitted into the environment and transferred off site as waste; encourage industry to use cleaner production techniques to reduce emissions and waste generation; track environmental progress; meet community right-to-know obligations; and assist government in identifying priorities for environmental decision making (Australian Government 2009a).

The total releases reported in the NPI 2007–2008 reporting year are shown in Table 4.2. In contrast to many of the PRTR programs, the NPI incorporates emissions of what the U.S. Environmental Protection Agency (USEPA) calls *criteria pollutant emissions* (carbon monoxide, volatile organic compounds, sulfur dioxide, oxides of nitrogen, and particulate matter), as well as total nitrogen, all of which are at the top of the mass emissions list.

EUROPEAN UNION

In addition to the PRTR programs in various European Union countries, the EU has established a PRTR that consolidates and centralizes pollutant release and transfer data from the member countries. The first Europe-wide register of industrial releases was called the European Pollutant Emission Register (EPER). This program has now

TABLE 4.2
NPI Emissions Report for Australia, Reporting Year 2007–2008

Substance	Totals (kg/year)
Carbon monoxide	5,600,000,000
Total volatile organic compounds	3,100,000,000
Sulfur dioxide	1,400,000,000
Oxides of nitrogen	1,400,000,000
Particulate matter 10.0 μm	1,200,000,000
Total nitrogen	240,000,000
Ammonia (total)	120,000,000
Hydrochloric acid	46,000,000
Particulate matter 2.5 μm	34,000,000
Toluene (methylbenzene)	33,000,000
Total phosphorus	30,000,000
Ethanol	29,000,000
Xylenes (individual or mixed isomers)	23,000,000
Benzene	15,000,000
n-Hexane	9,700,000
Formaldehyde (methyl aldehyde)	8,500,000
Acetone	7,900,000
Fluoride compounds	7,600,000
Methanol	6,500,000
Sulfuric acid	6,400,000
Methyl ethyl ketone	5,600,000
Acetaldehyde	5,300,000
Manganese and compounds	4,500,000
Cyclohexane	3,800,000
Boron and compounds	3,200,000
Ethylbenzene	3,100,000
1,3-Butadiene (vinyl ethylene)	2,800,000
Chlorine	2,500,000
Zinc and compounds	2,400,000
Methyl isobutyl ketone	2,400,000
Cyanide (inorganic) compounds	2,000,000
Ethyl acetate	1,900,000
Ethylene glycol (1,2-ethanediol)	1,600,000
Acetic acid (ethanoic acid)	1,600,000
Dichloromethane	1,400,000
Tetrachloroethylene	1,300,000
Lead and compounds	1,300,000
Polycyclic aromatic hydrocarbons	1,300,000
Copper and compounds	1,200,000
Hydrogen sulfide	1,000,000
Trichloroethylene	920,000

continued

TABLE 4.2 (continued)
NPI Emissions Report for Australia, Reporting Year 2007–2008

Substance	Totals (kg/year)
Styrene (ethenylbenzene)	900,000
Nickel and compounds	560,000
Carbon disulfide	420,000
Nitric acid	370,000
Chromium (III) compounds	300,000
Dibutyl phthalate	290,000
Cobalt and compounds	270,000
2-Ethoxyethanol acetate	190,000
Acrylamide	160,000
Arsenic and compounds	160,000
Phenol	150,000
Ethylene oxide	130,000
Antimony and compounds	130,000
1,2-Dibromoethane	82,000
Cumene (1-methylethylbenzene)	76,000
Cadmium and compounds	69,000
Glutaraldehyde	60,000
Acetonitrile	58,000
Acrylonitrile (2-propenenitrile)	55,000
Chloroethane (ethyl chloride)	27,000
Magnesium oxide fume	26,000
Vinyl chloride monomer	25,000
Mercury and compounds	25,000
Chloroform (trichloromethane)	23,000
1,2-Dichloroethane	22,000
2-Methoxyethanol	15,000
Selenium and compounds	13,000
Aniline (benzenamine)	13,000
Methyl methacrylate	11,000
2-Ethoxyethanol	9,100
Chromium (VI) compounds	8,800
Acrylic acid	7,700
1,1,2-Trichloroethane	6,000
Biphenyl (1,1-biphenyl)	5,300
Beryllium and compounds	5,300
Phosphoric acid	4,300
Di-(2-ethylhexyl) phthalate (DEHP)	3,600
Organo-tin compounds	2,900
Nickel subsulfide	2,600
Toluene-2,4-diisocyanate	930
Ethyl butyl ketone	750

TABLE 4.2 (continued)
NPI Emissions Report for Australia, Reporting Year 2007–2008

Substance	Totals (kg/year)
Chlorine dioxide	640
Chlorophenols (di, tri, tetra)	390
Methylenebis (phenylisocyanate)	300
1,1,1,2-Tetrachloroethane	87
Nickel carbonyl	39
Acrolein	24
2-Methoxyethanol acetate	2.8
Polychlorinated dioxins and furans	0.4

Source: Australian Government 2009b.

been replaced with the European Pollutant Release and Transfer Register (E-PRTR) (European Union 2009).

European Pollutant Emission Register

The EPER was the first European-wide register of industrial emissions into air and water and was established by a commission decision of July 17, 2000. The EPER required triennial reporting of 50 chemicals by member states. EPER chemicals included criteria pollutants, greenhouse gases, ozone-depleting compounds, pollutants harmful to water, metals, chlorinated organic compounds, nonchlorinated organic compounds, and inorganic compounds.

The EPER data were reported in 2001 and 2004 and are published on the Internet (European Union 2008).

European Pollutant Release and Transfer Register

The E-PRTR replaced the EPER. The E-PRTR introduced a number of changes to the program, including

- Increasing the reporting frequency to annual starting with reporting year 2007
- Expanding the list of facilities required to report
- Increasing the number of chemicals to 91
- Requiring reporting of releases to land, off-site transfer of waste, and fugitive emissions (called *diffuse emissions* by the EU)

The E-PRTR was established to improve public access to environmental information and contribute to long-term prevention and reduction of pollution.

The E-PRTR consists of an EU-level publicly accessible electronic database which is intended to meet the requirements of the OECD Protocol on Pollutant Release and Transfer Registers described in this chapter.

The register requires reporting of releases of pollutants to air, water, and land, as well as transfers of waste and pollutants, where emissions exceed certain threshold values and result from specific activities. The register will also cover releases of pollutants from fugitive sources, including transportation.

The pollutants reported under the E-PRTR include greenhouse gases, acid rain pollutants, ozone-depleting substances, heavy metals, and certain carcinogens, such as dioxins. Sources required to report releases include industrial sources such as power-generating facilities, mining, quarrying and metalworking industries, chemical plants, paper and timber industries, and waste and wastewater treatment plants.

Each country is responsible for collecting its industrial and diffuse emissions data. The data are then consolidated and reported to the European Commission within 12–15 months of the end of the reporting year. The European Commission then publishes the data within 16–21 months of the end of the reporting year (European Union 2008).

The full list of chemicals and reporting thresholds is shown below in Table 4.3. Facilities that emit more than the threshold values stated in the table are required to report under E-PRTR.

JAPAN

Japan established a PRTR that has required reporting of listed chemicals since 2001. The stated purposes of the Japanese PRTR are to collect basic environmental data, determine priorities for regulating chemical substances, promote voluntary improvement in chemical substance management by businesses, provide information and foster understanding of chemical substances, and understand the effect of environmental conservation measures on chemical substance release.

The Japanese program requires the reporting of 354 chemicals called *Class I designated chemical substances*, which are considered to be environmentally persistent over a substantial area and meet at least one of the following criteria:

- May be hazardous to human health or may adversely affect the ecosystem
- May easily form hazardous chemical substances through a naturally occurring chemical transformation
- Are ozone-depleting compounds

For most chemicals, if more than 1 ton of the material is used or manufactured per year, it is considered reportable with the exception of the following 12 substances called *specific Class I designated substances,* which have a reporting threshold of 0.5 tons per year:

Asbestos
Ethylene oxide
Cadmium
Chromium (VI)
Vinyl chloride
Dioxin
Nickel
Arsenic

TABLE 4.3
List of Reporting Thresholds for E-PRTR (European Commission, 2006)

Pollutant	Releases to Air (kg/year)	Releases to Water (kg/year)	Releases to Land (kg/year)
1,1,1-trichloroethane	100	—	—
1,1,2,2-tetrachloroethane	50	—	—
1,2,3,4,5,6-hexachlorocyclohexane (HCH)	10	1	1
1,2-dichloroethane (EDC)	1000	10	10
Alachlor	—	1	1
Aldrin	1	1	1
Ammonia (NH3)	10000	—	—
Anthracene	50	1	1
Arsenic and compounds (as As)	20	5	5
Asbestos	1	1	1
Atrazine	—	1	1
Benzene	1,000	200 (as BTEX)	200 (as BTEX)
Benzo(g,h,i)perylene		1	
Brominated diphenylethers (PBDE)	—	1	1
Cadmium and compounds (as Cd)	10	5	5
Carbon dioxide (CO2)	100 million	—	—
Carbon monoxide (CO)	500000	—	—
Chlordane	1	1	1
Chlordecone	1	1	1
Chlorfenvinphos	—	1	1
Chlorides (as total Cl)	—	2 million	2 million
Chlorine and inorganic compounds (as HCl)	10000	—	—
Chloro-alkanes, C10-C13	—	1	1
Chlorofluorocarbons (CFCs)	1	—	—
Chlorpyrifos	—	1	1
Chromium and compounds (as Cr)	100	50	50
Copper and compounds (as Cu)	100	50	50
Cyanides (as total CN)	—	50	50
DDT	1	1	1
Di-(2-ethyl hexyl) phthalate (DEHP)	10	1	1
Dichloromethane (DCM)	1000	10	10
Dieldrin	1	1	1
Diuron	—	1	1
Endosulphan	—	1	1
Endrin	1	1	1
Ethyl benzene	—	200 (as BTEX)	200 (as BTEX)
Ethylene oxide	1,000	10	10
Fluoranthene	—	1	—
Fluorides (as total F)	—	2000	2000

continued

TABLE 4.3 (continued)
List of Reporting Thresholds for E-PRTR (European Commission, 2006)

Pollutant	Releases to Air (kg/year)	Releases to Water (kg/year)	Releases to Land (kg/year)
Fluorine and inorganic compounds (as HF)	5000	—	—
Halogenated organic compounds (as AOX)	—	1000	1000
Halons	1	—	—
Heptachlor	1	1	1
Hexabromobiphenyl	0,1	0,1	0,1
Hexachlorobenzene (HCB)	10	1	1
Hexachlorobutadiene (HCBD)	—	1	1
Hydrochlorofluorocarbons (HCFCs)	1	—	—
Hydro-fluorocarbons (HFCs)	100	—	—
Hydrogen cyanide (HCN)	200	—	—
Isodrin	—	1	—
Isoproturon	—	1	1
Lead and compounds (as Pb)	200	20	20
Lindane	1	1	1
Mercury and compounds (as Hg)	10	1	1
Methane (CH4)	100000	—	—
Mirex	1	1	1
Naphthalene	100	10	10
Nickel and compounds (as Ni)	50	20	20
Nitrogen oxides (NOx/NO2)	100000	—	—
Nitrous oxide (N2O)	10000	—	—
Non-methane volatile organic compounds (NMVOC)	100000	—	—
Nonylphenol and Nonylphenol ethoxylates (NP/NPEs)	—	1	1
Octylphenols and Octylphenol ethoxylates	—	1	—
Organotin compounds(as total Sn)	—	50	50
Particulate matter (PM10)	50000	—	—
PCDD + PCDF (dioxins + furans) (as Teq)	0.0001	0.0001	0.0001
Pentachlorobenzene	1	1	1
Pentachlorophenol (PCP)	10	1	1
Perfluorocarbons (PFCs)	100	—	—
Phenols (as total C)	—	20	20
Polychlorinated biphenyls (PCBs)	0,1	0,1	0,1
Polycyclic aromatic hydrocarbons (PAHs)	50	5	5
Simazine	—	1	1
Sulphur hexafluoride (SF6)	50	—	—
Sulphur oxides (SOx/SO2)	150000	—	—
Tetrachloroethylene (PER)	2,000	10	—
Tetrachloromethane (TCM)	100	1	—

TABLE 4.3 (continued)
List of Reporting Thresholds for E-PRTR (European Commission, 2006)

Pollutant	Releases to Air (kg/year)	Releases to Water (kg/year)	Releases to Land (kg/year)
Toluene	—	200 (as BTEX)	200 (as BTEX)
Total nitrogen	—	50000	50000
Total organic carbon (TOC) (as total C or COD/3)	—	50000	—
Total phosphorus	—	5000	5000
Toxaphene	1	1	1
Tributyltin and compounds	—	1	1
Trichlorobenzenes (TCBs) (all isomers)	10	1	—
Trichloroethylene	2,000	10	—
Trichloromethane	500	10	—
Trifluralin	—	1	1
Triphenyltin and compounds	—	1	1
Vinyl chloride	1,000	10	10
Xylenes	—	200 (as BTEX)	200 (as BTEX)

Source: European Commission. *Regulation (EC) No. 166/2006 of the European Parliament and of the Council of 18 January 2006 Concerning the Establishment of a European Pollutant Release and Transfer Register and Amending Council Directives 91/689/EEC and 96/61/EC*, Brussels, BE: Commission for the European Communities, 2006.

Beryllium
Benzylidyne trichloride
Benzene
9-Methoxy-7H-furo[3,2-g][1]benzopyran-7-one; methoxsalen

Reporting is required by businesses that fall into 1 of 23 industrial categories and have at least 21 regular employees. In addition, facilities that fall under the mine safety law, sewage disposal facilities, municipal or industrial waste disposal facilities, and certain facilities regulated under the Act on Special Measures against Dioxins are required to report.

Facilities subject to reporting are required to quantify and report releases to air, public water bodies, and land on site, landfill disposal, transfers to sewage, and transfers off site (Government of Japan, Ministry of the Environment 2009).

ACCESSING PRTR DATA

North American Data

The Commission for Environmental Cooperation (CEC, 2010) is an international organization created by Canada, Mexico, and the United States under the North American Agreement on Environmental Cooperation (NAAEC). The CEC was

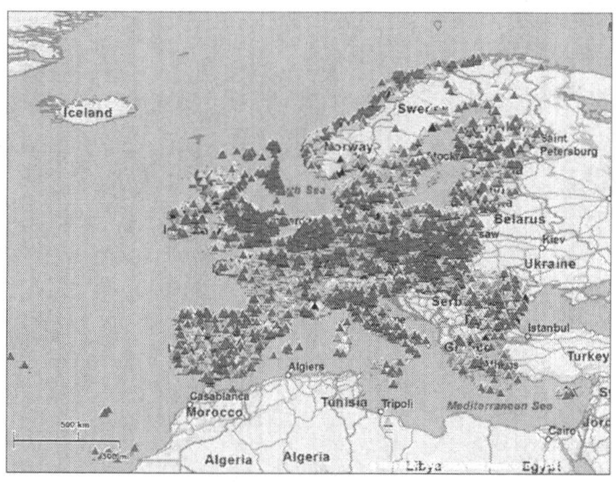

FIGURE 4.2 Graphical representation of PRTR releases in the European Union. (From European Environment Agency, Map Search, 2010. http://prtr.ec.europa.eu/MapSearch.aspx.)

established to address regional environmental concerns, help prevent potential trade and environmental conflicts, and promote the effective enforcement of environmental law. The CEC developed the North American Environmental Atlas, which provides a variety of geographic information through maps, GIS data, metadata, and interactive map layers, which depict the status and trends of environmental conditions across North America.

A portion of this database contains release information from about 35,000 industrial facilities in North America. This database can be queried in a number of different ways, including by year, pollutant, facility, industry sector, state/province, or country.

The data can also be searched via a graphic interface using Google Earth. The map layer of CEC plots the North American industrial facilities that reported releases and transfers of pollutants (currently with 2005 data). This tool allows the user to map any location in North America, locate nearby industrial facilities, and learn about the pollution profile of each facility, including which pollutants are generated and how the facility handles them. Users can also compare the performance of various facilities to other sources across North America (CEC 2010).

European Data from E-PRTR

The E-PRTR has a searchable database that includes releases and transfers throughout the European Union. Users can search data by facility, industrial activity, pollutant, and release or transfer type. The E-PRTR also provides a geographical map interface to allow the user to search for facilities near any point in Europe. The map interface allows the user to view one or more industrial sectors (Figure 4.2).

International PRTR Data from OECD

Some international PRTR data can also be accessed through the OECD Center for PRTR data. Users can create a report of PRTR data according to years, countries, regions, industry sectors, chemicals, types of release sources, and types of releases and transfers. The database does not provide information regarding releases from individual facilities. At the time this book was written, data from the following countries could be accessed: Australia, Belgium, Canada, Chile, Spain, European Union, England and Wales, Hungary, Japan, the Netherlands, Scotland, Sweden, and the United States (OECD 2010).

BIBLIOGRAPHY

Australian Government. 2009a. About the NPI. http://www.npi.gov.au/npi/index.html (accessed November 14, 2009).

Australian Government. 2009b. NPI location report—All Sources: Australia. http://www.npi.gov.au/data/overview/reports/national-location-report.html (accessed November 14, 2009).

Commission for Environmental Cooperation. 2010. Mapping Industrial Pollutants. http://www.cec.org/Page.asp?PageID=122&ContentID=2587 (accessed January 6, 2010).

Environment Canada. 2009a. About the National Pollutant Release Inventory (NPRI). http://www.ec.gc.ca/inrp-npri/default.asp?lang=En&n=4A577BB9-1 (accessed November 14, 2009).

Environment Canada. 2009b. National Pollutant Release Inventory (NPRI) 2008 Facility Data Summary. http://www.ec.gc.ca/inrp-npri/default.asp?lang=en&n=BF14CADF-1#part1a and http://www.ec.gc.ca/inrp-npri/default.asp?lang=en&n=DA8BFC79-1#part1b (accessed January 4, 2010).

European Commission. 2000. Commission Decision of 17 July 2000 on the Implementation of a European Pollutant Emission Register (EPER) According to Article 15 of Council Directive 96/61/EC Concerning Integrated Pollution Prevention and Control (IPPC). *Official Journal of the European Communities*, L 192/36, 28.7.2000.

European Commission. 2006. *Regulation (EC) No 166/2006 of the European Parliament and of the Council of 18 January 2006 Concerning the Establishment of a European Pollutant Release and Transfer Register and Amending Council Directives 91/689/EEC and 96/61/EC.*

European Environment Agency. 2010. Map Search. http://prtr.ec.europa.eu/MapSearch.aspx (accessed January 6, 2010).

European Union. 2008. The European Pollutant Emission Register (EPER) and the European Pollutant Release and Transfer Register (E-PRTR). http://ec.europa.eu/environment/air/pollutants/stationary/eper/index.htm (accessed November 14, 2009).

European Union. 2009. European Pollutant Release and Transfer Register (PRTR). http://europa.eu/legislation_summaries/environment/general_provisions/l28149_en.htm (accessed November 14, 2009).

Government of Japan, Ministry of the Environment. 2009. Background to Japanese PRTR, Japan's Efforts Toward Introducing PRTR. http://www.env.go.jp/en/chemi/prtr/about/index.html (accessed November 21, 2009).

Organization for Economic Co-operation and Development. 2009. Task Force on Pollutant Release and Transfer Registers. http://www.prtr.net/links_e.cfm (accessed November 14, 2009).

Organization for Economic Co-operation and Development. 2010. Centre for PRTR Data. http://www.oecd.org/env_prtr_data/ (accessed January 6, 2010).

U.S. Environmental Protection Agency. 2009. U.S. TRI Program a Leader in International Chemical Release Reporting. http://www.epa.gov/TRI/programs/international/#h1 (accessed January 4, 2010).

5 TRI Program Impacts on Reducing Toxic Chemical Releases

INTRODUCTION

Toxics Release Inventory (TRI) reporting has limitations; however, the program has been successful in reducing releases of chemicals included in the program. After the first reporting year, 1988, the releases declined steadily through 1996 (Figure 5.1). This decline was despite the addition of 286 new chemicals and the addition of federal facilities in 1994. The 1998 report included seven new industry sectors to the reports, which caused the reported releases to more than double. Following this expansion, the release volumes have again declined.

There are numerous factors that contributed to emission reduction over the time period since the TRI was introduced. Many who have analyzed the effects of TRI agree that the introduction of the program has led to substantial emission reductions (Stephan, Kraft, and Abel 2005).

Many give the TRI program credit for reductions as a direct result of the public disclosure aspect of the program and the pressure it puts on facilities. This pressure comes from a number of sources. For example, once information is published, the public has access to the data and can put direct pressure on facilities to reduce releases. This can come in the form of a community-based group, an environmental activist group, or many individuals. In addition, shareholders have been demonstrated to have tremendous power over the decisions a company makes regarding its environmental releases. According to one study, companies that showed the greatest drop in stock price following the initial release of TRI data in 1989 reduced their emissions over the following 3 years more than their industry peers (Konar and Cohen 1996).

The TRI program also can result in reductions of emissions from the mere potential for public pressure (Stephan, Kraft, and Abel 2005). The TRI data are published in a format that allows the public to view the top emitters by facility, industry type, and location. Facilities can monitor their own position in the rankings of top emitters to benchmark environmental performance based on the total mass of toxic chemicals released. The simple perception that top emitters will or could be in the spotlight could be enough to motivate some facilities to reduce releases of reported chemicals.

Another key factor was the desire by facilities to reduce their reporting burden. By reducing the use of a chemical targeted by the program, the facility is able to

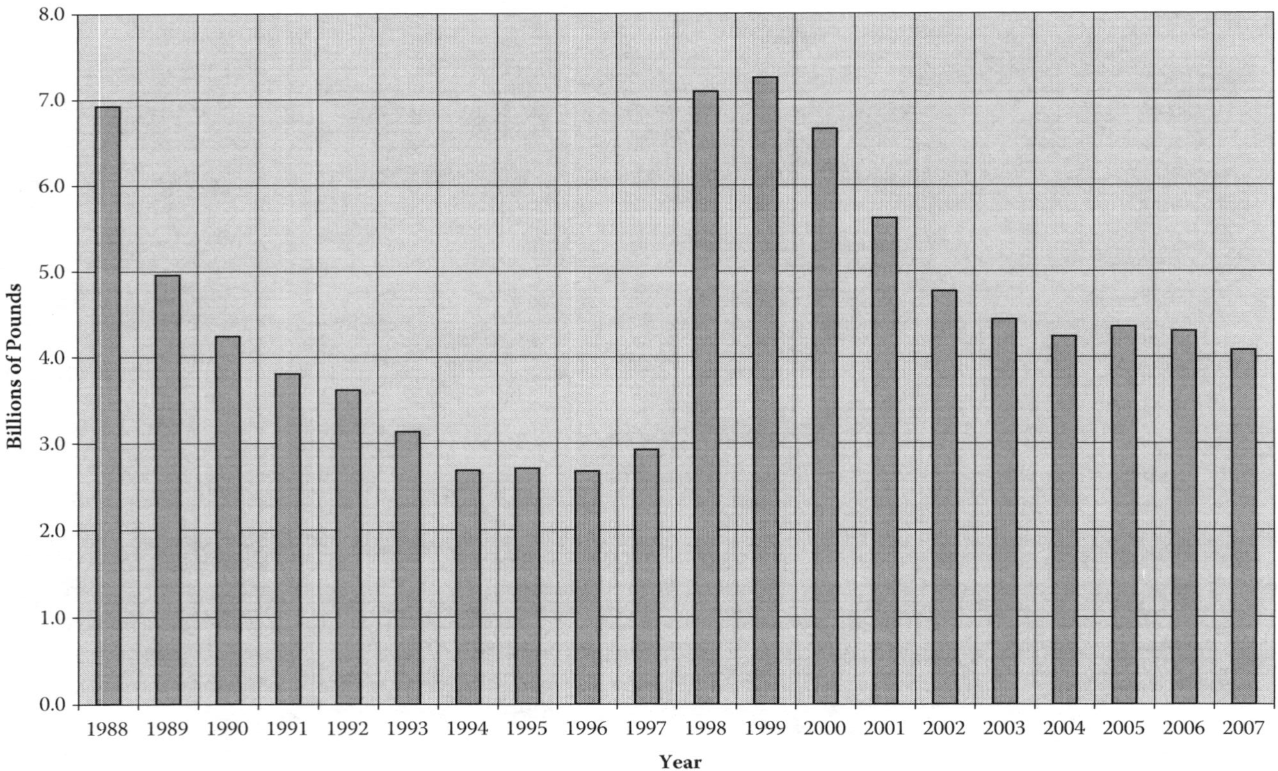

FIGURE 5.1 TRI releases over the years.

reduce or eliminate reporting requirements for this chemical. It has been our experience that regulated facilities will select products that do not contain TRI chemicals if an alternative is available, even if the alternate is more costly. For example, a recent client, a federal facility for which we are upgrading their treatment plant, is subject to TRI reporting. In municipal treatment plants, those not subject to TRI reporting, we typically provide methanol as the treatment chemical for removing nutrients. At the federal facility, the client chose to use acetic acid, even though it was twice the cost of methanol, to avoid the requirement of reporting methanol releases under TRI. In performing pollution prevention opportunity assessments at a series of military installations, one of the goals was to reduce the use of TRI chemicals below the reporting threshold.

A number of other factors not directly related to TRI may also have had an effect on release reductions (Stephan, Kraft, and Abel 2005). For example, facilities may choose to reduce releases to forestall mandatory regulations. The same reductions could also be implemented in anticipation of upcoming pollution reduction legislation.

Another mechanism by which companies could be inspired to reduce releases is through cost reduction. Through internal studies, mandatory reporting under TRI, or voluntary or mandatory pollution prevention planning, companies have found in a number of instances that projects that result in emission reductions may also lead to cost reductions through reduced energy use, reduced water use, reduced cost or quantity of chemicals used, or eliminated hazardous waste-related costs. This is discussed in more detail in Chapter 13.

In one survey, users of the TRI program were surveyed to determine the impacts of the program on environmental performance (Stephan, Kraft, and Abel 2009). In this survey, 74 percent of the respondents either agreed or were neutral to the assertion that the program helped identify needs and opportunities for source reduction at the facility level. Furthermore, 73 percent of the respondents agreed or were neutral to the assertion that the TRI program allowed facilities to set goals or demonstrate commitment to emission reductions. The results of this survey point to factors that can contribute to emission reductions.

According to a study conducted in 2005 that assessed trends in releases and estimated toxic risk from individual facilities between 1991 and 2002, a small group of large facilities contributed significantly to the national trend in toxic emission reductions. In 1999, in fact, 31 percent of the emission reductions came from 50 of the 21,000 facilities reporting under the program. The study also determined that, of the facilities reporting under the program, 43 percent of the facilities nationwide reduced both releases and public health risk between 1995 and 2000. Over the same time period, the study determined that 48 percent of facilities increased releases and public health risk (Stephan, Kraft, and Abel 2005). Although the national trends for toxic chemical release reduction would suggest that the TRI program has been effective in reducing the total amount of releases nationwide, the program is not uniform in its success. Many facilities do not seem to be affected in the same way by the program.

TRI SUCCESS STORIES

TRI success stories include the following:

- The Haartz Corporation, located in Acton, Massachusetts, makes coated fabrics used in automobiles. The firm once used 800,000 lb/yr of methyl ethyl ketone (MEK), a solvent that can cause dizziness, nausea, or unconsciousness when inhaled. In 1987, when Haartz was preparing its first TRI report, the company installed a new emissions control system to capture and recycle MEK. TRI data enabled Haartz to track the association between reduced toxic chemical releases and reduced costs. According to the Haartz environmental manager, the company's "emissions have stayed pretty flat" despite its "double-digit sales growth" between 1993 and 1998. In addition, reducing its MEK releases saved Haartz an estimated $200,000 annually (USEPA 2007).
- Marathon Oil installed a thermal desorption unit to process oily waste and recovered over 120,000 barrels of oil; Georgia Gulf Corporation relocated a methanol stripper purge line that resulted in the recovery of 9,300 gallons of methanol that previously underwent biological waste treatment.
- Attendees of the 1997 EPA Region III TRI workshop provided reasons for undertaking waste reduction activities. The most frequent reason given (98 percent of respondents) was cost reduction.

BIBLIOGRAPHY

Arora, S., and T. N. Cason. 1996. Why Do Firms Volunteer to Exceed Environmental Regulations? Understanding Participation in EPA's 33/50 Program. *Land Economics* 72(4): 413–432.

Konar, S., and M. Cohen. 1996. Information as Regulation: The Effect of Community Right to Know Laws on Toxic Emissions. http://www.vanderbilt.edu/vcems/papers/tri.pdf (accessed January 1, 2010).

Stephan, M., M. E. Kraft, and T. D. Abel. 2005. Information Politics and Environmental Performance: The Impact of the Toxics Release Inventory on Corporate Decision Making. Paper prepared for delivery at the 2005 annual meeting of the American Political Science Association, September 1–4, Washington, DC.

Stephan, M., M. E. Kraft., and T. D. Abel. 2009. Facility Level Perspectives on Toxics Release Inventory and Environmental Performance. Paper prepared for the 2009 TRI National Training Conference, March 30–April 2, Bethesda, MD.

U.S. Environmental Protection Agency. 1999. *Office of Pollution Prevention and Toxics, 33/50 Program: The Final Record.* EPA-745-R-99-004.

U.S. Environmental Protection Agency. 2007. How are the Data Used by Industry? *Chemical Emergency Prevention and Planning Newsletter* (June–July): 5.

U.S. Environmental Protection Agency. 2009a. Class I Ozone-Depleting Substances. http://www.epa.gov/docs/ozone/ods.html (accessed October 31, 2006).

U.S. Environmental Protection Agency. 2009b. Class II Ozone-Depleting Substances. http://www.epa.gov/docs/ozone/ods2.html (accessed October 31, 2006).

6 Quantifying Toxicity

INTRODUCTION

Generally, the emphasis of the Toxics Release Inventory (TRI) program to date has been on total pounds of releases, not on the toxicity of those releases. Reporting on releases has generally stressed total pounds released and not been concentrated on the toxicity of the individual chemicals. This in essence assumes that chemicals are either toxic and included in the inventory or they are not. All pounds are equal. Toxicity varies widely for chemicals that are reported. It is intuitively obvious that a pound of arsenic has more potential impact than a pound of saccharin, yet both are included in the TRI, and generally both are reported equally. The U.S. Environmental Protection Agency (USEPA) and state agencies have recognized a need to concentrate on chemicals that have greater impact, with an increased emphasis on chemicals that are persistent, bioaccumulative, and toxic (PBT). The purpose of this chapter is to develop a single toxicity factor (TF) for the chemicals in the TRI inventory, which would be useful in ranking chemicals by their potential toxic impact. The purpose of the TF would be to evaluate the relative toxic impacts of the annual total release of a chemical.

This is an initial proposed approach; we welcome suggestions on how to improve our analysis or use the same or additional toxicity data to relatively rank chemicals differently. The important point to keep in mind is that the objective of this approach was not to come up with an absolute number that we claim to represent the actual and fixed toxic impact of a chemical. Rather, the proposed methodology was developed solely for the purposes of coming up with a logical means of applying published and routinely used toxicological data to then develop chemical-specific "toxicity" factors to conduct an apples-to-apples comparison across chemicals. This would relatively rank the potential effect among various toxic chemicals, such that reduction in the use of toxic chemicals can be targeted and managed accordingly.

Toxicity values represent either acute or chronic impacts. Acute toxicity is the measure of how a single or short-term dose of a compound can cause death or other major impact. Chronic toxicity is the result of long-term ingestion or inhalation of a much smaller dose of the chemical.

Acute exposure is handled well by current laws, such as those of the Occupational Safety and Health Administration (OSHA). It is rare that the current public is exposed to acute levels of toxins. Chronic levels are much lower and therefore likely to be encountered by environmental releases or by the use of products containing toxic chemicals.

There are four measures of chronic toxicity used in our analysis, based on method of exposure and associated effect:

- Ingestion carcinogenicity
- Inhalation carcinogenicity
- Ingestion noncarcinogenic toxicity
- Inhalation noncarcinogenic toxicity

The toxicity information was downloaded from the USEPA human health-based risk assessment Web address at the time this book was written (USEPA 2009).

INGESTION CARCINOGENICITY

We based our assessment of ingestion carcinogenicity, that is, cancer risk due to ingestion of a chemical, on the oral slope factor (SFO) of the USEPA. The SFO is expressed as inverse of dose (milligram of compound per kilogram of body mass per day)$^{-1}$ and represents the proportion of a population that is estimated to develop increased cancer risk if each individual were to ingest 1 mg of a chemical per kilogram of their body weight each day over a lifetime (upper-bound estimate). The higher the SFO is, the higher will be the cancer risk from a particular chemical relative to a chemical that has a lower SFO. The SFO is generally reserved for use in the low-dose region of the dose-response relationship (i.e., for exposures corresponding to risks less than 1 in 100).

For example, a chemical with an SFO of 11 (mg/kg-day)$^{-1}$ represents 11 extra cancer cases per person ingesting 1 mg of a chemical per kilogram body mass per day. An SFO of 1×10^{-2} (mg/kg-day)$^{-1}$ represents 1 excess cancer case per 100 people ingesting 1 mg of the chemical per kilogram body mass per day.

INHALATION CARCINOGENICITY

We based our assessment of inhalation carcinogenicity, that is, cancer risk due to inhalation of a chemical, on the inhalation unit risk (IUR) number of the USEPA. The IUR, expressed in units of (microgram/cubic meter)$^{-1}$, represents an upper-bound excess lifetime cancer risk estimated to result from continuous exposure to a chemical at a concentration of 1 μg/m^3 of air. The higher the IUR is, the higher the cancer risk from a particular chemical will be relative to a chemical that has a lower IUR.

For example, if the IUR of a chemical equals 3×10^{-6} (μg/m^3)$^{-1}$, 3 excess cancer cases are expected to develop per 1 million people if exposed daily for a lifetime to 1 μg of the chemical per cubic meter of air. An IUR of 38 (μg/m^3)$^{-1}$ represents 38 extra cancer cases per person inhaling 1 μg of the chemical per cubic meter of air for a lifetime.

INGESTION NONCARCINOGENIC TOXICITY

We based our assessment of ingestion noncarcinogenic toxicity, that is, noncancer risk due to ingestion of a chemical, on the oral reference dose (RfDo) of the USEPA, which is an estimate (with uncertainty spanning perhaps an order of magnitude) of a continuous oral exposure to the human population (including sensitive subgroups) that is likely to be without an appreciable risk of deleterious effects during a lifetime.

The RfDo is expressed as milligrams of chemical per kilogram of body weight per day. It can be thought of as the daily dose that would have no impact or the dose above which there would be a toxic impact. The higher the RfDo is, the lower the noncarcinogenic toxicity from a particular chemical will be (i.e., a higher RfDo represents a higher "acceptable" threshold) relative to a chemical that has a lower RfDo.

For example, a chemical with an RfDo of 2 mg/kg-day represents the fact that 2 mg/kg can be ingested per day and have no noncarcinogenic impact, whereas a dose above 2 mg/kg ingested per day could pose a noncarcinogenic toxic impact.

INHALATION NONCARCINOGENIC TOXICITY

We based our assessment of inhalation noncarcinogenic toxicity (noncancer risk due to inhalation of a chemical) on the inhalation reference concentration (RfCi) of the USEPA, which is an estimate (with uncertainty spanning perhaps an order of magnitude) of a continuous inhalation exposure to the human population (including sensitive subgroups) that is likely to be without an appreciable risk of deleterious effects during a lifetime. The RfCi is expressed as milligrams of chemical per cubic meters of air. It can be thought of as the concentration of a chemical in air that can be breathed over a lifetime without experiencing a toxic impact. The RfCi is not expressed relative to body weight because an individual's breathing rate is proportional to his or her body weight. The higher the RfCi is, the lower the noncarcinogenic toxicity from a particular chemical will be (i.e., a higher RfCi represents a higher "acceptable" threshold) relative to a chemical that has a lower RfCi.

For example, a chemical with an RfCi of 3 mg/m^3 represents the fact that 3 mg/m^3 can be inhaled per day and have no noncarcinogenic impact, whereas a dose inhaled above 3 mg/m^3 can pose a noncarcinogenic toxic impact.

DEVELOPMENT OF A SINGLE TOXICITY FACTOR

These four individual chemical-specific measures of chronic toxicity were used to develop a relative single TF for TRI chemicals. The TF is based on the concept of how many individuals could equally share (ingest or inhale) a pound of chemical released per year to have no adverse impact, either from excess cancers or from noncarcinogenic effects (toxicity threshold).

The analysis of specific toxicity of any given chemical does require a more detailed analysis of exposure pathways and individuals affected. We have simplified the analysis to come up with a single, relative chemical-specific TF to rank chemicals relatively by their overall toxic effect. In addition, the measures of toxicity obtained from the USEPA Web site do get updated periodically; therefore, our calculations of TFs as presented in this book are indeed intended solely for relative ranking versus representing absolute numbers.

We welcome suggestions on how to improve our analysis or come up with additional toxicity information to better rank chemicals. Not to rank chemicals by some toxicity measure and therefore not prioritize which ones to focus on eliminating is unacceptable. Without some type of toxicity ranking, all chemicals will be considered "bad," the undertaking to reduce or eliminate use of all (vs. targeted) chemicals will be overwhelming, and ultimately nothing will be done to reach the next step of reducing overall toxicity.

Ingestion Carcinogenic Toxicity Factor Development

For the ingestion carcinogenicity TF, we took a pound of a chemical per year, converted it to milligrams per day using standard unit conversion factors, and divided by a typical average body weight (70 kg) and 365 days per year:

$$\frac{1 \text{ lb chemical}}{\text{yr}} \times \frac{454{,}000 \text{ mg}}{\text{lb}} \times \frac{1 \text{ yr}}{365 \text{ days}} \times \frac{1}{70 \text{ kg}} = \frac{18 \text{ mg chemical}}{\text{kg-day}}$$

Subsequently, this value was multiplied by the SFO $(\text{mg/kg-day})^{-1}$ to yield the portion of the population that could develop excess cancers due to exposure of 1 lb of chemical released:

$$\frac{18 \text{ mg chemical}}{\text{kg-day}} \times \frac{\text{SFO (kg-day)}}{\text{mg}}$$

However, this is not an acceptable risk. While no cancer is desirable, there is a threshold of cancer risk that is generally acceptable, and according to the USEPA, this is usually from 1 excess cancer case per 100,000 people to 1 excess cancer case per million people. For this analysis, we factored in 1 excess cancer risk per million people and hence multiplied the result by 1 million to obtain the estimated number of individuals who could share/ingest (number of doses) a pound of the chemical released and still result in an "acceptable" level of cancers:

$$\text{Ingestion Carcinogenic TF (doses/lb)} = \frac{18 \text{ mg chemical}}{\text{kg-day}} \times \frac{\text{SFO (kg-day)}}{\text{mg}} \times 10^6 \text{ people}$$

Inhalation Carcinogenic Toxicity Factor Development

According to the USEPA, the adult male inhalation rate is 15.2 m^3/day, and the adult female inhalation rate is 11.3 m^3/day. Hence, the average inhalation rate for an adult is 13.25 m^3/day, or 4,836 m^3/yr, using the standard conversion factor of 365 days per year. For the inhalation carcinogenicity TF, we took a pound of a chemical per year and converted it to micrograms per cubic meters using standard unit conversion factors and the average annual inhalation of air (4,836 m^3/yr):

$$\frac{1 \text{ lb chemical}}{\text{yr}} \times \frac{454{,}000{,}000\ \mu\text{g}}{\text{lb}} \times \frac{1 \text{ yr}}{4{,}836 \text{ m}^3} = \frac{93{,}879\ \mu\text{g chemical}}{\text{m}^3}$$

Subsequently, this value was multiplied by the IUR $(\mu g/m^3)^{-1}$. As noted with the ingestion carcinogenicity TF, we then multiplied this result by 1 million people to obtain the estimated number of individuals who could equally share/inhale (number of doses) a pound of the chemical and still result in an acceptable level of cancers:

$$\text{Inhalation Carcinogenic TF (doses/lb)} = \frac{93{,}879\ \mu\text{g chemical}}{\text{m}^3} \times \frac{1}{\text{IUR } (\mu\text{g/m}^3)} \times 10^6 \text{ people}$$

Ingestion Noncarcinogenic Toxicity Factor Development

For the ingestion noncarcinogenic TF, we took a pound of a chemical released per year, converted it to milligrams per day using standard unit conversion factors, and divided by a typical 70-kg body weight and 365 days per year:

$$\frac{1 \text{ lb chemical}}{\text{yr}} \times \frac{454{,}000 \text{ mg}}{\text{lb}} \times \frac{\text{yr}}{365 \text{ days}} \times \frac{1}{70 \text{ kg}} = \frac{18 \text{ mg chemical}}{\text{kg-day}}$$

Subsequently, this value was divided by the RfDo (mg/kg-day) to obtain the ingestion noncarcinogenic TF:

$$\text{Ingestion Noncarcinogenic TF (doses/lb)} = \frac{18 \text{ mg chemical}}{\text{kg-day}} \times \frac{1}{\text{RfDo (mg/kg-day)}}$$

This results in the number of individuals who could equally share/ingest (number of doses) a pound of the chemical released and for whom there would be no adverse noncarcinogenic impact from the chemical.

Inhalation Noncarcinogenic Toxicity Factor Development

According to the USEPA, the adult male inhalation rate is 15.2 m^3/day and the adult female inhalation rate is 11.3 m^3/day. Hence, the average inhalation rate for an adult is 13.25 m^3/day, or 4,836 m^3/yr, using the standard conversion factor of 365 days per year. For the inhalation noncarcinogenic TF, we took a pound of a chemical released per year and converted it to micrograms per cubic meter released by using standard unit conversion factors and the average annual inhalation of air (4,836 m^3/yr):

$$\frac{1 \text{ lb chemical}}{\text{yr}} \times \frac{454{,}000 \text{ mg}}{\text{lb}} \times \frac{1 \text{ yr}}{4{,}836 \text{ m}^3} = \frac{93.879 \text{ mg chemical}}{\text{m}^3}$$

Subsequently, this value was divided by a standard unit conversion factor and the RfCi (mg/m^3) to obtain the inhalation noncarcinogenic TF:

$$\text{Inhalation Noncarcinogenic TF (doses/lb)} = \frac{93.879 \text{ mg chemical}}{\text{m}^3} \times \frac{1}{\text{RfCi (mg/m}^3)}$$

This results in the number of individuals who could equally share/inhale (number of doses) a pound of the chemical released and for whom there would be no adverse noncarcinogenic impact from the chemical.

Single Combined Relative Toxicity Factor Development

What follows are the combination of the individual oral and inhalation TFs developed for cancer and noncancer risks to calculate a single, combined relative TF for TRI chemicals. This chemical-specific single relative TF represents how many individuals could ingest or inhale each pound of chemical per year released to have no adverse carcinogenic or noncarcinogenic impact from the chemical (i.e., the threshold beyond which an adverse effect of some type could occur) in the unit of doses/pound. As noted, the higher the measures of cancer risk (based on USEPA cancer toxicity values) are, the higher the cancer toxicity associated with the chemical will be, whereas the higher the measures of noncancer risk (based on USEPA noncancer toxicity values) are, the lower the noncancer toxicity will be. However, as was described in detail, as part of developing the single relative chemical-specific TFs, all four of the measures of chronic toxicity were converted into a dose relationship (doses/pound), which resulted in a direct relationship to toxicity (the higher the number of doses per pound for a chemical, the higher the toxicity); hence, the carcinogenic and noncarcinogenic effects become additive:

Single Chemical-Specific TF (doses/lb) = Σ (Ingestion Carcinogenic TF, Inhalation Carcinogenic TF, Ingestion Noncarcinogenic TF, Inhalation Noncarcinogenic TF)

For example, a chemical with a calculated TF of 57 doses/lb represents the fact that 1 lb of this chemical can be "shared" equally across 57 individuals (1/57 lb per person) without having an adverse carcinogenic/noncarcinogenic toxic effect, whereas in comparison, another chemical with a higher calculated TF of 129 doses/lb represents the fact that 1 lb of this chemical can be shared equally across 129 individuals (1/129 lb per person) without having an adverse carcinogenic/noncarcinogenic toxic effect. In essence, the higher the TF is, the higher the toxicity will be because each individual can only ingest or inhale a smaller fraction of the pound of chemical released before reaching the toxicity threshold.

Subsequently, each chemical-specific TF was then divided by the current (as of the time of writing this book) population of the United States (306 million) to determine the chemical-specific TF in units of doses/capita-pound:

Single Chemical-Specific TF (doses/capita-lb) = Single TF (doses/lb) ÷ 306,000,000 people in the United States

The TF in doses/capita-pound, as presented in Table 6.1, can be interpreted as follows:

- A chemical-specific TF of 1 dose/capita-lb represents the fact that each of the 306 million people in the United States has the potential to reach the threshold beyond which an adverse effect can occur (carcinogenic or noncarcinogenic) from sharing 1 lb of that chemical, assuming exposure and intake (ingestion or inhalation) are equal across the population.

- A chemical-specific TF of greater than 1 dose/capita-lb (e.g., 500 doses/capita-lb) represents the fact that each of the 306 million people in the United States has 500 times the potential to reach the threshold beyond which an adverse effect can occur from sharing 1 lb of that chemical (carcinogenic or noncarcinogenic), assuming exposure and intake (ingestion or inhalation) are equal across the population.
- A chemical-specific TF of less than 1 dose/capita-lb (e.g., 0.06 doses/capita-lb) represents the fact that less than 306 million people in the United States will be exposed to the threshold beyond which an adverse effect (carcinogenic or noncarcinogenic) can occur from sharing 1 lb of that chemical, assuming exposure and intake (ingestion or inhalation) are equal across the population.

The TRI chemicals for which measures of toxicity were not available are not included in Table 6.1.

In the next few chapters of this book, factors for mobility (Chapter 7), persistence (Chapter 8), and bioconcentration (Chapter 9) are developed for the TRI chemicals. Subsequently, in Chapter 10, these factors are integrated into the chemical-specific TFs developed in this chapter to come up with effective toxicity factors (ETFs) that can then be used to evaluate the relative impacts of the various TRI chemical releases. The purpose of this is to develop a better overall relative toxicity index of releases rather than volume alone to guide policy in reducing these releases, as compared to the present system of reporting on and reducing the total volume of these releases regardless of relative toxicity.

TABLE 6.1
Toxicity Factors of TRI Chemicals

	USEPA Toxicity Values: Carcinogenic		USEPA Toxicity Values: Noncarcinogenic		Toxicity Factor: Carcinogenic		Toxicity Factor: Noncarcinogenic		Toxicity Factor: Combined (TF)
Chemical	**Ingestion (SFO) (mg/kg-day)$^{-1}$**	**Inhalation (IUR) (μg/m^3)$^{-1}$**	**Ingestion (RfDo) (mg/kg-day)**	**Inhalation (RfCi) (mg/m^3)**	**Ingestion (Doses/lb)**	**Inhalation (Doses/lb)**	**Ingestion (Doses/lb)**	**Inhalation (Doses/lb)**	**(Doses/ Capita-lb)**
(1,1′-Biphenyl)-4,4′-diamine,3,3′-dimethyl-	1.10E+01	—	—	—	1.95E+08	—	—	—	6.39E–01
1-(3-Chloroallyl)-3,5,7-triaza-1-azoniaadamantane chloride	—	—	—	—	—	—	—	—	—
1,1,1,2-Tetrachloro-2-fluoroethane (Hcfc-121a)	—	—	—	—	—	—	—	—	—
1,1,1,2-Tetrachloroethane	2.60E–02	7.40E–06	3.00E–02	—	4.62E+05	6.95E+05	5.92E+02	—	3.78E–03
1,1,1-Trichloroethane	—	—	2.00E+00	5.00E+00	—	—	8.88E+00	1.88E+01	9.04E–08
1,1,1-Trifluoro-2,2-dichloroethane	—	—	—	—	—	—	—	—	—
1,1,2,2-Tetrachloro-1-fluoroethane	—	—	—	—	—	—	—	—	—
1,1,2,2-Tetrachloroethane	2.00E–01	5.80E–05	4.00E–03	—	3.55E+06	5.44E+06	4.44E+03	—	2.94E–02
1,1,2-Trichloroethane	5.70E–02	1.60E–05	4.00E–03	—	1.01E+06	1.50E+06	4.44E+03	—	8.23E–03
1,1′-Bi(ethylene oxide)	—	—	—	—	—	—	—	—	—
1,1-Dichloro-1-fluoroethane	—	—	—	—	—	—	—	—	—
1,1-Dichloroethane	5.70E–03	1.60E–06	2.00E–01		1.01E+05	1.50E+05	8.88E+01		8.22E–04
1,1-Dichloroethylene	—	—	5.00E–02	2.00E–01	—	—	3.55E+02	4.69E+02	2.70E–06
1,1-Dimethyl hydrazine	6.10E+00	—	—	—	1.08E+08	—	—	—	3.54E–01
1,1′-Methylenebis(4-isocyanatobenzene)	—	—	—	6.00E–04	-	—	—	1.56E+05	5.11E–04
1,2,3-Trichloropropane	7.00E+00	—	6.00E–03	—	1.24E+08	—	2.96E+03		4.06E–01
1,2,4-Trichlorobenzene	3.60E–03	—	1.00E–02	4.00E–03	6.40E+04	—	1.78E+03	2.35E+04	2.92E–04
1,2,4-Trimethylbenzene	—	—	—	7.00E–03	—	—	—	1.34E+04	4.38E–05

1,2-Butylene oxide	—	—	—	2.00E–02	—	—	—	4.69E+03	1.53E–05
1,2-Dibromo-3-chloropropane (Dbcp)	8.00E–01	6.00E–03	2.00E–04	2.00E–04	1.42E+07	5.63E+08	8.88E+04	4.69E+05	1.89E+00
1,2-Dibromoethane	2.00E+00	6.00E–04	9.00E–03	9.00E–03	3.55E+07	5.63E+07	1.97E+03	1.04E+04	3.00E–01
1,2-Dichloro-1,1,2-trifluoroethane	—	—	—	—	—	—	—	—	—
1,2-Dichloro-1,1-difluoroethane	—	—	—	—	—	—	—	—	—
1,2-Dichlorobenzene	—	—	9.00E–02	2.00E–01	—	—	1.97E+02	4.69E+02	2.18E–06
1,2-Dichloroethane	9.10E–02	2.60E–05	2.00E–02	2.40E+00	1.62E+06	2.44E+06	8.88E+02	3.91E+01	1.33E–02
1,2-Dichloroethylene	—	—	9.00E–03	—	—	—	1.97E+03	—	6.45E–06
1,2-Dichloropropane	3.60E–02	1.00E–05	9.00E–02	4.00E–03	6.40E+05	9.39E+05	1.97E+02	2.35E+04	5.24E–03
1,2-Diphenylhydrazine	8.00E–01	2.20E–04	—	—	1.42E+07	2.07E+07	—	—	1.14E–01
1,3-Butadiene	3.40E+00	3.00E–05	—	2.00E–03	6.04E+07	2.82E+06	—	4.69E+04	2.07E–01
1,3-Dichloro-1,1,2,2,3-pentafluoropropane	—	—	—	—	—	—	—	—	—
1,3-Dichlorobenzene	—	—	—	—	—	—	—	—	—
1,3-Dichloropropene (mixed isomers)	1.00E–01	4.00E–06	3.00E–02	2.00E–02	1.78E+06	3.76E+05	5.92E+02	4.69E+03	7.05E–03
1,4-Dichloro-2-butene	—	4.20E–03	—	—	—	3.94E+08	—	—	1.29E+00
1,4-Dichlorobenzene	5.40E–03	1.10E–05	7.00E–02	8.00E–01	9.60E+04	1.03E+06	2.54E+02	1.17E+02	3.69E–03
1,4-Dioxane	1.10E–02	7.70E–06	1.00E–01	3.60E+00	1.95E+05	7.23E+05	1.78E+02	2.61E+01	3.00E–03
1-Bromo-1-(bromomethyl)-1,3-propanedicarbonitrile	—	—	—	—	—	—	—	—	—
1-Chloro-1,1,2,2-tetrafluoroethane	1.20E+00	—	—	—	2.13E+07	—	—	—	6.97E–02
1-Chloro-1,1-difluoroethane	—	—	—	5.00E+01	—	—	—	1.88E+00	6.14E–09
2,2′,6,6′-Tetrabromo-4,4′-isopropylidenediphenol	1.00E+00	—	—	—	1.78E+07	—	—	—	5.81E–02
2,3-Dichloropropene	—	—	2.00E–02	—	—	—	8.88E+02	—	2.90E–06
2,4,5-Trichlorophenol	—	—	1.00E–01	—	—	—	1.78E+02	—	5.81E–07
2,4,6-Trichlorophenol	1.10E–02	3.10E–06	1.00E–03		1.95E+05	2.91E+05	1.78E+04	—	1.65E–03
2,4,6-Trinitrophenol	—	—	—	—	—	—	—	—	—
2,4-D	—	—	1.00E–02	—	—	—	1.78E+03	—	5.81E–06

continued

TABLE 6.1 (continued)
Toxicity Factors of TRI Chemicals

Chemical	USEPA Toxicity Values: Carcinogenic		USEPA Toxicity Values: Noncarcinogenic		Toxicity Factor: Carcinogenic		Toxicity Factor: Noncarcinogenic		Toxicity Factor: Combined (TF)
	Ingestion (SFO) (mg/kg-day)$^{-1}$	Inhalation (IUR) (μg/m^3)$^{-1}$	Ingestion (RfDo) (mg/kg-day)	Inhalation (RfCi) (mg/m^3)	Ingestion (Doses/lb)	Inhalation (Doses/lb)	Ingestion (Doses/lb)	Inhalation (Doses/lb)	(Doses/ Capita-lb)
2,4-D 2-Ethylhexyl ester	4.90E–01	—	—	—	8.71E+06	—	—	—	2.85E–02
2,4-D Butoxyethyl ester	4.60E–01	—	—	1.00E–01	8.17E+06	—	—	9.39E+02	2.67E–02
2,4-D Butyl ester	—	—	1.00E–02	—	—	—	1.78E+03	—	5.81E–06
2,4-D Sodium salt	1.98E–01	—	—	—	3.52E+06		—	—	1.15E–02
2,4-Db	—	—	8.00E–03	—			2.22E+03	—	7.26E–06
2,4-Diaminoanisole	—	—	—	—			—	—	—
2,4-Diaminotoluene	3.80E+00	1.10E–03	—	—	6.75E+07	1.03E+08	—	—	5.58E–01
2,4-Dichlorophenol	—	—	3.00E–03	—			5.92E+03	—	1.94E–05
2,4-Dimethylphenol	—	—	2.00E–02	—			8.88E+02	—	2.90E–06
2,4-Dinitrophenol	—	—	2.00E–03	—			8.88E+03	—	2.90E–05
2,4-Dinitrotoluene	3.10E–01	8.90E–05	2.00E–03	—	5.51E+06	8.36E+06	8.88E+03	—	4.53E–02
2,6-Dinitrotoluene	—	—	1.00E–03	—			1.78E+04	—	5.81E–05
2,6-Xylidine	1.70E–01	—	—	—	3.02E+06		—	—	9.87E–03
2-Acetylaminofluorene	3.80E+00	1.30E—03	—	—	6.75E+07	1.22E+08	—	—	6.19E–01
2-Aminonaphthalene	1.80E+00	0.00E+00	—	—	3.20E+07	0.00E+00	—	—	1.05E–01
2-Chlor-1,3-butadiene	—	—	2.00E–02	7.00E–03	—	—	8.88E+02	1.34E+04	4.67E–05
2-Chloro-1,1,1,2-tetrafluoroethane	1.70E–01	—	—	—	3.02E+06	—	—	—	9.87E–03
2-Chloro-1,1,1-trifluoroethane					0.00E+00				0.00E+00
2-Chloroacetophenone				3.00E–05				3.13E+06	1.02E–02
2-Mercaptobenzothiazole	1.40E–01	—	—	—	2.49E+06	—	—	—	8.13E–03

2-Methyllactonitrile	—	—	3.00E–03	6.00E–02		—	5.92E+03	1.56E+03	2.45E–05
2-Methylpyridine	1.40E–01	—	—	—	2.49E+06	—	—	—	8.13E–03
2-Nitrophenol	2.20E–01	—	9.00E–04	—	3.91E+06	—	1.97E+04	—	1.28E–02
2-Nitropropane	—	2.70E–03	—	2.00E–02		2.53E+08	—	4.69E+03	8.28E–01
2-Phenylphenol	1.90E–03	—	—	—	3.38E+04	—	—	—	1.10E–04
3,3-Dichloro-1,1,1,2,2-pentafluoropropane	—	—	—	—	—	—	—	—	—
3,3′-Dichlorobenzidine	4.50E–01	3.40E–04	—	—	8.00E+06	3.19E+07	—	—	1.30E–01
3,3′-Dichlorobenzidine dihydrochloride	7.20E–02		—	—	1.28E+06		—	—	4.18E–03
3,3′-Dichlorobenzidine sulfate	4.50E–01	3.40E–04	—	—	8.00E+06	3.19E+07	—	—	1.30E–01
3,3′-Dimethoxybenzidine	1.40E–02	—	—	—	2.49E+05	—	—	—	8.13E–04
3,3′-Dimethoxybenzidine dihydrochloride	7.00E–02	—	—	—	1.24E+06	—	—	—	4.06E–03
3-Chloro-2-methyl-1-propene	4.25E–02	—	—	5.00E–01	7.55E+05	—	—	1.88E+02	2.47E–03
3-Chloropropionitrile	—	—	—	—	—	—	—	—	—
3-Iodo-2-propynyl butylcarbamate	—	—	1.00E–02	—	—	—	1.78E+03	—	5.81E–06
4,4′-Diaminodiphenyl ether	2.00E–02	—	—	—	3.55E+05	—	—	—	1.16E–03
4,4′-Isopropylidenediphenol	—	—	5.00E–02	—	—	—	3.55E+02	—	1.16E–06
4,4′-Methylenebis(2-chloroaniline)	1.00E–01	4.30E–04	2.00E–03	—	1.78E+06	4.04E+07	8.88E+03	—	1.38E–01
4,4′-Methylenebis(N,N-dimethyl) benzenamine	4.60E–02	1.30E–05	—	—	8.17E+05	1.22E+06	—	—	6.66E–03
4,4′-Methylenedianiline	1.60E+00	4.60E–04	—	2.00E–02	2.84E+07	4.32E+07	—	4.69E+03	2.34E–01
4,6-Dinitro-O-cresol	—	—	1.00E–04	—	—	—	1.78E+05	—	5.81E–04
4-Aminoazobenzene	1.88E–02	—	—	—	3.34E+05	—	—	—	1.09E–03
4-Aminobiphenyl	2.10E+01	6.00E–03	—	—	3.73E+08	5.63E+08	—	—	3.06E+00
4-Dimethylaminoazobenzene	4.60E+00	1.30E–03	—	—	8.17E+07	1.22E+08	—	—	6.66E–01
4-Nitrophenol	—	—	2.00E–03	—	—	—	8.88E+03	—	2.90E–05
5-Nitro-O-anisidine	4.90E–02	1.40E–05	—	—	8.71E+05	1.31E+06	—	—	7.14E–03

continued

TABLE 6.1 (continued)
Toxicity Factors of TRI Chemicals

Chemical	USEPA Toxicity Values: Carcinogenic		USEPA Toxicity Values: Noncarcinogenic		Toxicity Factor: Carcinogenic		Toxicity Factor: Noncarcinogenic		Toxicity Factor: Combined (TF)
	Ingestion (SFO) (mg/kg-day)$^{-1}$	Inhalation (IUR) (μg/m^3)$^{-1}$	Ingestion (RfDo) (mg/kg-day)	Inhalation (RfCi) (mg/m^3)	Ingestion (Doses/lb)	Inhalation (Doses/lb)	Ingestion (Doses/lb)	Inhalation (Doses/lb)	(Doses/ Capita-lb)
5-Nitro-O-toluidine	3.30E–02	—	—	—	5.86E+05	—	—	—	1.92E–03
Abamectin	—	—	5.00E–05	—	—	—	3.55E+05	—	1.16E–03
Acephate	8.70E–03	—	4.00E–03	—	1.55E+05	—	4.44E+03	—	5.20E–04
Acetaldehyde	—	2.20E–06	—	9.00E–03	—	2.07E+05	—	1.04E+04	7.09E–04
Acetamide	—	—	—	5.00E–03	—	—	—	1.88E+04	6.14E–05
Acetone	—	—	9.00E–01	3.10E+01	—	—	1.97E+01	3.03E+00	7.44E–08
Acetonitrile	—	—	—	6.00E–02	—	—	—	1.56E+03	5.11E–06
Acetophenone	—	—	1.00E–01	—	—	—	1.78E+02	—	5.81E–07
Acifluorfen, sodium salt	—	—	1.00E–04	—	—	—	1.78E+05	—	5.81E–04
Acrolein	—	—	5.00E–04	2.00E–05	—	—	3.55E+04	4.69E+06	1.55E–02
Acrylamide	4.50E+00	1.30E–03	2.00E–04	—	8.00E+07	1.22E+08	8.88E+04	—	6.60E–01
Acrylic acid	—	—	5.00E–01	1.00E–03	—		3.55E+01	9.39E+04	3.07E–04
Acrylonitrile	5.40E–01	6.80E–05	4.00E–02	2.00E–03	9.60E+06	6.38E+06	4.44E+02	4.69E+04	5.24E–02
Alachlor	5.60E–02	—	1.00E–02	—	9.95E+05	—	1.78E+03	—	3.26E–03
Aldicarb	—	—	1.00E–03	—	—	—	1.78E+04	—	5.81E–05
Aldrin	1.70E+01	4.90E–03	3.00E–05	—	3.02E+08	4.60E+08	5.92E+05	—	2.49E+00
Allyl alcohol	—	—	5.00E–03	3.00E–04	—	—	3.55E+03	3.13E+05	1.03E–03
Allyl amine	—	—	—	2.00E–02	—	—	—	4.69E+03	1.53E–05
Allyl chloride	2.10E–02	6.00E–06	—	1.00E–03	3.73E+05	5.63E+05	—	9.39E+04	3.37E–03
alpha-Naphthylamine	1.80E+00	–	—	—	3.20E+07	—	—	—	1.05E–01

Aluminum	—	—	1.00E+00	5.00E–03	—	—	1.78E+01	1.88E+04	6.14E–05
Aluminum oxide (fibrous forms)	—	—	—	—	—	—	—	—	—
Aluminum phosphide	—	—	4.00E–04	—	—	—	4.44E+04	—	1.45E–04
Ametryn	—	—	9.00E–03	—	—	—	1.97E+03	—	6.45E–06
Amitraz	—	—	2.50E–03	—	—	—	7.11E+03	—	2.32E–05
Amitrole	—	—	—	4.00E–02	0.00E+00	—	—	2.35E+03	7.67E–06
Ammonia	—	—	—	1.00E–01	—	—	—	9.39E+02	3.07E–06
Ammonium nitrate (solution)	—	—	—	—	—	—	—	—	—
Ammonium sulfate (solution)	—	—	—	—	—	—	—	—	—
Aniline	5.70E–03	1.60E–06	7.00E–03	1.00E–03	1.01E+05	1.50E+05	2.54E+03	9.39E+04	1.14E–03
Anthracene	—	—	3.00E–01	—	—	—	5.92E+01	—	1.94E–07
Antimony and antimony compounds	—	—	4.00E–04	—	—	—	4.44E+04	—	1.45E–04
Arsenic and arsenic compounds	1.50E+00	4.30E–03	3.00E–04	1.50E–05	2.67E+07	4.04E+08	5.92E+04	6.26E+06	1.43E+00
Asbestos (friable)	—	—	—	8.00E–02	—	—	—	1.17E+03	3.83E–06
Atrazine	2.30E–01	—	3.50E–02	—	4.09E+06	—	5.08E+02	—	1.34E–02
Auramine	—	—	—	—	—	—	—	—	—
Barium and barium compounds	—	—	2.00E–01	5.00E–04	—	—	8.88E+01	1.88E+05	6.14E–04
Bendiocarb	—	—	—	—	—	—	—	—	—
Benfluralin	—	—	3.00E–01	—	—	—	5.92E+01	—	1.94E–07
Benomyl	—	—	5.00E–02	—	—	—	3.55E+02	—	1.16E–06
Benzal chloride	—	—	—	5.00E–01	0.00E+00	—	—	1.88E+02	6.14E–07
Benzene	5.50E–02	7.80E–06	4.00E–03	3.00E–02	9.77E+05	7.32E+05	4.44E+03	3.13E+03	5.61E–03
Benzidine	2.30E+02	6.70E–02	3.00E–03	—	4.09E+09	6.29E+09	5.92E+03	—	3.39E+01
Benzo(Ghi)perylene	—	—	—	5.00E–01	—	—	—	1.88E+02	6.14E–07
Benzoic trichloride	1.30E+01	—	—	—	2.31E+08	—	—	—	7.55E–01
Benzoyl chloride	—	—	—	1.00E+00	—	—	—	9.39E+01	3.07E–07
Benzoyl peroxide	—	—	—	1.00E+00	—	—	—	9.39E+01	3.07E–07
Benzyl chloride	1.70E–01	4.90E–05	2.00E–03	1.00E–03	3.02E+06	4.60E+06	8.88E+03	9.39E+04	2.52E–02

continued

TABLE 6.1 (continued)
Toxicity Factors of TRI Chemicals

	USEPA Toxicity Values: Carcinogenic		USEPA Toxicity Values: Noncarcinogenic		Toxicity Factor: Carcinogenic		Toxicity Factor: Noncarcinogenic		Toxicity Factor: Combined (TF)
Chemical	Ingestion (SFO) (mg/kg-day)$^{-1}$	Inhalation (IUR) (μg/m^3)$^{-1}$	Ingestion (RfDo) (mg/kg-day)	Inhalation (RfCi) (mg/m^3)	Ingestion (Doses/lb)	Inhalation (Doses/lb)	Ingestion (Doses/lb)	Inhalation (Doses/lb)	(Doses/ Capita-lb)
Beryllium and beryllium compounds	—	2.40E–03	2.00E–03	2.00E–05	—	2.25E+08	8.88E+03	4.69E+06	7.52E–01
Bifenthrin	—	—	1.00E–02	—	—	—	1.78E+03	—	5.81E–06
Biphenyl	—	—	5.00E–02	—	—	—	3.55E+02	—	1.16E–06
Bis(2-chloro-1-methylethyl) ether	7.00E–02	1.00E–05	4.00E–02	—	1.24E+06	9.39E+05	4.44E+02	—	7.13E–03
Bis(2-chloroethoxy)methane	—	—	3.00E–03	—	—	—	5.92E+03	—	1.94E–05
Bis(2-chloroethyl) ether	1.10E+00	3.30E–04	—	—	1.95E+07	3.10E+07		—	1.65E–01
Bis(2-ethylhexyl) adipate	1.20E–03	—	6.00E–01	—	2.13E+04		2.96E+01	—	6.98E–05
Bis(2-ethylhexyl)phthalate	1.40E–02	2.40E–06	2.00E–02	—	2.49E+05	2.25E+05	8.88E+02	—	1.55E–03
Bis(chloromethyl) ether	2.20E+02	6.20E–02	—	—	3.91E+09	5.82E+09	—	—	3.18E+01
Bis(tributyltin) oxide	—	—	3.00E–04	—	—	—	5.92E+04	—	1.94E–04
Boron trichloride	—	—	—	—	—	—	—	—	—
Boron trifluoride	—	—	—	7.00E–04	—	—	—	1.34E+05	4.38E–04
Bromacil	—	—	—	1.00E+00	—	—	—	9.39E+01	3.07E–07
Bromine	—	—	—	—	—	—	—	—	—
Bromochlorodifluoromethane	—	—	—	1.00E+00	—	—	—	9.39E+01	3.07E–07
Bromotrifluoromethane	—	—	—	1.50E+00	—	—	—	6.26E+01	2.05E–07
Bromoxynil	—	—	2.00E–02	—	—	—	8.88E+02	—	2.90E–06
Bromoxynil octanoate	—	—	2.00E–02	—	—	—	8.88E+02	—	2.90E–06
Brucine	—	—	—	—	—	—	—	—	—
Butyl acrylate	—	—	—	1.80E+00	0.00E+00	—	—	5.22E+01	1.70E–07

Butyl benzyl phthalate	1.90E–03	—	2.00E–01	—	3.38E+04	—	8.88E+01	—	1.11E–04
Butyraldehyde	—	—	—	2.00E+01	0.00E+00	—	—	4.69E+00	1.53E–08
C.I. acid red 114	—	—	—	—	—	—	—	—	—
C.I. basic green 4	—	—	—	—	—	—	—	—	—
C.I. basic red 1	—	—	—	—	—	—	—	—	—
C.I. direct blue 218	—	—	—	1.50E+02	—	—	—	6.26E–01	2.05E–09
C.I. disperse yellow 3	—	—	—	—	—	—	—	—	—
C.I. food red 15	—	—	—	—	—	—	—	—	—
C.I. solvent orange 7	—	—	—	—	—	—	—	—	—
C.I. solvent yellow 14	—	—	—	—	—	—	—	—	—
C.I. solvent yellow 3	—	—	—	—	—	—	—	—	—
Cadmium and cadmium compounds	—	4.20E–03	1.00E–03	1.00E–05	—	3.94E+08	1.78E+04	9.39E+06	1.32E+00
Calcium cyanamide	—	—	—	—	—	—	—	—	—
Camphechlor	1.10E+00	3.20E–04	—	—	1.95E+07	3.00E+07	—	—	1.62E–01
Captan	2.30E–03	6.60E–07	1.30E–01	—	4.09E+04	6.20E+04	1.37E+02	—	3.36E–04
Carbaryl	—	—	1.00E–01	—	—	—	1.78E+02	—	5.81E–07
Carbofuran	—	—	5.00E–03	—	—	—	3.55E+03	—	1.16E–05
Carbon disulfide	—	—	1.00E–01	7.00E–01			1.78E+02	1.34E+02	1.02E–06
Carbon tetrachloride	1.30E–01	1.50E–05	7.00E–04	1.90E–01	2.31E+06	1.41E+06	2.54E+04	4.94E+02	1.22E–02
Carbonyl sulfide	—	—	—	4.17E+02	—	—	—	2.25E—01	7.36E–10
Carboxin	—	—	1.00E–01	—	—	—	1.78E+02	—	5.81E–07
Catechol	—	—	—	4.17E+02	—	—	—	2.25E–01	7.36E–10
Chloramben	—	—	1.50E–02	—	—	—	1.18E+03	—	3.87E–06
Chlordane	—	—	2.00E+01	—	—	—	8.88E–01	—	2.90E–09
Chlorendic acid	—	—	—	4.17E+02	—	—	—	2.25E–01	7.36E–10
Chlorimuron ethyl	—	—	7.00E–04	—	—	—	2.54E+04	—	8.30E–05
Chlorine	—	—	1.00E–01	1.50E–04	—	—	1.78E+02	6.26E+05	2.05E–03
Chlorine dioxide	—	—	3.00E–02	2.00E–04	—	—	5.92E+02	4.69E+05	1.54E–03

continued

TABLE 6.1 (continued)
Toxicity Factors of TRI Chemicals

Chemical	USEPA Toxicity Values: Carcinogenic		USEPA Toxicity Values: Noncarcinogenic		Toxicity Factor: Carcinogenic		Toxicity Factor: Noncarcinogenic		Toxicity Factor: Combined (TF)
	Ingestion (SFO) (mg/kg-day)$^{-1}$	Inhalation (IUR) (μg/m^3)$^{-1}$	Ingestion (RfDo) (mg/kg-day)	Inhalation (RfCi) (mg/m^3)	Ingestion (Doses/lb)	Inhalation (Doses/lb)	Ingestion (Doses/lb)	Inhalation (Doses/lb)	(Doses/ Capita-lb)
Chloroacetic acid	—	—	2.00E–03	—	—	—	8.88E+03	—	2.90E–05
Chlorobenzene	—	—	2.00E–02	5.00E–02	—	—	8.88E+02	1.88E+03	9.04E–06
Chlorobenzilate	1.10E–01	3.10E–05	2.00E–02	—	1.95E+06	2.91E+06	8.88E+02	—	1.59E–02
Chlorodifluoromethane	—	—	—	5.00E+01	—	—	—	1.88E+00	6.14E–09
Chloroethane	—	—	—	1.00E+01	—	—	—	9.39E+00	3.07E–08
Chloroform	3.10E–02	2.30E–05	1.00E–02	9.80E–02	5.51E+05	2.16E+06	1.78E+03	9.58E+02	8.87E–03
Chloromethane	—	—	—	9.00E–02	—	—	—	1.04E+03	3.41E–06
Chloromethyl methyl ether	2.40E+00	6.90E–04	—	—	4.26E+07	6.48E+07	—	—	3.51E–01
Chlorophenols	1.20E–01	4.60E–06	3.00E–02	—	2.13E+06	4.32E+05	5.92E+02	—	8.38E–03
Chloropicrin	—	—	—	4.00E–04	—	—	—	2.35E+05	7.67E–04
Chlorotetrafluoroethane	3.00E–02	8.60E–06	—	—	5.33E+05	8.07E+05	—	—	4.38E–03
Chlorothalonil	3.10E–03	8.90E–07	1.50E–02	—	5.51E+04	8.36E+04	1.18E+03	—	4.57E–04
Chlorotrifluoromethane	—	—	—	4.17E+02	—	—	—	2.25E–01	7.36E–10
Chlorpyrifos methyl	—	—	1.00E–02	—	—	—	1.78E+03	—	5.81E–06
Chlorsulfuron	—	—	1.30E–02	—	—	—	1.37E+03	—	4.47E–06
Chromium and chromium compounds	—	1.20E–02	—	—	—	1.13E+09	—	—	3.68E+00
Cobalt and cobalt compounds	—	9.00E–03	3.00E–04	6.00E–06	—	8.45E+08	5.92E+04	1.56E+07	2.81E+00
Copper and copper compounds	—	—	4.00E–02		—		4.44E+02	—	1.45E–06
Creosotes	—	—	—	4.17E+02	—		—	2.25E–01	7.36E–10
Cresol (mixed isomers)	—	—	1.00E–01	6.00E–01	—		1.78E+02	1.56E+02	1.09E–06

Crotonaldehyde	1.90E+00	—	—	—	3.38E+07		—	—	1.10E–01
Cumene	—	—	1.00E–01	4.00E–01	—		1.78E+02	2.35E+02	1.35E–06
Cumene hydroperoxide	—	—	1.00E–01	4.00E–01	—		1.78E+02	2.35E+02	1.35E–06
Cupferron	—	—		—	—		—	—	0.00E+00
Cyanazine	8.40E–01	—	2.00E–03	—	1.49E+07		8.88E+03	—	4.88E–02
Cyanide compounds	—	—	2.00E–02	3.00E–03	—	—	8.88E+02	3.13E+04	1.05E–04
Cycloate	—	—	—	—	—	—	—	—	0.00E+00
Cyclohexane	—	—	—	6.00E+00	—	—	—	1.56E+01	5.11E–08
Cyclohexanol	—	—	2.00E–01	—	—	—	8.88E+01	—	2.90E–07
Cyfluthrin	1.50E–01	—	9.00E–03	—	2.67E+06		1.97E+03	—	8.72E–03
Cyhalothrin	—	—	3.00E–04	—	—	—	5.92E+04	—	1.94E–04
Dazomet	—	—	—	4.17E+02	—	—	—	2.25E–01	7.36E–10
Dazomet, sodium salt	—	—	—	4.17E+02	—	—	—	2.25E–01	7.36E–10
Decabromodiphenyl oxide	7.00E–04	—	7.00E–03	—	1.24E+04		2.54E+03	—	4.89E–05
Desmedipham	—	—	—	4.17E+02	—	—	—	2.25E–01	7.36E–10
Diallate	6.10E–02	—	—	—	1.08E+06		—	—	3.54E–03
Diaminotoluene (mixed isomers)	—	—	—	—	—	—	—	—	0.00E+00
Diazinon	—	—	7.00E–04	—	—	—	2.54E+04	—	8.30E–05
Dibenzofuran	—	—	—	—	—	—	—	—	—
Dibromotetrafluoroethane (Halon 2402)	—	—	—	—	—	—	—	—	—
Dibutyl phthalate	—	—	1.00E–01	—	—	—	1.78E+02	—	5.81E–07
Dicamba	—	—	3.00E–02	—	—	—	5.92E+02	—	1.94E–06
Dichloran	—	—	—	—	—	—	—	—	—
Dichlorobenzene (mixed isomers)	—	—	9.00E–02	2.00E–01	—	—	1.97E+02	4.69E+02	2.18E–06
Dichlorobromomethane	6.20E–02	3.70E–05	2.00E–02	—	1.10E+06	3.47E+06	8.88E+02	—	1.50E–02
Dichlorodifluoromethane	—	—	2.00E–01	2.00E–01	—	—	8.88E+01	4.69E+02	1.82E–06
Dichlorofluoromethane	—	—	—	4.00E+01	—	—	—	2.35E+00	7.67E–09
Dichloromethane	7.50E–03	4.70E–07	6.00E–02	1.00E+00	1.33E+05	4.41E+04	2.96E+02	9.39E+01	5.81E–04

continued

TABLE 6.1 (continued)
Toxicity Factors of TRI Chemicals

	USEPA Toxicity Values: Carcinogenic		USEPA Toxicity Values: Noncarcinogenic		Toxicity Factor: Carcinogenic		Toxicity Factor: Noncarcinogenic		Toxicity Factor: Combined (TF)
Chemical	Ingestion (SFO) (mg/kg-day)$^{-1}$	Inhalation (IUR) (μg/m^3)$^{-1}$	Ingestion (RfDo) (mg/kg-day)	Inhalation (RfCi) (mg/m^3)	Ingestion (Doses/lb)	Inhalation (Doses/lb)	Ingestion (Doses/lb)	Inhalation (Doses/lb)	(Doses/ Capita-lb)
Dichlorotetrafluoroethane (Cfc-114)	—	—	2.00E–01	2.00E–01	—	—	8.88E+01	4.69E+02	1.82E–06
Dichlorotrifluoroethane	—	—	2.00E–01	2.00E–01	—	—	8.88E+01	4.69E+02	1.82E–06
Dichlorpentafluoro-propane	—	—	2.00E–01	2.00E–01	—	—	8.88E+01	4.69E+02	1.82E–06
Dichlorvos	2.90E–01	8.30E–05	5.00E–04	5.00E–04	5.15E+06	7.79E+06	3.55E+04	1.88E+05	4.30E–02
Dicofol	—	—	—	—	—	—	—	—	—
Dicyclopentadiene	—	—	8.00E–03	7.00E–03	—	—	2.22E+03	1.34E+04	5.11E–05
Diethanolamine	—	—		4.17E+02	—	—	—	2.25E–01	7.36E–10
Diethyl phthalate	—	—	8.00E–01	—	—	—	2.22E+01	—	7.26E–08
Diethyl sulfate	—	—		4.17E+02	0.00E+00	—	—	2.25E–01	7.36E–10
Diflubenzuron	—	—	2.00E–02	—	—	—	8.88E+02	—	2.90E–06
Diglycidyl resorcinol ether (Dgre)	—	—	—	4.17E+02	—	—	—	2.25E–01	7.36E–10
Dihydrosafrole	—	—	—	—	—	—	—	—	—
Diisocyanates	—	—	—	6.00E–04	—	—	—	1.56E+05	5.11E–04
Dimethipin	—	—	2.00E–02	—	—	—	8.88E+02	—	2.90E–06
Dimethoate	—	—	2.00E–04	—	—	—	8.88E+04	—	2.90E–04
Dimethyl chlorothiophosphate	—	—	-	—	—	—	—	—	—
Dimethyl phthalate	—	—	8.00E–01	—	—	—	2.22E+01	—	7.26E–08
Dimethyl sulfate	—	—	—	—	—	—	—	—	—
Dimethylamine	—	—	—	7.00E–03	—	—	—	1.34E+04	4.38E–05
Dimethylamine dicamba	—	—	3.00E–02	—	—	—	5.92E+02	—	1.94E–06

Dimethylcarbamoyl chloride	—	—	—	—	—	—	—	—	—
Dinitrobutyl phenol	—	—	1.00E–03	—	—	—	1.78E+04	—	5.81E–05
Dinitrotoluene (mixed isomers)	6.80E–01	—	—	—	1.21E+07	—	—	—	3.95E–02
Dinocap		—	—	—	—	—	—	—	—
Di-N-propylnitrosamine	7.00E+00	2.00E–03	—	—	1.24E+08	1.88E+08	—	—	1.02E+00
Dioxin and dioxin-like compounds	1.30E+05	3.80E+01	1.00E–09	4.00E–08	2.31E+12	3.57E+12	1.78E+10	2.35E+09	1.93E+04
Diphenylamine	—	—	2.50E–02	—	—	—	7.11E+02	—	2.32E–06
Dipotassium endothall	—	—	2.00E–02	—	—	—	8.88E+02	—	2.90E–06
Dipropyl isocinchomeronate	—	—	—	—	—	—	—	—	—
Direct black 38	7.40E+00	2.10E–03	—	—	1.31E+08	1.97E+08	—	—	1.07E+00
Disodium cyanodithioimidocarbonate	—	—	—	—	—	—	—	—	—
Dithiobiuret	—	—	—	—	—	—	—	—	—
Diuron	—	—	2.00E–03	—	—	—	8.88E+03	—v	2.90E–05
Dodine	—	—	4.00E–03	—	—	—	4.44E+03	—	1.45E–05
D-Trans-allethrin	—	—	—	—	—	—	—	—	0.00E+00
Epichlorohydrin	9.90E–03	1.20E–06	6.00E–03	1.00E–03	1.76E+05	1.13E+05	2.96E+03	9.39E+04	1.26E–03
Ethoprop	—	—	—	—	—	—	—	—	—
Ethyl acrylate	4.80E–02	—	—	—	8.53E+05	—	—	—	2.79E–03
Ethyl chloroformate	—	—	—	—	—	—	—	—	—
Ethyl dipropylthiocarbamate	—	—	2.50E–02	—	—	—	7.11E+02	—	2.32E–06
Ethylbenzene	1.10E–02	2.50E–06	1.00E–01	1.00E+00	1.95E+05	2.35E+05	1.78E+02	9.39E+01	1.41E–03
Ethylene	—	—	—	—	—	—	—	—	—
Ethylene glycol	—	—	2.00E+00	4.00E–01	—	—	8.88E+00	2.35E+02	7.96E–07
Ethylene glycol monoethyl ether	—	—	4.00E–01	2.00E–01	—	—	4.44E+01	4.69E+02	1.68E–06
Ethylene glycol monomethyl ether	—	—	3.00E–03	2.00E–02	—	—	5.92E+03	4.69E+03	3.47E–05
Ethylene oxide	3.10E–01	8.80E–05	—	3.00E–02	5.51E+06	8.26E+06	—	3.13E+03	4.50E–02
Ethylene thiourea	4.50E–02	1.30E–05	8.00E–05	–	8.00E+05	1.22E+06	2.22E+05	—	7.33E–03

continued

TABLE 6.1 (continued)
Toxicity Factors of TRI Chemicals

Chemical	USEPA Toxicity Values: Carcinogenic		USEPA Toxicity Values: Noncarcinogenic		Toxicity Factor: Carcinogenic		Toxicity Factor: Noncarcinogenic		Toxicity Factor: Combined (TF)
	Ingestion (SFO) $(mg/kg\text{-}day)^{-1}$	Inhalation (IUR) $(\mu g/m^3)^{-1}$	Ingestion (RfDo) (mg/kg-day)	Inhalation (RfCi) (mg/m^3)	Ingestion (Doses/lb)	Inhalation (Doses/lb)	Ingestion (Doses/lb)	Inhalation (Doses/lb)	(Doses/ Capita-lb)
Ethylenebisdithiocarbamic acid, salts, and esters	—	—	—	—	—	—	—	—	—
Ethyleneimine	—	—	—	—	—	—	—	—	—
Famphur	—	—	—	—	—	—	—	—	—
Fenarimol	3.00E+01	8.60E–03	7.00E–06	—	5.33E+08	8.07E+08	2.54E+06	—	4.39E+00
Fenbutatin oxide	—	—	—	—	—	—	—	—	0.00E+00
Fenoxycarb	9.00E–03	—	—	—	1.60E+05	—	—	—	5.23E–04
Fenpropathrin	—	—	2.50E–02	—	—	—	7.11E+02	—	2.32E–06
Fenthion	—	—	—	—	—	—	—	—	—
Fenvalerate	—	—	2.50E–02	—	—	—	7.11E+02	—	2.32E–06
Fluazifop-butyl	—	—	—	—	—	—	—	—	—
Fluometuron	—	—	1.30E–02	—	—	—	1.37E+03	—	4.47E–06
Fluorine	—	—	6.00E–02	—	—	—	2.96E+02	—	9.68E–07
Fluoroacetic acid, sodium salt	—	—	2.00E–05	—	—	—	8.88E+05	—	2.90E–03
Fluorouracil	—	—	—	—	—	—	—	—	—
Fluvalinate	—	—	2.50E–02	—	—	—	7.11E+02	—	2.32E–06
Folpet	3.50E–03	—	1.00E–01	—	6.22E+04	—	1.78E+02	—	2.04E–04
Fomesafen	1.30E+01	3.80E–03	—	—	2.31E+08	3.57E+08	—	—	1.92E+00
Formaldehyde	—	1.30E–05	2.00E–01	9.80E–03	—	1.22E+06	8.88E+01	9.58E+03	4.02E–03
Formic acid	—	—	2.00E+00	3.00E–03	—	—	8.88E+00	3.13E+04	1.02E–04

Freon 113	—	—	3.00E+01	3.00E+01	—	—	5.92E–01	3.13E+00	1.22E–08
gamma-Lindane	1.10E+00	3.10E–04	3.00E–04	—	1.95E+07	2.91E+07	5.92E+04	—	1.59E–01
Glycol ethers	—	—	5.00E–01	1.30E+01	—	—	3.55E+01	7.22E+00	1.40E–07
Heptachlor	4.50E+00	1.30E–03	5.00E–04	—	8.00E+07	1.22E+08	3.55E+04	—	6.60E–01
Hexachloro-1,3-butadiene	7.80E–02	2.20E–05	1.00E–03	—	1.39E+06	2.07E+06	1.78E+04	—	1.13E–02
Hexachlorobenzene	1.60E+00	4.60E–04	8.00E–04	—	2.84E+07	4.32E+07	2.22E+04	—	2.34E–01
Hexachlorocyclopentadiene	—	—	6.00E–03	2.00E–04	—	—	2.96E+03	4.69E+05	1.54E–03
Hexachloroethane	1.40E–02	4.00E–06	1.00E–03	—	2.49E+05	3.76E+05	1.78E+04	—	2.10E–03
Hexachlorophene (Hcp)	—	—	3.00E–04	—	—	—	5.92E+04	—	1.94E–04
Hexamethylphosphoramide	—	—	—	—	—	—	—	—	—
Hexazinone	—	—	3.30E–02	—	—	—	5.38E+02	—	1.76E–06
Hydramethylnon	—	—	6.00E–02	—	—	—	2.96E+02	—	9.68E–07
Hydrazine	3.00E+00	4.90E–03	—	2.00E–04	5.33E+07	4.60E+08	—	4.69E+05	1.68E+00
Hydrazine sulfate	3.00E+00	4.90E–03	—	—	5.33E+07	4.60E+08	—	—	1.68E+00
Hydrochloric acid	—	—	—	2.00E–02	—	—	—	4.69E+03	1.53E–05
Hydrofluoric acid	—	—	4.00E–02	1.40E–02	—	—	4.44E+02	6.71E+03	2.34E–05
Hydrogen cyanide	—	—	2.00E–02	3.00E–03	—	—	8.88E+02	3.13E+04	1.05E–04
Hydroquinone	5.60E–02	—	4.00E–02	—	9.95E+05	—	4.44E+02	—	3.25E–03
Iron pentacarbonyl	—	—	—	—	—	—	—	—	—
Isobutyraldehyde	—	—	—	—	—	—	—	—	—
Isodrin	—	—	—	—	—	—	—	—	—
Isofenphos	—	—	—	—	—	—	—	—	—
Isopropyl alcohol	—	—	—	7.00E+00	—	—	—	1.34E+01	4.38E–08
Isosafrole	—	—	—	—	—	—	—	—	—
Lactofen	—	—	1.30E–02	—	—	—	1.37E+03	—	4.47E–06
Lead and lead compounds	—	—	1.00E–07	—	—	—	1.78E+08	—	5.81E–01
Linuron	—	—	2.00E–03	—	—	—	8.88E+03	—	2.90E–05
Lithium carbonate	—	—	5.00E–03	—	—	—	3.55E+03	—	1.16E–05

continued

TABLE 6.1 (continued)
Toxicity Factors of TRI Chemicals

	USEPA Toxicity Values: Carcinogenic		USEPA Toxicity Values: Noncarcinogenic		Toxicity Factor: Carcinogenic		Toxicity Factor: Noncarcinogenic		Toxicity Factor: Combined (TF)
Chemical	**Ingestion (SFO) (mg/kg-day)$^{-1}$**	**Inhalation (IUR) (μg/m^3)$^{-1}$**	**Ingestion (RfDo) (mg/kg-day)**	**Inhalation (RfCi) (mg/m^3)**	**Ingestion (Doses/lb)**	**Inhalation (Doses/lb)**	**Ingestion (Doses/lb)**	**Inhalation (Doses/lb)**	**(Doses/ Capita-lb)**
Malathion	—	—	2.00E–02	—	—	—	8.88E+02	—	2.90E–06
Maleic anhydride	—	—	1.00E–01	7.00E–04	—	—	1.78E+02	1.34E+05	4.39E–04
Malononitrile	—	—	1.00E–04	—	—	—	1.78E+05	—	5.81E–04
Maneb	—	—	5.00E–03	—	—	—	3.55E+03	—	1.16E–05
Manganese and manganese compounds	—	—	1.40E–01	5.00E–05	—	—	1.27E+02	1.88E+06	6.14E–03
M-Cresol	—	—	5.00E–02	—	—	—	3.55E+02	—	1.16E–06
M-Dinitrobenzene	—	—	1.00E–04	—	—	—	1.78E+05	—	5.81E–04
Mecoprop	—	—	1.00E–03	—	—	—	1.78E+04	—	5.81E–05
Mercury and mercury compounds	—	—	1.60E–03	3.00E–04	—	—	1.11E+04	3.13E+05	1.06E–03
Merphos	—	—	3.00E–05	—	—	—	5.92E+05	—	1.94E–03
Methacrylonitrile	—	—	1.00E–04	7.00E–04	—	—	1.78E+05	1.34E+05	1.02E–03
Metham sodium	—	—	—	—	—	—	—	—	—
Methanamine, N-methyl-N-nitroso	5.10E+01	1.40E–02	8.00E–06	—	9.06E+08	1.31E+09	2.22E+06	—	7.26E+00
Methanol	—	—	5.00E–01	4.00E+00	—	—	3.55E+01	2.35E+01	1.93E–07
Methoxone	—	—	5.00E–04	—	—	—	3.55E+04	—	1.16E–04
Methoxychlor	—	—	5.00E–03	—	—	—	3.55E+03	—	1.16E–05
Methyl acrylate	—	—	3.00E–02	—	—	—	5.92E+02	—	1.94E–06
Methyl bromide	—	—	1.40E–03	5.00E–03	—	—	1.27E+04	1.88E+04	1.03E–04
Methyl chlorocarbonate	—	—	—	—	—	—	—	—	—
Methyl ethyl ketone	—	—	6.00E–01	5.00E+00	—	—	2.96E+01	1.88E+01	1.58E–07

Methyl hydrazine	3.00E+00	4.90E–03	—	—	5.33E+07	4.60E+08	—	—	1.68E+00
Methyl iodide	—	—	—	—	—	—	—	—	—
Methyl isobutyl ketone	—	—	8.00E–02	3.00E+00	—	—	2.22E+02	3.13E+01	8.28E–07
Methyl isocyanate	3.90E–02	1.10E–05	—	7.00E–05	6.93E+05	1.03E+06	—	1.34E+06	1.00E–02
Methyl isothiocyanate	—	—	1.00E–01	—	—	—	1.78E+02	—	5.81E–07
Methyl methacrylate	—	—	1.40E+00	7.00E–01	—	—	1.27E+01	1.34E+02	4.80E–07
Methyl parathion	—	—	2.50E–04	—	—	—	7.11E+04	—	2.32E–04
Methyl tert-butyl ether	1.80E–03	2.60E–07	—	3.00E+00	3.20E+04	2.44E+04	—	3.13E+01	1.84E–04
Methylene bromide	—	—	1.00E–02	—	—	—	1.78E+03	—	5.81E–06
Metribuzin	—	—	2.50E–02	—	—	—	7.11E+02	—	2.32E–06
Michler's ketone	—	—	—	—	—	—	—	—	—
Molinate	—	—	2.00E–03	—	—	—	8.88E+03	—	2.90E–05
Molybdenum trioxide	—	—	—	—	—	—	—	—	—
Monochloropentafluoroethane	—	—	—	—	—	—	—	—	—
M-Phenylenediamine	—	—	6.00E–03	—	—	—	2.96E+03	—	9.68E–06
M-Xylene	—	—	2.00E+00	7.00E–01	—	—	8.88E+00	1.34E+02	4.67E–07
Myclobutanil	—	—	2.00E–01	—	—	—	8.88E+01	—	2.90E–07
N,N-Dimethylaniline	—	—	2.00E–03	—	—	—	8.88E+03	—	2.90E–05
N,N-Dimethylformamide	—	—	1.00E–01	3.00E–02	—	—	1.78E+02	3.13E+03	1.08E–05
Nabam	—	—	-	—	—	—	—	—	—
Naled	—	—	2.00E–03	—	—	—	8.88E+03	—	2.90E–05
Naphthalene	—	3.40E–05	2.00E–02	3.00E–03	—	3.19E+06	8.88E+02	3.13E+04	1.05E–02
N-Butyl alcohol	—	—	1.00E–01	—	—	—	1.78E+02	—	5.81E–07
N-Dioctyl phthalate	—	—	—	—	—	—	—	—	—
N-Ethyl-N-nitrosourea	2.70E+01	7.70E–03	—	—	4.80E+08	7.23E+08	—	—	3.93E+00
N-Hexane	—	—	6.00E–02	7.00E–01	—	—	2.96E+02	1.34E+02	1.41E–06
Nickel and nickel compounds	—	2.60E–04	2.00E–02	9.00E–05	—	2.44E+07	8.88E+02	1.04E+06	8.32E–02
Nicotine and salts	—	—	—	—	—	—	—	—	—

continued

TABLE 6.1 (continued)
Toxicity Factors of TRI Chemicals

	USEPA Toxicity Values: Carcinogenic		USEPA Toxicity Values: Noncarcinogenic		Toxicity Factor: Carcinogenic		Toxicity Factor: Noncarcinogenic		Toxicity Factor: Combined (TF)
Chemical	Ingestion (SFO) (mg/kg-day)$^{-1}$	Inhalation (IUR) (μg/m^3)$^{-1}$	Ingestion (RfDo) (mg/kg-day)	Inhalation (RfCi) (mg/m^3)	Ingestion (Doses/lb)	Inhalation (Doses/lb)	Ingestion (Doses/lb)	Inhalation (Doses/lb)	(Doses/ Capita-lb)
Nitrapyrin	—	—	—	—	—	—	—	—	—
Nitrate compounds	—	—	—	—	—	—	—	—	4.52E–08
Nitric acid	—	—	—	—	—	—	—	—	—
Nitrilotriacetic acid	—	—	—	—	—	—	—	—	—
Nitrobenzene	—	4.00E–05	2.00E–03	9.00E–03	—	3.76E+06	8.88E+03	1.04E+04	1.23E–02
Nitrofen	—	—	—	—	—	—	—	—	—
Nitroglycerin	1.70E–02	—	1.00E–04	—	3.02E+05	—	1.78E+05	—	1.57E–03
N-Methyl-2-pyrrolidone	—	—	—	—	—	—	—	—	0.00E+00
N-Methylolacrylamide	4.50E+00	1.30E–03	2.00E–04	—	8.00E+07	1.22E+08	8.88E+04	—	6.60E–01
N-Nitrosodiethylamine	1.50E+02	4.30E–02	—	—	2.67E+09	4.04E+09	—	—	2.19E+01
N-Nitrosodi-N-butylamine	5.40E+00	1.60E–03	—	—	9.60E+07	1.50E+08	—	—	8.04E–01
N-Nitrosodiphenylamine	4.90E–03	2.60E–06	—	—	8.71E+04	2.44E+05	—	—	1.08E–03
N-Nitrosomethylvinylamine	—	—	—	—			—	—	—
N-Nitroso-N-methylurea	1.20E+02	3.40E–02	—	—	2.13E+09	3.19E+09	—	—	1.74E+01
N-Nitrosopiperidine	9.40E+00	2.70E–03	—	—	1.67E+08	2.53E+08	—	—	1.37E+00
Norflurazon	—	—	4.00E–02	—	—	—	4.44E+02	—	1.45E–06
O-Anisidine	4.90E–02	1.40E–05	—	—	8.71E+05	1.31E+06	—	—	7.14E–03
O-Cresol	—	—	5.00E–02	—	—	—	3.55E+02	—	1.16E–06
O-Dinitrobenzene	—	—	1.00E–04	—	—	—	1.78E+05	—	5.81E–04
O-Phenylenediamine	4.70E–02	—	—	—	8.35E+05	—	—	—	2.73E–03

O-Phenylphenate, sodium	—	—	—	—	—	—	—	—	—
Oryzalin	—	—	5.00E–02	—	—	—	3.55E+02	—	1.16E–06
Osmium oxide oso4 (T-4)	—	—	—	—	—	—	—	—	—
O-Toluidine	1.80E–01	5.10E–05	—	—	3.20E+06	4.79E+06	—	—	2.61E–02
O-Toluidine hydrochloride	1.30E–01	3.70E–05	—	—	2.31E+06	3.47E+06	—	—	1.89E–02
Oxydemeton methyl	—	—	—	—	—	—	—	—	0.00E+00
Oxydiazon	—	—	5.00E–03	—	—	—	3.55E+03		1.16E–05
Oxyfluorfen	—	—	3.00E–03	—	—	—	5.92E+03		1.94E–05
O-Xylene	—	—	2.00E+00	7.00E–01	—	—	8.88E+00	1.34E+02	4.67E–07
Ozone	—	—	—	—	—	—	—	—	—
P-Anisidine	4.90E–02	1.40E–05	—	—	8.71E+05	1.31E+06	—	—	7.14E–03
Paraldehyde	—	—	—	—	—	—	—	—	—
Paraquat	—	—	4.50E–03	—	—	—	3.95E+03	—	1.29E–05
Parathion	—	—	6.00E–03	—	—	—	2.96E+03	—	9.68E–06
P-Chloroaniline	2.00E–01	—	4.00E–03	—	3.55E+06	—	4.44E+03	—	1.16E–02
P-Cresidine	—	—	—	—	—	—	—	—	—
P-Cresol	—	—	5.00E–03	—	—	—	3.55E+03	—	1.16E–05
P-Dinitrobenzene	—	—	1.00E–04	—	—	—	1.78E+05	—	5.81E–04
Pebulate	—	—	5.00E–02	—	—	—	3.55E+02	—	1.16E–06
Pendimethalin	—	—	4.00E–02	—	—	—	4.44E+02	—	1.45E–06
Pentachlorobenzene	—	—	8.00E–04	—	—	—	2.22E+04	—	7.26E–05
Pentachloroethane	9.00E–02	—	—	—	1.60E+06	—	—	—	5.23E–03
Pentachlorophenol	1.20E–01	4.60E–06	3.00E–02	—	2.13E+06	4.32E+05	5.92E+02	—	8.38E–03
Peracetic acid	—	—	—	—	—	—	—	—	—
Perchloromethyl mercaptan	—	—	—	—	—	—	—	—	—
Permethrin	—	—	5.00E–02	—	—	—	3.55E+02	—	1.16E–06
Phenanthrene	—	—	—	—	—	—	—	—	—
Phenol	—	—	3.00E–01	2.00E–01	—	—	5.92E+01	4.69E+02	1.73E–06

continued

TABLE 6.1 (continued)
Toxicity Factors of TRI Chemicals

Chemical	USEPA Toxicity Values: Carcinogenic		USEPA Toxicity Values: Noncarcinogenic		Toxicity Factor: Carcinogenic		Toxicity Factor: Noncarcinogenic		Toxicity Factor: Combined (TF)
	Ingestion (SFO) (mg/kg-day)$^{-1}$	Inhalation (IUR) (μg/m^3)$^{-1}$	Ingestion (RfDo) (mg/kg-day)	Inhalation (RfCi) (mg/m^3)	Ingestion (Doses/lb)	Inhalation (Doses/lb)	Ingestion (Doses/lb)	Inhalation (Doses/lb)	(Doses/ Capita-lb)
Phenothrin	—	—	—	—	—	—	—	—	—
Phenytoin	—	—	—	—	—	—	—	—	—
Phosgene	—	—	—	3.00E–04	—	—	—	3.13E+05	1.02E–03
Phosphine	—	—	3.00E–04	3.00E–04	—	—	5.92E+04	3.13E+05	1.22E–03
Phosphorus (yellow or white)	—	—	2.00E–05	—	—	—	8.88E+05	—	2.90E–03
Phosphoric acid	—	—	—	1.00E–02	—	—	—	9.39E+03	3.07E–05
Phthalic anhydride	—	—	2.00E+00	2.00E–02	—	—	8.88E+00	4.69E+03	1.54E–05
Picloram	—	—	7.00E–02	—	—	—	2.54E+02	—	8.30E–07
Piperonyl butoxide	—	—	—	—	—	—	—	—	—
Pirimiphos methyl	—	—	1.00E–02	—	—	—	1.78E+03	—	5.81E–06
P-Nitroaniline	2.00E–02	—	4.00E–03	6.00E–03	3.55E+05	—	4.44E+03	1.56E+04	1.23E–03
P-Nitrosodiphenylamine	4.90E–03	2.60E–06	—	—	8.71E+04	2.44E+05	—	—	1.08E–03
Polychlorinated alkanes (C10–C13)	—	—	8.00E–03	—	—	—	2.22E+03	—	7.26E–06
Polychlorinated biphenyls	2.00E+00	5.70E–04	—		3.55E+07	5.35E+07	—	—	2.91E–01
Polycyclic aromatic compounds	7.30E+00	1.10E–03	—	—	1.30E+08	1.03E+08	—	—	7.61E–01
Potassium bromate	—	—	—	—	—	—	—	—	—
Potassium dimethyldithiocarbamate	—	—	—	—	—	—	—	—	—
Potassium N-methyldithiocarbamate	—	—	—	—	—	—	—	—	—
P-Phenylenediamine	—	—	1.90E–01	—	—	—	9.35E+01	—	3.06E–07
Profenofos	—	—	—	—	—	—	—	—	—

Prometryn	—	—	4.00E–03	—	—	—	4.44E+03	—	1.45E–05
Pronamide	—	—	7.50E–02	—	—	—	2.37E+02	—	7.74E–07
Propachlor	—	—	1.30E–02	—	—	—	1.37E+03	—	4.47E–06
Propane sultone	—	—	—	—	—	—	—	—	—
Propanil	—	—	5.00E–03	—	—	—	3.55E+03	—	1.16E–05
Propargite	—	—	2.00E–02	—	—	—	8.88E+02	—	2.90E–06
Propargyl alcohol	—	—	2.00E–03	—	—	—	8.88E+03	—	2.90E–05
Propetamphos	—	—	—	—	—	—	—	—	—
Propiconazole	—	—	1.30E–02	—	—	—	1.37E+03	—	4.47E–06
Propionaldehyde	—	—	—	8.00E–03	—	—	—	1.17E+04	3.83E–05
Propoxur	—	—	4.00E–03	—	—	—	4.44E+03	—	1.45E–05
Propylene	—	—	—	—	—	—	—	—	—
Propylene oxide	2.40E–01	3.70E–06	—	3.00E–02	4.26E+06	3.47E+05	—	3.13E+03	1.51E–02
Propyleneimine	—	—	—	—	—	—	—	—	—
P-Xylene	—	—	—	7.00E–01	—	—	—	1.34E+02	4.38E–07
Pyridine	—	—	1.00E–03	—	—	—	1.78E+04	—	5.81E–05
Quinoline	3.00E+00	—	—	—	5.33E+07	—	—	—	1.74E–01
Quinone	—	—	—	—	—	—	—	—	—
Quintozene	2.60E–01	—	3.00E–03	—	4.62E+06	—	5.92E+03	—	1.51E–02
Quizalofop-ethyl	—	—	2.50E–01	—	—	—	7.11E+01	—	2.32E–07
Resmethrin	—	—	3.00E–02	—	—	—	5.92E+02	—	1.94E–06
S,S,S-Tributyltrithiophosphate	—	—	3.00E–05	—	—	—	5.92E+05	—	1.94E–03
Saccharin	—	—	–	—	—	—	—	—	—
Safrole	2.20E–01	6.30E–05	–	—	3.91E+06	5.91E+06	—	—	3.21E–02
sec-Butyl alcohol	—	—	2.00E+00	3.00E+01	—	—	8.88E+00	3.13E+00	3.93E–08
Selenium and selenium compounds	—	—	5.00E–03	2.00E–02	—	—	3.55E+03	4.69E+03	2.70E–05
Sethoxydim	1.90E–01	—	—	—	3.38E+06	—	—	—	1.10E–02
Silver and silver compounds	—	—	5.00E–03	—	—	—	3.55E+03	—	1.16E–05

continued

TABLE 6.1 (continued)
Toxicity Factors of TRI Chemicals

	USEPA Toxicity Values: Carcinogenic		USEPA Toxicity Values: Noncarcinogenic		Toxicity Factor: Carcinogenic		Toxicity Factor: Noncarcinogenic		Toxicity Factor: Combined (TF)
Chemical	**Ingestion (SFO) (mg/kg-day)$^{-1}$**	**Inhalation (IUR) (μg/m^3)$^{-1}$**	**Ingestion (RfDo) (mg/kg-day)**	**Inhalation (RfCi) (mg/m^3)**	**Ingestion (Doses/lb)**	**Inhalation (Doses/lb)**	**Ingestion (Doses/lb)**	**Inhalation (Doses/lb)**	**(Doses/ Capita-lb)**
Simazine	1.20E–01	—	5.00E–03	—	2.13E+06	—	3.55E+03	—	6.98E–03
Sodium azide	—	—	4.00E–03	—	—	—	4.44E+03	—	1.45E–05
Sodium dicamba	—	—	3.00E–02	—	—	—	5.92E+02	—	1.94E–06
Sodium dimethyldithiocarbamate	—	—	—	—	—	—	—	—	—
Sodium hydroxide (solution)	—	—	—	—	—	—	—	—	—
Sodium nitrite	—	—	—	—	—	—	—	—	—
Strychnine	—	—	3.00E–04	—	—	—	5.92E+04	—	1.94E–04
Styrene	—	—	2.00E–01	1.00E+00	—	—	8.88E+01	9.39E+01	5.97E–07
Styrene oxide	—	—	—	—	—	—	—	—	—
Sulfuric acid	—	—	—	—	—	—	—	—	8.13E–09
Sulfuryl fluoride	—	—	—	—	—	—	—	—	—
Sulprofos	—	—	—	—	—	—	—	—	—
Tebuthiuron	—	—	7.00E–02	—	—	—	2.54E+02	—	8.30E–07
Temephos	—	—	2.00E–02	—	—	—	8.88E+02	—	2.90E–06
Terbacil	—	—	1.30E–02	—	—	—	1.37E+03	—	4.47E–06
Terephthalic acid	—	—	1.00E+00	—	—	—	1.78E+01	—	5.81E–08
tert-Butyl alcohol	—	—	2.00E+00	3.00E+01	—	—	8.88E+00	3.13E+00	3.93E–08
Tetrachloroethylene	5.40E–01	5.90E–06	1.00E–02	2.70E–01	9.60E+06	5.54E+05	1.78E+03	3.48E+02	3.32E–02
Tetrachlorvinphos	2.40E–02	—	3.00E–02	—	4.26E+05	—	5.92E+02	—	1.40E–03
Tetracycline hydrochloride	—	—	—	—	—	—	—	—	—

Tetramethrin	—	—	—	—	—	—	—	—	—
Thallium and thallium compounds	—	—	6.50E–05	—	—	—	2.73E+05	—	8.93E–04
Thiabendazole	—	—	—	—	—	—	—	—	—
Thioacetamide	—	—	—	—	—	—	—	—	—
Thiobencarb	—	—	1.00E–02	—	—	—	1.78E+03	—	5.81E–06
Thiodicarb	—	—	3.00E–03	—	—	—	5.92E+03	—	1.94E–05
Thiophanate-methyl	—	—	8.00E–02	—	—	—	2.22E+02	—	7.26E–07
Thiosemicarbazide	—	—	—	—	—	—	—	—	—
Thiourea	4.50E–02	1.30E–05	8.00E–05		8.00E+05	1.22E+06	2.22E+05		7.33E–03
Thiram	—	—	5.00E–03	—	—	—	3.55E+03	—	1.16E–05
Thorium dioxide	—	—	—	—	—	—	—	—	—
Titanium tetrachloride	—	—	—	1.00E–04	—	—	—	9.39E+05	3.07E–03
Toluene	—	—	8.00E–02	5.00E+00	—	—	2.22E+02	1.88E+01	7.87E–07
Toluene diisocyanate (mixed isomers)	3.90E–02	1.10E–05	—	7.00E–05	6.93E+05	1.03E+06	—	1.34E+06	1.00E–02
Toluene-2,4-diisocyanate	3.90E–02	1.10E–05	—	7.00E–05	6.93E+05	1.03E+06	—	1.34E+06	1.00E–02
Toluene-2,6-diisocyanate	3.90E–02	1.10E–05	—	7.00E–05	6.93E+05	1.03E+06	—	1.34E+06	1.00E–02
trans-1,3-Dichloropropene	1.00E–01	4.00E–06	3.00E–02	2.00E–02	1.78E+06	3.76E+05	5.92E+02	4.69E+03	7.05E–03
trans-1,4-Dichloro-2-butene	—	4.20E–03	—	—	—	3.94E+08	—	—	1.29E+00
Triadimefon	—	—	3.00E–02	—	—	—	5.92E+02	—	1.94E–06
Triallate	—	—	1.30E–02	—	—	—	1.37E+03	—	4.47E–06
Tribenuron methyl	—	—	-	—	—	—	—	—	—
Tribromomethane	7.90E–03	1.10E–06	2.00E–02	—	1.40E+05	1.03E+05	8.88E+02	—	7.99E–04
Tributyltin methacrylate	—	—	3.00E–04	—	—	—	5.92E+04	—	1.94E–04
Trichlorfon	—	—	—	—	—	—	—	—	—
Trichloroacetyl chloride	—	—	—	—	—	—	—	—	—
Trichloroethylene	1.30E–02	2.00E–06	—	—	2.31E+05	1.88E+05	—	—	1.37E–03
Trichlorofluoromethane	—	—	3.00E–01	7.00E–01	—	—	5.92E+01	1.34E+02	6.32E–07
Triclopyr triethylammonium salt	—	—	2.00E–02	—	—	—	8.88E+02		2.90E–06

continued

TABLE 6.1 (continued)
Toxicity Factors of TRI Chemicals

	USEPA Toxicity Values: Carcinogenic		USEPA Toxicity Values: Noncarcinogenic		Toxicity Factor: Carcinogenic		Toxicity Factor: Noncarcinogenic		Toxicity Factor: Combined (TF)
Chemical	**Ingestion (SFO) (mg/kg-day)$^{-1}$**	**Inhalation (IUR) (μg/m^3)$^{-1}$**	**Ingestion (RfDo) (mg/kg-day)**	**Inhalation (RfCi) (mg/m^3)**	**Ingestion (Doses/lb)**	**Inhalation (Doses/lb)**	**Ingestion (Doses/lb)**	**Inhalation (Doses/lb)**	**(Doses/ Capita-lb)**
Triethylamine	—	—	—	7.00E–03	—	—		1.34E+04	4.38E–05
Trifluralin	7.70E–03	—	7.50E–03	—	1.37E+05	—	2.37E+03		4.55E–04
Triforine	—	—	—	—	—	—	—	—	—
Triphenyltin chloride	—	—	—	—	—	—	—	—	—
Triphenyltin hydroxide	—	—	—	—	—	—	—	—	—
tris(2,3-Dibromopropyl) phosphate	—	—	—	—	—	—	—	—	—
Trypan blue	—	—	—	—	—	—	—	—	—
Urethane	—	—	—	—	—	—	—	—	—
Vanadium and vanadium compounds	—	—	7.00E–03	—	—	—	2.54E+03	—	8.30E–06
Vinclozolin	—	—	2.50E–02	—	—	—	7.11E+02		2.32E–06
Vinyl acetate	—	—	1.00E+00	2.00E–01	—	—	1.78E+01	4.69E+02	1.59E–06
Vinyl bromide	—	3.20E–05	—	3.00E–03	—	3.00E+06	—	3.13E+04	9.92E–03
Vinyl chloride	7.20E–01	4.40E–06	3.00E–03	1.00E–01	1.28E+07	4.13E+05	5.92E+03	9.39E+02	4.32E–02
Warfarin and salts	—	—	3.00E–04	—	—	—	5.92E+04	—	1.94E–04
Xylene (mixed isomers)	—	—	2.00E–01	1.00E–01	—	—	8.88E+01	9.39E+02	3.36E–06
Zinc and zinc compounds	—	—	3.00E–01	—	—	—	5.92E+01	—	1.94E–07
Zineb	—	—	5.00E–02	—	—	—	3.55E+02	—	1.16E–06

Source: U.S. Environmental Protection Agency, EPA Exposure Factors Handbook, 1997. http://cfpub.epa.gov/ncea/cfm/recordisplay.cfm?deid=12464. U.S. Environmental Protection Agency, Guidelines for Carcinogen Risk Assessment, 2005. http://cfpub.epa.gov/ncea/cfm/recordisplay.cfm?deid=116283.

BIBLIOGRAPHY

U.S. Environmental Protection Agency. 1997. EPA Exposure Factors Handbook. http://cfpub.epa.gov/ncea/cfm/recordisplay.cfm?deid=12464 (accessed January 20, 2010).

U.S. Environmental Protection Agency. 2005. Guidelines for Carcinogen Risk Assessment (2005). http://cfpub.epa.gov/ncea/cfm/recordisplay.cfm?deid=116283 (accessed January 20, 2010).

U.S. Environmental Protection Agency. 2009. Mid-Atlantic Risk Assessment. http://www.epa.gov/reg3hwmd/risk/human/rb-concentration_table/Generic_Tables/index.htm (accessed January 20, 2010).

7 Quantifying Mobility

INTRODUCTION

The impact of a chemical release is dependent not only on the toxicity of the compound but also on the likelihood that we will ingest or inhale the toxin. The impact of a particular chemical increases proportionally to the mobility of the chemical, that is, its tendency to dissolve and enter the water we drink or evaporate and enter the air that we breathe.

The tendency of a chemical to dissolve in a liquid solvent (e.g., water), to form a homogeneous solution is its *solubility*. Solubility is dependent on the chemical itself as well as temperature and pressures. The extent of the solubility of a specific chemical in a specific solvent is measured as the saturation concentration, for which adding more chemical does not increase the concentration of the solution. The tendency of a chemical to evaporate is measured by its vapor pressure (VP) at typical temperature or approximately 70°F. VP is the pressure of a vapor in equilibrium with its nonvapor phases or the pressure at which the gas of that substance is in dynamic equilibrium with its liquid or solid forms, an indication of the evaporation rate of a liquid. A substance with a high VP at normal temperatures is referred to as *volatile*. Water, as all liquids, starts to boil when its VP reaches its surrounding pressure. At higher elevations, the atmospheric pressure is lower, and water will boil at a lower temperature.

This chapter presents the development of a chemical-specific mobility factor (MF) (or a factor that represents the tendency of a chemical to dissolve in water and evaporate into air) based on the solubility of the chemical (in milligrams per liter) and VP (in atmospheres).

AIR MOBILITY FACTOR

The VP (in atmospheres) of the Toxics Release Inventory (TRI) chemicals was obtained from the Pennsylvania Department of Environmental Protection. For the purposes of this analysis, the VPs were set such that none of the VPs was set at more than 1 atm (anything greater than 1 atm does not have an additional impact with respect to release to the air), and the minimum was 1×10^{-9} atm (such that some dissipation affect is always taken into account). The chemical-specific air MFs, that is, the tendency for the chemical to evaporate and enter the air that we breathe, were derived by taking the square root of each chemical-specific VP (in atmospheres) to narrow the range of the values. Without this adjustment, the VPs would adjust the mobility of chemicals by a range of over 1 million to 1, which was too much of a weighting factor for this parameter. By taking the square root of the VPs, MFs ranged over approximately 1,000 to 1 across TRI chemicals, as shown in Table 7.1.

WATER MOBILITY FACTOR

The solubility (in milligrams per liter) of the TRI chemicals was obtained from the Risk-Screening Environmental Indicators (RSEI) Database of the U.S. Environmental Protection Agency (USEPA). For the purposes of this analysis, the solubilities were set such that the maximum was set at 100 percent. Anything greater than this value does not have an additional impact with respect to release in water. The minimum was set at 1 part per billion, such that some dissolution effect is always taken into account. Subsequently, chemical-specific water MFs, or the tendency for the chemical to dissolve and enter the water we drink, were derived by taking the square root of each chemical-specific water solubility (after converting the water solubility to kilograms per liter using a standard unit conversion factor) to narrow the range of the values (just as with the VP values), as shown in Table 7.1.

COMBINED MOBILITY FACTOR

The chemical-specific combined MF was derived by averaging the respective chemical-specific air and water MFs. This resulted in combined MFs spanning about five orders of magnitude across TRI chemicals, as shown in Table 7.1. TRI chemicals for which data were not available are not included in this table.

TABLE 7.1
Mobility Factors of TRI Chemicals

Chemical Name	VP (atm)	Air Mobility Factor	Solubility (mg/L)	Water Mobility Factor	Combined Mobility Factor
(1,1′-Biphenyl)-4,4′-diamine, 3,3′-dimethyl-	1.00E–09	3.16E–05	1.30E+03	3.61E–02	1.80E–02
1-(3-Chloroallyl)-3,5,7-triaza-1-azoniaadamantane chloride	1.00E–09	3.16E–05	1.00E+06	1.00E+00	5.00E–01
1,1,1,2-Tetrachloro-2-fluoroethane (Hcfc-121a)	1.00E–09	3.16E–05	2.09E+02	1.45E–02	7.24E–03
1,1,1,2-Tetrachloroethane	1.58E–02	1.26E–01	1.10E+03	3.32E–02	7.94E–02
1,1,1-Trichloroethane	1.32E–01	3.63E–01	1.49E+03	3.86E–02	2.01E–01
1,1,1-Trifluoro-2,2-dichloroethane	9.29E–01	9.64E–01	1.49E+03	3.86E–02	5.01E–01
1,1,2,2-Tetrachloro-1-fluoroethane	1.00E–09	3.16E–05	2.09E+02	1.45E–02	7.24E–03
1,1,2,2-Tetrachloroethane	6.58E–03	8.11E–02	2.96E+03	5.44E–02	6.78E–02
1,1,2-Trichloroethane	2.50E–02	1.58E–01	4.42E+03	6.65E–02	1.12E–01
1,1-Dichloro-1-fluoroethane	5.42E–01	7.36E–01	2.63E+03	5.13E–02	3.94E–01
1,1-Dichloroethane	3.01E–01	5.49E–01	5.06E+03	7.11E–02	3.10E–01
1,1-Dichloroethylene	7.78E–01	8.82E–01	2.25E+03	4.74E–02	4.65E–01
1,1-Dimethyl hydrazine	2.07E–01	4.55E–01	1.00E+06	1.00E+00	7.27E–01

TABLE 7.1 (continued)
Mobility Factors of TRI Chemicals

Chemical Name	VP (atm)	Air Mobility Factor	Solubility (mg/L)	Water Mobility Factor	Combined Mobility Factor
1,2,3-Trichloropropane	2.63E–03	5.13E–02	1.75E+03	4.18E–02	4.66E–02
1,2,4-Trichlorobenzene	4.50E–04	2.12E–02	4.90E+01	7.00E–03	1.41E–02
1,2,4-Trimethylbenzene	1.36E–03	3.69E–02	5.70E+01	7.55E–03	2.22E–02
1,2-Butylene oxide	2.32E–01	4.81E–01	9.50E+04	3.08E–01	3.95E–01
1,2-Dibromo-3-chloropropane (Dbcp)	1.05E–03	3.24E–02	1.23E+03	3.51E–02	3.38E–02
1,2-Dibromoethane	1.45E–02	1.20E–01	4.15E+03	6.44E–02	9.24E–02
1,2-Dichloro-1,1,2-trifluoroethane	1.00E–09	3.16E–05	1.86E+02	1.36E–02	6.83E–03
1,2-Dichloro-1,1-difluoroethane	1.00E–09	3.16E–05	9.99E+02	3.16E–02	1.58E–02
1,2-Dichlorobenzene	1.84E–03	4.29E–02	1.56E+02	1.25E–02	2.77E–02
1,2-Dichloroethane	8.03E–02	2.83E–01	8.52E+03	9.23E–02	1.88E–01
1,2-Dichloroethylene	4.26E–01	6.53E–01	3.50E+03	5.92E–02	3.56E–01
1,2-Dichloropropane	5.53E–02	2.35E–01	2.80E+03	5.29E–02	1.44E–01
1,2-Diphenylhydrazine	3.42E–08	1.85E–04	2.21E+02	1.49E–02	7.53E–03
1,3-Butadiene	1.00E+00	1.00E+00	7.35E+02	2.71E–02	5.14E–01
1,3-Dichloro-1,1,2,2,3-pentafluoropropane	1.00E–09	3.16E–05	1.59E+01	3.99E–03	2.01E–03
1,3-Dichlorobenzene	3.03E–03	5.50E–02	1.25E+02	1.12E–02	3.31E–02
1,3-Dichloropropene (mixed isomers)	3.29E–02	1.81E–01	2.80E+03	5.29E–02	1.17E–01
1,4-Dichloro-2-butene	4.51E–03	6.72E–02	5.80E+02	2.41E–02	4.56E–02
1,4-Dichlorobenzene	7.90E–04	2.81E–02	8.13E+01	9.02E–03	1.86E–02
1,4-Dioxane	4.88E–02	2.21E–01	1.00E+06	1.00E+00	6.10E–01
1-Chloro-1,1,2,2-Tetrafluoroethane	1.00E+00	1.00E+00	4.04E+02	2.01E–02	5.10E–01
1-Chloro-1,1-difluoroethane	1.00E+00	1.00E+00	1.40E+03	3.74E–02	5.19E–01
2,2′,6,6′-Tetrabromo-4,4′-isopropylidenediphenol	1.00E–09	3.16E–05	1.00E–03	3.16E–05	3.16E–05
2,3-Dichloropropene	6.97E–02	2.64E–01	2.15E+03	4.64E–02	1.55E–01
2,4,5-Trichlorophenol	2.90E–05	5.38E–03	1.20E+03	3.46E–02	2.00E–02
2,4,6-Trichlorophenol	1.11E–05	3.32E–03	8.00E+02	2.83E–02	1.58E–02
2,4,6-Trinitrophenol	1.00E–09	3.16E–05	1.27E+04	1.13E–01	5.64E–02
2,4-D	6.19E–06	2.49E–03	6.77E+02	2.60E–02	1.43E–02
2,4-D 2-Ethylhexyl ester	1.00E–12	1.00E–06	3.47E–02	1.86E–04	9.36E–05
2,4-D Butoxyethyl ester	5.92E–09	7.70E–05	1.20E+01	3.46E–03	1.77E–03
2,4-D Sodium salt	1.00E–09	3.16E–05	3.35E+05	5.79E–01	2.89E–01
2,4-Diaminotoluene	7.26E–08	2.70E–04	7.48E+04	2.74E–01	1.37E–01
2,4-Dichlorophenol	1.17E–04	1.08E–02	4.50E+03	6.71E–02	3.90E–02
2,4-Dimethylphenol	1.29E–04	1.14E–02	7.87E+03	8.87E–02	5.00E–02
2,4-Dinitrophenol	5.13E–07	7.16E–04	2.79E+03	5.28E–02	2.68E–02
2,4-Dinitrotoluene	1.45E–07	3.80E–04	2.70E+02	1.64E–02	8.41E–03
2,6-Dinitrotoluene	4.61E–07	6.79E–04	1.82E+02	1.35E–02	7.08E–03

continued

TABLE 7.1 (continued)
Mobility Factors of TRI Chemicals

Chemical Name	VP (atm)	Air Mobility Factor	Solubility (mg/L)	Water Mobility Factor	Combined Mobility Factor
2,6-Xylidine	1.00E–09	3.16E–05	8.24E+03	9.08E–02	4.54E–02
2-Acetylaminofluorene	1.00E–09	3.16E–05	5.53E+00	2.35E–03	1.19E–03
2-Chlor-1,3-butadiene	2.29E–01	4.79E–01	8.75E+02	2.96E–02	2.54E–01
2-Chloro-1,1,1,2-tetrafluoroethane	1.00E+00	1.00E+00	4.04E+02	2.01E–02	5.10E–01
2-Chloro-1,1,1-trifluoroethane	1.00E–09	3.16E–05	9.20E+03	9.59E–02	4.80E–02
2-Mercaptobenzothiazole	1.00E–09	3.16E–05	1.20E+02	1.10E–02	5.49E–03
2-Methyllactonitrile	1.05E–03	3.24E–02	1.00E+06	1.00E+00	5.16E–01
2-Methylpyridine	1.05E–02	1.03E–01	1.00E+06	1.00E+00	5.51E–01
2-Nitrophenol	1.99E–04	1.41E–02	2.50E+03	5.00E–02	3.20E–02
2-Nitropropane	1.71E–02	1.31E–01	1.70E+04	1.30E–01	1.31E–01
2-Phenylphenol	2.63E–06	1.62E–03	7.00E+02	2.65E–02	1.40E–02
3,3-Dichloro-1,1,1,2,2-pentafluoropropane	1.00E–09	3.16E–05	1.59E+01	3.99E–03	2.01E–03
3,3′-Dichlorobenzidine dihydrochloride	1.00E–09	3.16E–05	3.59E+03	5.99E–02	3.00E–02
3,3′-Dimethoxybenzidine	1.00E–09	3.16E–05	6.00E+01	7.75E–03	3.89E–03
3,3′-Dimethoxybenzidine dihydrochloride	1.00E–09	3.16E–05	7.24E+02	2.69E–02	1.35E–02
3-Chloro-2-methyl-1-propene	1.34E–01	3.66E–01	1.40E+03	3.74E–02	2.02E–01
3-Iodo-2-propynyl butylcarbamate	1.00E–09	3.16E–05	1.27E+02	1.13E–02	5.65E–03
4,4′-Diaminodiphenyl ether	1.00E–09	3.16E–05	5.60E+02	2.37E–02	1.18E–02
4,4′-Isopropylidenediphenol	1.00E–09	3.16E–05	1.20E+02	1.10E–02	5.49E–03
4,4′-Methylenebis(2-chloroaniline)	1.00E–09	3.16E–05	1.39E+01	3.73E–03	1.88E–03
4,4′-Methylenedianiline	3.91E–03	6.25E–02	1.00E+03	3.16E–02	4.71E–02
4,6-Dinitro-O-cresol	1.09E–07	3.30E–04	1.98E+02	1.41E–02	7.20E–03
4-Aminoazobenzene	1.00E–09	3.16E–05	3.20E+01	5.66E–03	2.84E–03
4-Aminobiphenyl	7.90E–08	2.81E–04	2.24E+02	1.50E–02	7.62E–03
4-Dimethylaminoazobenzene	1.00E–09	3.16E–05	2.30E–01	4.80E–04	2.56E–04
4-Nitrophenol	5.33E–08	2.31E–04	1.16E+04	1.08E–01	5.40E–02
5-Nitro-O-toluidine	3.68E–04	1.92E–02	1.88E+03	4.33E–02	3.13E–02
Abamectin	1.00E–09	3.16E–05	5.00E+00	2.24E–03	1.13E–03
Acephate	2.24E–09	4.73E–05	8.18E+05	9.04E–01	4.52E–01
Acetaldehyde	9.74E–01	9.87E–01	1.00E+06	1.00E+00	9.93E–01
Acetamide	1.45E–04	1.20E–02	2.25E+06	1.00E+00	5.06E–01
Acetonitrile	9.74E–02	3.12E–01	1.00E+06	1.00E+00	6.56E–01
Acetophenone	5.22E–04	2.29E–02	6.13E+03	7.83E–02	5.06E–02
Acrolein	2.90E–01	5.38E–01	2.12E+05	4.60E–01	4.99E–01
Acrylamide	9.21E–06	3.04E–03	6.40E+05	8.00E–01	4.02E–01
Acrylic acid	4.21E–03	6.49E–02	1.00E+06	1.00E+00	5.32E–01
Acrylonitrile	1.09E–01	3.30E–01	7.45E+04	2.73E–01	3.02E–01
Alachlor	4.08E–08	2.02E–04	2.40E+02	1.55E–02	7.85E–03

TABLE 7.1 (continued)
Mobility Factors of TRI Chemicals

Chemical Name	VP (atm)	Air Mobility Factor	Solubility (mg/L)	Water Mobility Factor	Combined Mobility Factor
Aldicarb	4.61E–08	2.15E–04	6.03E+03	7.77E–02	3.89E–02
Aldrin	9.87E–08	3.14E–04	1.70E–02	1.30E–04	2.22E–04
Allyl alcohol	2.63E–02	1.62E–01	1.00E+06	1.00E+00	5.81E–01
Allyl amine	1.00E–09	3.16E–05	1.00E+06	1.00E+00	5.00E–01
Allyl chloride	4.47E–01	6.69E–01	3.37E+03	5.81E–02	3.63E–01
Aluminum	1.32E–03	3.63E–02	1.00E+01	3.16E–03	1.97E–02
Aluminum oxide (fibrous forms)	1.00E–09	3.16E–05	1.00E–03	3.16E–05	3.16E–05
Aluminum phosphide	1.00E–09	3.16E–05	1.92E+05	4.38E–01	2.19E–01
Ametryn	1.11E–09	3.32E–05	2.09E+02	1.45E–02	7.25E–03
Amitrole	1.00E–09	3.16E–05	2.80E+05	5.29E–01	2.65E–01
Ammonia	1.00E+00	1.00E+00	4.82E+05	6.94E–01	8.47E–01
Aniline	3.95E–04	1.99E–02	3.60E+04	1.90E–01	1.05E–01
Anthracene	3.51E–09	5.93E–05	4.34E–02	2.08E–04	1.34E–04
Antimony	1.00E–09	3.16E–05	1.00E+00	1.00E–03	5.16E–04
Arsenic	1.00E–09	3.16E–05	1.00E+00	1.00E–03	5.16E–04
Asbestos	1.32E–08	1.15E–04	1.00E–02	1.00E–04	1.07E–04
Atrazine	1.00E–09	3.16E–05	3.47E+01	5.89E–03	2.96E–03
Barium	1.00E–09	3.16E–05	1.00E+00	1.00E–03	5.16E–04
Benfluralin	8.69E–08	2.95E–04	1.00E–01	3.16E–04	3.05E–04
Benzal chloride	3.95E–04	1.99E–02	2.50E+02	1.58E–02	1.78E–02
Benzene	1.00E–01	3.16E–01	1.79E+03	4.23E–02	1.79E–01
Benzidine	1.32E–08	1.15E–04	3.22E+02	1.79E–02	9.03E–03
Benzo(Ghi)perylene	1.00E–09	3.16E–05	2.60E–04	1.61E–05	2.39E–05
Benzoic trichloride	1.97E–04	1.40E–02	5.30E+01	7.28E–03	1.07E–02
Benzoyl chloride	5.26E–04	2.29E–02	4.94E+03	7.03E–02	4.66E–02
Benzoyl peroxide	1.00E–09	3.16E–05	9.10E+00	3.02E–03	1.52E–03
Benzyl chloride	1.32E–03	3.63E–02	5.25E+02	2.29E–02	2.96E–02
Beryllium	1.00E–09	3.16E–05	1.00E–02	1.00E–04	6.58E–05
Bifenthrin	1.00E–09	3.16E–05	1.00E–01	3.16E–04	1.74E–04
Biphenyl	7.69E–07	8.77E–04	6.94E+00	2.63E–03	1.76E–03
Bis(2-chloro-1-methylethyl) ether	1.12E–03	3.34E–02	1.70E+03	4.12E–02	3.73E–02
Bis(2-chloroethoxy)methane	1.84E–07	4.29E–04	7.80E+03	8.83E–02	4.44E–02
Bis(2-chloroethyl) ether	9.34E–04	3.06E–02	1.72E+04	1.31E–01	8.09E–02
Bis(2-ethylhexyl)phthalate	1.00E–09	3.16E–05	2.70E–01	5.20E–04	2.76E–04
Bis(tributyltin) oxide	1.00E–09	3.16E–05	1.00E+02	1.00E–02	5.02E–03
Boron trichloride	4.06E–01	6.37E–01	1.05E+04	1.03E–01	3.70E–01
Boron trifluoride	1.00E+00	1.00E+00	3.32E+06	1.00E+00	1.00E+00
Bromacil	1.00E–09	3.16E–05	8.15E+02	2.85E–02	1.43E–02
Bromine	2.26E–01	4.76E–01	3.50E+04	1.87E–01	3.31E–01
Bromochlorodifluoromethane	1.00E–09	3.16E–05	2.77E+02	1.66E–02	8.33E–03

continued

TABLE 7.1 (continued)
Mobility Factors of TRI Chemicals

Chemical Name	VP (atm)	Air Mobility Factor	Solubility (mg/L)	Water Mobility Factor	Combined Mobility Factor
Bromotrifluoromethane	1.00E–09	3.16E–05	3.20E+02	1.79E–02	8.96E–03
Bromoxynil	6.32E–09	7.95E–05	1.30E+02	1.14E–02	5.74E–03
Bromoxynil octanoate	6.32E–09	7.95E–05	8.00E–02	2.83E–04	1.81E–04
Butyl acrylate	5.26E–03	7.26E–02	2.00E+03	4.47E–02	5.86E–02
Butyraldehyde	1.00E–09	3.16E–05	7.10E+04	2.66E–01	1.33E–01
C.I. direct blue 218	1.00E–09	3.16E–05	3.15E–06	1.77E–06	1.67E–05
Cadmium	1.00E–09	3.16E–05	1.00E–02	1.00E–04	6.58E–05
Calcium cyanamide	2.11E–01	4.59E–01	1.93E+05	4.40E–01	4.49E–01
Camphechlor	6.62E–09	8.14E–05	5.50E–01	7.42E–04	4.11E–04
Captan	9.87E–09	9.93E–05	5.10E+00	2.26E–03	1.18E–03
Carbaryl	1.79E–09	4.23E–05	1.10E+02	1.05E–02	5.27E–03
Carbofuran	6.38E–09	7.99E–05	3.20E+02	1.79E–02	8.98E–03
Carbon disulfide	3.42E–01	5.85E–01	1.18E+03	3.44E–02	3.10E–01
Carbon tetrachloride	1.18E–01	3.44E–01	7.93E+02	2.82E–02	1.86E–01
Carbonyl sulfide	1.00E+00	1.00E+00	1.22E+03	3.49E–02	5.17E–01
Carboxin	1.00E–09	3.16E–05	1.99E+02	1.41E–02	7.07E–03
Catechol	3.95E–05	6.28E–03	4.61E+05	6.79E–01	3.43E–01
Chlordane	1.32E–08	1.15E–04	5.60E–02	2.37E–04	1.76E–04
Chlorendic acid	1.00E–09	3.16E–05	3.50E+03	5.92E–02	2.96E–02
Chlorimuron ethyl	1.00E–09	3.16E–05	1.20E+03	3.46E–02	1.73E–02
Chlorine	1.00E+00	1.00E+00	6.30E+03	7.94E–02	5.40E–01
Chlorine dioxide	1.00E+00	1.00E+00	1.00E+06	1.00E+00	1.00E+00
Chloroacetic acid	1.97E–04	1.40E–02	8.58E+05	9.26E–01	4.70E–01
Chlorobenzene	1.56E–02	1.25E–01	4.98E+02	2.23E–02	7.36E–02
Chlorobenzilate	2.90E–09	5.38E–05	1.30E+01	3.61E–03	1.83E–03
Chlorodifluoromethane	1.00E+00	1.00E+00	2.77E+03	5.26E–02	5.26E–01
Chloroethane	1.00E+00	1.00E+00	5.68E+03	7.54E–02	5.38E–01
Chloroform	2.11E–01	4.59E–01	7.95E+03	8.92E–02	2.74E–01
Chloromethane	1.00E+00	1.00E+00	5.32E+03	7.29E–02	5.36E–01
Chloromethyl methyl ether	2.83E–01	5.32E–01	6.94E+04	2.64E–01	3.98E–01
Chlorophenols	1.00E–09	3.16E–05	2.40E+03	4.90E–02	2.45E–02
Chloropicrin	2.22E–02	1.49E–01	1.62E+03	4.02E–02	9.47E–02
Chlorotetrafluoroethane	1.00E+00	1.00E+00	9.17E–03	9.58E–05	5.00E–01
Chlorothalonil	1.00E–09	3.16E–05	6.00E–01	7.75E–04	4.03E–04
Chlorotrifluoromethane	1.00E+00	1.00E+00	9.00E+01	9.49E–03	5.05E–01
Chlorpyrifos methyl	5.53E–08	2.35E–04	4.76E+00	2.18E–03	1.21E–03
Chlorsulfuron	5.92E–09	7.70E–05	2.80E+04	1.67E–01	8.37E–02
Chromium	1.32E–08	1.15E–04	8.67E+04	2.94E–01	1.47E–01
Cobalt	1.00E–09	3.16E–05	1.00E–02	1.00E–04	6.58E–05
Copper	1.00E–09	3.16E–05	1.00E–02	1.00E–04	6.58E–05
Creosotes	1.00E–09	3.16E–05	1.00E–03	3.16E–05	3.16E–05

TABLE 7.1 (continued)
Mobility Factors of TRI Chemicals

Chemical Name	VP (atm)	Air Mobility Factor	Solubility (mg/L)	Water Mobility Factor	Combined Mobility Factor
Cresol (mixed isomers)	5.66E–04	2.38E–02	9.07E+03	9.52E–02	5.95E–02
Crotonaldehyde	2.50E–02	1.58E–01	1.81E+05	4.25E–01	2.92E–01
Cumene	4.21E–03	6.49E–02	6.13E+01	7.83E–03	3.64E–02
Cumene hydroperoxide	1.45E–05	3.80E–03	1.39E+04	1.18E–01	6.09E–02
Cyanazine	1.00E–09	3.16E–05	1.70E+02	1.30E–02	6.54E–03
Cyanide compounds	8.16E–01	9.03E–01	1.00E+05	3.16E–01	6.10E–01
Cyclohexane	1.01E–01	3.18E–01	5.50E+01	7.42E–03	1.63E–01
Cyclohexanol	9.87E–04	3.14E–02	4.20E+04	2.05E–01	1.18E–01
Cyfluthrin	4.34E–08	2.08E–04	3.00E–03	5.48E–05	1.32E–04
Dazomet	1.00E–09	3.16E–05	3.00E+03	5.48E–02	2.74E–02
Decabromodiphenyl oxide	1.00E–09	3.16E–05	2.50E–02	1.58E–04	9.49E–05
Desmedipham	1.00E–09	3.16E–05	7.00E+00	2.65E–03	1.34E–03
Diallate	1.97E–07	4.44E–04	1.40E+01	3.74E–03	2.09E–03
Diaminotoluene (mixed isomers)	2.33E–06	1.53E–03	7.48E+04	2.74E–01	1.38E–01
Diazinon	1.84E–07	4.29E–04	4.00E+01	6.32E–03	3.38E–03
Dibenzofuran	3.26E–06	1.81E–03	3.10E+00	1.76E–03	1.78E–03
Dibutyl phthalate	5.53E–08	2.35E–04	1.12E+01	3.35E–03	1.79E–03
Dicamba	4.47E–08	2.12E–04	8.31E+03	9.12E–02	4.57E–02
Dichlorobenzene (mixed isomers)	1.97E–03	4.44E–02	8.00E+01	8.94E–03	2.67E–02
Dichlorobromomethane	6.58E–02	2.57E–01	3.03E+03	5.51E–02	1.56E–01
Dichlorodifluoromethane	1.00E+00	1.00E+00	2.80E+02	1.67E–02	5.08E–01
Dichlorofluoromethane	1.00E+00	1.00E+00	1.88E+04	1.37E–01	5.69E–01
Dichloromethane	4.59E–01	6.78E–01	1.30E+04	1.14E–01	3.96E–01
Dichlorotetrafluoroethane (Cfc-114)	1.00E+00	1.00E+00	1.30E+02	1.14E–02	5.06E–01
Dichlorotrifluoroethane	1.00E–09	3.16E–05	1.00E–03	3.16E–05	3.16E–05
Dichlorpentafluoro-propane	1.00E–09	3.16E–05	1.00E–03	3.16E–05	3.16E–05
Dichlorvos	1.58E–05	3.97E–03	8.00E+03	8.94E–02	4.67E–02
Dicyclopentadiene	1.84E–03	4.29E–02	2.65E+01	5.14E–03	2.40E–02
Diethanolamine	3.66E–07	6.05E–04	1.00E+06	1.00E+00	5.00E–01
Diethyl sulfate	4.85E–04	2.20E–02	7.00E+03	8.37E–02	5.28E–02
Diglycidyl resorcinol ether (Dgre)	5.36E–02	2.31E–01	2.96E+03	5.44E–02	1.43E–01
Dihydrosafrole	7.37E–05	8.58E–03	5.69E+01	7.54E–03	8.06E–03
Diisocyanates	1.00E–09	3.16E–05	1.00E+06	1.00E+00	5.00E–01
Dimethipin	1.00E–09	3.16E–05	4.60E+03	6.78E–02	3.39E–02
Dimethoate	1.09E–08	1.04E–04	2.50E+04	1.58E–01	7.91E–02
Dimethyl chlorothiophosphate	1.00E–09	3.16E–05	4.40E+03	6.63E–02	3.32E–02
Dimethyl phthalate	2.17E–06	1.47E–03	4.00E+03	6.32E–02	3.24E–02
Dimethyl sulfate	6.42E–04	2.53E–02	2.80E+04	1.67E–01	9.63E–02
Dimethylamine	1.00E+00	1.00E+00	1.63E+06	1.00E+00	1.00E+00
Dimethylamine dicamba	1.00E–09	3.16E–05	7.20E+05	8.49E–01	4.24E–01

continued

TABLE 7.1 (continued)
Mobility Factors of TRI Chemicals

Chemical Name	VP (atm)	Air Mobility Factor	Solubility (mg/L)	Water Mobility Factor	Combined Mobility Factor
Dimethylcarbamoyl chloride	2.57E–03	5.07E–02	4.59E+05	6.77E–01	3.64E–01
Dinitrobutyl phenol	9.87E–05	9.93E–03	5.20E+01	7.21E–03	8.57E–03
Dinitrotoluene (mixed isomers)	2.37E–05	4.87E–03	2.70E+02	1.64E–02	1.06E–02
Di-N-Propylnitrosamine	1.13E–04	1.06E–02	1.30E+04	1.14E–01	6.23E–02
Dioxin	1.00E–09	3.16E–05	1.90E–03	4.36E–05	3.76E–05
Diphenylamine	6.04E–07	7.77E–04	5.30E+01	7.28E–03	4.03E–03
Dipotassium endothall	1.00E–09	3.16E–05	1.10E+02	1.05E–02	5.26E–03
Diuron	1.00E–09	3.16E–05	4.20E+01	6.48E–03	3.26E–03
Epichlorohydrin	1.58E–02	1.26E–01	6.59E+04	2.57E–01	1.91E–01
Ethoprop	4.59E–07	6.78E–04	7.50E+02	2.74E–02	1.40E–02
Ethyl acrylate	3.82E–02	1.95E–01	1.50E+04	1.22E–01	1.59E–01
Ethyl chloroformate	1.00E–09	3.16E–05	3.21E+04	1.79E–01	8.95E–02
Ethyl dipropylthiocarbamate	4.61E–05	6.79E–03	3.75E+02	1.94E–02	1.31E–02
Ethylbenzene	9.21E–03	9.60E–02	1.69E+02	1.30E–02	5.45E–02
Ethylene	1.00E+00	1.00E+00	1.31E+02	1.14E–02	5.06E–01
Ethylene glycol	6.58E–05	8.11E–03	1.00E+06	1.00E+00	5.04E–01
Ethylene glycol monoethyl ether	7.01E–03	8.38E–02	1.00E+06	1.00E+00	5.42E–01
Ethylene glycol monomethyl ether	8.16E–03	9.03E–02	1.00E+06	1.00E+00	5.45E–01
Ethylene oxide	1.00E+00	1.00E+00	1.00E+06	1.00E+00	1.00E+00
Ethylene thiourea	1.84E–07	4.29E–04	2.00E+04	1.41E–01	7.09E–02
Ethylenebisdithiocarbamic acid, salts, and esters	4.83E–07	6.95E–04	1.00E+06	1.00E+00	5.00E–01
Ethyleneimine	2.11E–01	4.59E–01	1.00E+06	1.00E+00	7.29E–01
Fenoxycarb	1.00E–09	3.16E–05	6.00E+00	2.45E–03	1.24E–03
Ferbam	1.00E–09	3.16E–05	1.30E+02	1.14E–02	5.72E–03
Fluometuron	1.00E–09	3.16E–05	1.10E+02	1.05E–02	5.26E–03
Fluorine	1.00E+00	1.00E+00	1.69E+00	1.30E–03	5.01E–01
Fluorouracil	1.00E–09	3.16E–05	1.11E+04	1.05E–01	5.27E–02
Folpet	1.28E–08	1.13E–04	1.00E+00	1.00E–03	5.57E–04
Fomesafen	1.00E–09	3.16E–05	5.00E+01	7.07E–03	3.55E–03
Formaldehyde	1.00E+00	1.00E+00	4.00E+05	6.32E–01	8.16E–01
Formic acid	4.61E–02	2.15E–01	1.00E+06	1.00E+00	6.07E–01
Freon 113	3.75E–01	6.12E–01	1.70E+02	1.30E–02	3.13E–01
gamma-Lindane	1.24E–08	1.11E–04	7.30E+00	2.70E–03	1.41E–03
Glycol ethers	1.65E–02	1.28E–01	1.00E–03	3.16E–05	6.41E–02
Heptachlor	3.07E–07	5.54E–04	1.80E–01	4.24E–04	4.89E–04
Hexachloro-1,3-butadiene	1.97E–04	1.40E–02	3.20E+00	1.79E–03	7.92E–03
Hexachlorobenzene	1.43E–08	1.20E–04	6.20E–03	7.87E–05	9.93E–05
Hexachlorocyclopentadiene	1.05E–04	1.03E–02	1.80E+00	1.34E–03	5.80E–03
Hexachloroethane	2.76E–04	1.66E–02	5.00E+01	7.07E–03	1.18E–02
Hexachlorophene (Hcp)	1.00E–09	3.16E–05	1.40E+02	1.18E–02	5.93E–03

TABLE 7.1 (continued)
Mobility Factors of TRI Chemicals

Chemical Name	VP (atm)	Air Mobility Factor	Solubility (mg/L)	Water Mobility Factor	Combined Mobility Factor
Hexazinone	1.00E–09	3.16E–05	3.30E+04	1.82E–01	9.08E–02
Hydramethylnon	1.00E–09	3.16E–05	6.00E–03	7.75E–05	5.45E–05
Hydrazine	1.32E–02	1.15E–01	1.00E+06	1.00E+00	5.57E–01
Hydrochloric acid	1.00E+00	1.00E+00	8.23E+05	9.07E–01	9.54E–01
Hydrofluoric acid	1.00E–01	3.16E–01	9.22E+02	3.04E–02	1.73E–01
Hydrogen cyanide	8.16E–01	9.03E–01	1.00E+06	1.00E+00	9.52E–01
Hydroquinone	8.82E–07	9.39E–04	7.20E+04	2.68E–01	1.35E–01
Iron pentacarbonyl	1.00E–09	3.16E–05	0.00E+00	0.00E+00	1.58E–05
Isobutyraldehyde	3.98E–02	2.00E–01	8.90E+04	2.98E–01	2.49E–01
Isodrin	7.90E–09	8.89E–05	1.42E–02	1.19E–04	1.04E–04
Isosafrole	3.22E–05	5.68E–03	1.44E+02	1.20E–02	8.85E–03
Lactofen	1.00E–09	3.16E–05	1.00E–01	3.16E–04	1.74E–04
Lead	1.32E–08	1.15E–04	1.00E–02	1.00E–04	1.07E–04
Linuron	1.97E–08	1.40E–04	7.50E+01	8.66E–03	4.40E–03
Lithium carbonate	1.00E–09	3.16E–05	1.28E+04	1.13E–01	5.66E–02
Malathion	1.04E–08	1.02E–04	1.43E+02	1.20E–02	6.03E–03
Maleic anhydride	3.29E–04	1.81E–02	4.91E+03	7.01E–02	4.41E–02
Malononitrile	1.97E–05	4.44E–03	1.33E+05	3.65E–01	1.85E–01
Manganese	1.00E–09	3.16E–05	1.00E+01	3.16E–03	1.60E–03
M-Cresol	1.88E–04	1.37E–02	2.27E+04	1.51E–01	8.22E–02
M-Dinitrobenzene	1.71E–07	4.14E–04	1.15E+02	1.07E–02	5.57E–03
Mecoprop	1.00E–09	3.16E–05	6.20E+02	2.49E–02	1.25E–02
Mercury	2.63E–06	1.62E–03	1.00E–02	1.00E–04	8.61E–04
Merphos	1.00E–09	3.16E–05	3.50E–03	5.92E–05	4.54E–05
Methacrylonitrile	9.37E–02	3.06E–01	2.54E+04	1.59E–01	2.33E–01
Metham sodium	1.00E–09	3.16E–05	7.22E+05	8.50E–01	4.25E–01
Methanol	1.21E–01	3.48E–01	1.00E+06	1.00E+00	6.74E–01
Methazole	1.00E–09	3.16E–05	1.50E+00	1.22E–03	6.28E–04
Methoxone	1.97E–09	4.44E–05	6.30E+02	2.51E–02	1.26E–02
Methoxychlor	9.87E–09	9.93E–05	1.00E–01	3.16E–04	2.08E–04
Methyl acrylate	9.21E–02	3.04E–01	4.94E+04	2.22E–01	2.63E–01
Methyl bromide	1.00E+00	1.00E+00	1.52E+04	1.23E–01	5.62E–01
Methyl chlorocarbonate	1.85E–01	4.30E–01	9.28E+04	3.05E–01	3.67E–01
Methyl iodide	5.26E–01	7.26E–01	1.38E+04	1.17E–01	4.22E–01
Methyl isobutyl ketone	7.90E–03	8.89E–02	1.90E+04	1.38E–01	1.13E–01
Methyl isocyanate	4.58E–01	6.77E–01	2.92E+04	1.71E–01	4.24E–01
Methyl methacrylate	5.05E–02	2.25E–01	1.50E+04	1.22E–01	1.74E–01
Methyl parathion	1.00E–09	3.16E–05	3.77E+01	6.14E–03	3.09E–03
Methyl tert-butyl ether	3.28E–01	5.72E–01	5.10E+04	2.26E–01	3.99E–01
Methylene bromide	6.04E–02	2.46E–01	1.19E+04	1.09E–01	1.77E–01

continued

TABLE 7.1 (continued)
Mobility Factors of TRI Chemicals

Chemical Name	VP (atm)	Air Mobility Factor	Solubility (mg/L)	Water Mobility Factor	Combined Mobility Factor
Metribuzin	1.00E–09	3.16E–05	1.05E+03	3.24E–02	1.62E–02
Molybdenum trioxide	1.00E–09	3.16E–05	4.90E+02	2.21E–02	1.11E–02
Monochloropentafluoroethane	1.00E+00	1.00E+00	5.80E+01	7.62E–03	5.04E–01
M-Phenylenediamine	2.76E–06	1.66E–03	2.38E+05	4.88E–01	2.45E–01
M-Xylene	7.90E–03	8.89E–02	1.61E+02	1.27E–02	5.08E–02
Myclobutanil	2.11E–09	4.59E–05	1.42E+02	1.19E–02	5.98E–03
N,N-Dimethylaniline	6.58E–04	2.57E–02	1.45E+03	3.81E–02	3.19E–02
N,N-Dimethylformamide	4.87E–03	6.98E–02	1.00E+06	1.00E+00	5.35E–01
Nabam	1.00E–09	3.16E–05	2.00E+05	4.47E–01	2.24E–01
Naled	2.63E–07	5.13E–04	1.50E+00	1.22E–03	8.69E–04
Naphthalene	1.08E–04	1.04E–02	3.10E+01	5.57E–03	7.98E–03
N-Butyl alcohol	8.55E–03	9.25E–02	6.32E+04	2.51E–01	1.72E–01
N-Ethyl-N-nitrosourea	2.37E–05	4.87E–03	1.30E+04	1.14E–01	5.94E–02
N-Hexane	1.58E–01	3.97E–01	9.50E+00	3.08E–03	2.00E–01
Nickel	1.00E–09	3.16E–05	1.00E–02	1.00E–04	6.58E–05
Nicotine and salts	5.00E–05	7.07E–03	1.00E+06	1.00E+00	5.04E–01
Nitrapyrin	3.68E–06	1.92E–03	7.20E+01	8.49E–03	5.20E–03
Nitrate compounds	1.00E–09	3.16E–05	1.00E+06	1.00E+00	5.00E–01
Nitric acid	8.30E–02	2.88E–01	9.09E+04	3.01E–01	2.95E–01
Nitrilotriacetic acid	3.95E–08	1.99E–04	5.91E+04	2.43E–01	1.22E–01
Nitrobenzene	1.97E–04	1.40E–02	2.09E+03	4.57E–02	2.99E–02
Nitroglycerin	3.29E–07	5.74E–04	1.38E+03	3.71E–02	1.89E–02
N-Methyl-2-pyrrolidone	1.00E–09	3.16E–05	1.00E+06	1.00E+00	5.00E–01
N-Methylolacrylamide	1.00E–09	3.16E–05	1.00E+06	1.00E+00	5.00E–01
N-Nitrosodiethylamine	1.13E–03	3.36E–02	1.06E+05	3.26E–01	1.80E–01
N-Nitrosodi-N-butylamine	3.95E–05	6.28E–03	1.27E+03	3.56E–02	2.10E–02
N-Nitrosodiphenylamine	1.32E–04	1.15E–02	3.50E+01	5.92E–03	8.69E–03
N-Nitroso-N-methylurea	4.41E–02	2.10E–01	1.44E+04	1.20E–01	1.65E–01
N-Nitrosopiperidine	1.21E–04	1.10E–02	7.65E+04	2.77E–01	1.44E–01
Norflurazon	1.00E–09	3.16E–05	3.37E+01	5.81E–03	2.92E–03
O-Anisidine	1.97E–05	4.44E–03	9.60E+03	9.80E–02	5.12E–02
O-Cresol	3.16E–04	1.78E–02	2.59E+04	1.61E–01	8.94E–02
Octachlorostyrene	1.00E–09	3.16E–05	1.00E–03	3.16E–05	3.16E–05
O-Dinitrobenzene	5.13E–08	2.27E–04	1.33E+02	1.15E–02	5.88E–03
O-Phenylenediamine	1.28E–05	3.58E–03	4.04E+04	2.01E–01	1.02E–01
O-Toluidine	1.32E–04	1.15E–02	1.66E+04	1.29E–01	7.02E–02
O-Toluidine hydrochloride	1.57E–06	1.25E–03	8.29E+03	9.11E–02	4.62E–02
Oxydiazon	1.00E–09	3.16E–05	7.00E–01	8.37E–04	4.34E–04
Oxyfluorfen	1.00E–09	3.16E–05	1.16E–01	3.41E–04	1.86E–04
O-Xylene	6.58E–03	8.11E–02	1.78E+02	1.33E–02	4.72E–02
Ozone	1.00E+00	1.00E+00	7.44E+05	8.63E–01	9.31E–01

TABLE 7.1 (continued)
Mobility Factors of TRI Chemicals

Chemical Name	VP (atm)	Air Mobility Factor	Solubility (mg/L)	Water Mobility Factor	Combined Mobility Factor
Paraldehyde	3.33E–02	1.82E–01	1.12E+05	3.35E–01	2.59E–01
Paraquat	1.00E–09	3.16E–05	7.00E+05	8.37E–01	4.18E–01
Parathion	4.97E–08	2.23E–04	1.10E+01	3.32E–03	1.77E–03
P-Chloroaniline	1.97E–05	4.44E–03	3.90E+03	6.24E–02	3.34E–02
P-Cresidine	1.34E–05	3.66E–03	2.81E+03	5.30E–02	2.83E–02
P-Cresol	1.45E–04	1.20E–02	2.15E+04	1.47E–01	7.93E–02
P-Dinitrobenzene	2.96E–07	5.44E–04	6.90E+01	8.31E–03	4.43E–03
Pendimethalin	3.95E–08	1.99E–04	2.75E–01	5.24E–04	3.62E–04
Pentachlorobenzene	4.78E–05	6.91E–03	8.31E–01	9.12E–04	3.91E–03
Pentachloroethane	4.47E–03	6.69E–02	4.80E+02	2.19E–02	4.44E–02
Pentachlorophenol	1.45E–07	3.80E–04	1.40E+01	3.74E–03	2.06E–03
Pentobarbital sodium	1.00E–09	3.16E–05	5.13E+02	2.26E–02	1.13E–02
Peracetic acid	1.00E–09	3.16E–05	1.00E+06	1.00E+00	5.00E–01
Permethrin	1.00E–09	3.16E–05	6.00E–03	7.75E–05	5.45E–05
Phenanthrene	8.95E–07	9.46E–04	1.15E+00	1.07E–03	1.01E–03
Phenol	2.63E–04	1.62E–02	8.28E+04	2.88E–01	1.52E–01
Phenothrin	1.00E–09	3.16E–05	9.70E–03	9.85E–05	6.51E–05
Phenytoin	1.00E–09	3.16E–05	3.20E+01	5.66E–03	2.84E–03
Phosgene	1.00E+00	1.00E+00	4.75E+05	6.89E–01	8.45E–01
Phosphine	1.00E+00	1.00E+00	2.05E+05	4.53E–01	7.26E–01
Phosphorus (yellow or white)	2.38E–04	1.54E–02	2.05E+05	4.53E–01	2.34E–01
Phthalic anhydride	2.63E–07	5.13E–04	6.20E+03	7.87E–02	3.96E–02
Picloram	1.00E–09	3.16E–05	4.30E+02	2.07E–02	1.04E–02
Piperonyl butoxide	1.00E–09	3.16E–05	1.43E+01	3.78E–03	1.91E–03
P-Nitroaniline	1.97E–06	1.40E–03	7.28E+02	2.70E–02	1.42E–02
Polychlorinated alkanes (C10–C13)	1.00E–09	3.16E–05	1.00E–03	3.16E–05	3.16E–05
Polychlorinated biphenyls	1.01E–07	3.18E–04	7.00E–01	8.37E–04	5.77E–04
Polycyclic aromatic compounds	1.00E–09	3.16E–05	1.00E–03	3.16E–05	3.16E–05
Potassium bromate	1.00E–09	3.16E–05	6.90E+04	2.63E–01	1.31E–01
Potassium dimethyldithiocarbamate	1.00E–09	3.16E–05	2.90E+04	1.70E–01	8.52E–02
Potassium N-methyldithiocarbamate	1.00E–09	3.16E–05	4.40E+05	6.63E–01	3.32E–01
P-Phenylenediamine	6.58E–06	2.57E–03	3.70E+04	1.92E–01	9.75E–02
Prometryn	1.32E–09	3.63E–05	3.30E+01	5.74E–03	2.89E–03
Pronamide	1.12E–07	3.34E–04	1.50E+01	3.87E–03	2.10E–03
Propachlor	3.03E–07	5.50E–04	7.00E+02	2.65E–02	1.35E–02
Propane Sultone	8.38E–07	9.16E–04	1.71E+05	4.13E–01	2.07E–01
Propanil	1.00E–09	3.16E–05	1.52E+02	1.23E–02	6.18E–03
Propargite	3.95E–03	6.28E–02	5.00E–01	7.07E–04	3.18E–02
Propargyl alcohol	1.26E–02	1.12E–01	1.00E+06	1.00E+00	5.56E–01
Propiconazole	1.32E–09	3.63E–05	1.10E+02	1.05E–02	5.26E–03

continued

TABLE 7.1 (continued)
Mobility Factors of TRI Chemicals

Chemical Name	VP (atm)	Air Mobility Factor	Solubility (mg/L)	Water Mobility Factor	Combined Mobility Factor
Propionaldehyde	3.09E–01	5.56E–01	3.06E+05	5.53E–01	5.55E–01
Propoxur	3.95E–09	6.28E–05	1.86E+03	4.31E–02	2.16E–02
Propylene	1.00E+00	1.00E+00	2.00E+02	1.41E–02	5.07E–01
Propylene oxide	5.86E–01	7.65E–01	5.90E+05	7.68E–01	7.67E–01
Propyleneimine	1.47E–01	3.84E–01	1.00E+06	1.00E+00	6.92E–01
P-Xylene	8.55E–03	9.25E–02	1.62E+02	1.27E–02	5.26E–02
Pyridine	2.63E–02	1.62E–01	1.00E+06	1.00E+00	5.81E–01
Quinoline	1.26E–04	1.12E–02	6.11E+03	7.82E–02	4.47E–02
Quinone	1.18E–04	1.09E–02	1.11E+04	1.05E–01	5.81E–02
Quintozene	6.58E–08	2.57E–04	4.40E–01	6.63E–04	4.60E–04
S,S,S-Tributyltrithiophosphate	2.11E–06	1.45E–03	2.30E+00	1.52E–03	1.48E–03
Saccharin	1.20E–09	3.46E–05	4.00E+03	6.32E–02	3.16E–02
Safrole	7.90E–05	8.89E–03	1.21E+02	1.10E–02	9.94E–03
sec-Butyl alcohol	1.32E–02	1.15E–01	1.81E+05	4.25E–01	2.70E–01
Selenium	9.87E–07	9.93E–04	2.06E+03	4.54E–02	2.32E–02
Silver	1.00E–09	3.16E–05	1.00E–02	1.00E–04	6.58E–05
Simazine	1.00E–09	3.16E–05	6.20E+00	2.49E–03	1.26E–03
Sodium azide	1.00E–09	3.16E–05	3.67E+04	1.92E–01	9.58E–02
Sodium dicamba	1.00E–09	3.16E–05	3.60E+05	6.00E–01	3.00E–01
Sodium dimethyldithiocarbamate	1.00E–09	3.16E–05	1.00E+06	1.00E+00	5.00E–01
Sodium nitrite	1.00E–09	3.16E–05	1.00E+06	1.00E+00	5.00E–01
Strychnine	1.00E–09	3.16E–05	1.43E+02	1.20E–02	5.99E–03
Styrene	6.58E–03	8.11E–02	3.10E+02	1.76E–02	4.94E–02
Styrene oxide	3.95E–04	1.99E–02	3.00E+03	5.48E–02	3.73E–02
Sulfuric acid	7.81E–08	2.79E–04	1.00E+06	1.00E+00	5.00E–01
Sulfuryl fluoride	1.00E–09	3.16E–05	7.50E+02	2.74E–02	1.37E–02
Tebuthiuron	2.63E–09	5.13E–05	2.50E+03	5.00E–02	2.50E–02
tert-Butyl alcohol	5.45E–02	2.33E–01	1.00E+06	1.00E+00	6.17E–01
Tetrachloroethylene	1.84E–02	1.36E–01	2.00E+02	1.41E–02	7.49E–02
Tetrachlorvinphos	1.00E–09	3.16E–05	1.10E+01	3.32E–03	1.67E–03
Tetracycline hydrochloride	1.00E–09	3.16E–05	2.49E+05	4.99E–01	2.50E–01
Tetramethrin	1.00E–09	3.16E–05	1.83E+00	1.35E–03	6.92E–04
Thallium	1.00E–09	3.16E–05	1.00E–02	1.00E–04	6.58E–05
Thiabendazole	1.00E–09	3.16E–05	5.00E+01	7.07E–03	3.55E–03
Thioacetamide	1.00E–09	3.16E–05	1.63E+05	4.04E–01	2.02E–01
Thiodicarb	4.25E–08	2.06E–04	3.50E+01	5.92E–03	3.06E–03
Thiophanate-methyl	1.00E–09	3.16E–05	4.39E+02	2.09E–02	1.05E–02
Thiourea	1.00E–09	3.16E–05	1.42E+05	3.77E–01	1.88E–01
Thiram	1.05E–08	1.03E–04	3.00E+01	5.48E–03	2.79E–03
Titanium tetrachloride	1.32E–02	1.15E–01	2.74E+03	5.23E–02	8.35E–02
Toluene	2.90E–02	1.70E–01	5.26E+02	2.29E–02	9.65E–02

TABLE 7.1 (continued)
Mobility Factors of TRI Chemicals

Chemical Name	VP (atm)	Air Mobility Factor	Solubility (mg/L)	Water Mobility Factor	Combined Mobility Factor
Toluene diisocyanate (mixed isomers)	2.86E–05	5.34E–03	3.76E+01	6.13E–03	5.74E–03
Toluene-2,4-diisocyanate	1.32E–05	3.63E–03	3.76E+01	6.13E–03	4.88E–03
Toluene-2,6-diisocyanate	2.63E–05	5.13E–03	3.76E+01	6.13E–03	5.63E–03
trans-1,3-Dichloropropene	3.29E–02	1.81E–01	1.99E+03	4.47E–02	1.13E–01
trans-1,4-Dichloro-2-butene	4.51E–03	6.72E–02	8.50E+02	2.92E–02	4.82E–02
Triallate	1.58E–07	3.97E–04	4.00E+00	2.00E–03	1.20E–03
Tribenuron methyl	1.00E–09	3.16E–05	5.00E+01	7.07E–03	3.55E–03
Tribromomethane	7.11E–03	8.43E–02	3.10E+03	5.57E–02	7.00E–02
Trichlorfon	1.03E–08	1.01E–04	1.20E+05	3.46E–01	1.73E–01
Trichloroacetyl chloride	1.00E–09	3.16E–05	9.49E+03	9.74E–02	4.87E–02
Trichloroethylene	7.90E–02	2.81E–01	1.10E+03	3.32E–02	1.57E–01
Trichlorofluoromethane	9.04E–01	9.51E–01	1.10E+03	3.32E–02	4.92E–01
Triclopyr triethylammonium salt	1.00E–09	3.16E–05	2.99E+02	1.73E–02	8.66E–03
Triethylamine	6.58E–02	2.57E–01	7.37E+04	2.71E–01	2.64E–01
Trifluralin	1.45E–07	3.80E–04	1.84E–01	4.29E–04	4.05E–04
Tris(2,3-Dibromopropyl) phosphate	2.50E–07	5.00E–04	8.00E+00	2.83E–03	1.66E–03
Trypan blue	1.00E–09	3.16E–05	8.35E–02	2.89E–04	1.60E–04
Urethane	3.42E–04	1.85E–02	4.80E+05	6.93E–01	3.56E–01
Vanadium	1.00E–09	3.16E–05	1.00E–02	1.00E–04	6.58E–05
Vinyl acetate	1.12E–01	3.34E–01	2.00E+04	1.41E–01	2.38E–01
Vinyl chloride	1.00E+00	1.00E+00	8.80E+03	9.38E–02	5.47E–01
Warfarin and salts	1.00E–09	3.16E–05	1.70E+01	4.12E–03	2.08E–03
Xylene	1.05E–02	1.03E–01	1.06E+02	1.03E–02	5.64E–02
Zinc	1.32E–03	3.63E–02	1.00E+01	3.16E–03	1.97E–02

Source: U.S. Environmental Protection Agency, Risk-Screening Environmental Indicators (RSEI) Database, 2009. http://www.epa.gov/oppt/rsei/pubs/basic_information.html.

BIBLIOGRAPHY

Pennsylvania Department of Environmental Protection. 2010. Land Recycling Program, Chemical and Physical Properties Database. http://www.dep.state.pa.us/physicalproperties/CPP_Search.htm (accessed January 23, 2010).

U.S. Environmental Protection Agency. 2009. Risk-Screening Environmental Indicators (RSEI) Database. http://www.epa.gov/oppt/rsei/pubs/basic_information.html (accessed January 20, 2010).

8 Quantifying Persistence

INTRODUCTION

A chemical that is persistent, that is, that does not degrade or decompose, will accumulate in the environment over time, thereby increasing our exposure as compared with a chemical that rapidly degrades. A measure of how rapidly a chemical decomposes is the *half-life* (HL) or the time needed for half of the chemical to degrade. This Chapter presents data on the HLs of the Toxics Release Inventory (TRI) chemicals and a method for converting these to a persistence factor (PF).

PERSISTENCE FACTOR

The developed PF is based on how much of a chemical would accumulate in the environment over an average 70-yr life span or the residual if 1 lb of the chemical were released to the environment each year for 70 yrs. If none of the compound degrades (as indicated by an infinite HL or at least a HL much longer than 70 yrs), then, at the end of 70 yrs a residual of 70 lb would accumulate in the environment. If the HL were short, or about a day or less, then at the end of a year, although there would be an insignificant remaining residual concentration, there would still be exposure of the amount released at the initial time of release. Also, while a toxic chemical may degrade in the environment, the compound is not likely to degrade in use, either in a manufacturing process or in the home or workplace when using a product containing the compound. Therefore, minimum impact would be exposure from the chemical at the time of use or release, with no impact due to accumulation, equating to a minimum residual of 1 lb.

The developed PF is based on the amount of chemical initially released plus the residual amount that would accumulate through release of an additional pound each year over the remaining 69 yrs. Specifically, the chemical- and media-specific HLs (in days) were obtained for the organic (nonmetal) TRI chemicals for water, soil, sediment, and air from the U.S. Environmental Protection Agency (USEPA) Persistent, Bioaccumulative, and Toxic (PBT) Profiler (USEPA 2006). For inorganics (metals), HL data are not available; however, it is known that metals do have long HLs, so the HLs for inorganics were set at 10 million years. These HLs were converted into units of years by dividing the HLs (in days) by 365 days per year, as applicable. Subsequently, the media-specific HLs were averaged to obtain a chemical-specific average HL. Using these chemical-specific average HLs, the residual or remaining amount of chemical after 70 yrs was calculated using the following formula:

Residual (pounds) = (0.5^[Year 1/HL]) + (0.5^[Year 2/HL])

+ (0.5^[Year 3/HL]) + … (0.5^[Year 70/HL])

This equation yields the same results as the following equation, used to calculate how much of a chemical will remain after time t, or residual, based on frequency, time, and quantity due to both new releases and residual up to the point just before time t, based on the HL of a substance:

$$N_t = N_0 \times e^{-\frac{\ln 2}{t_{1/2}} \times t}$$

where N_t is the quantity after time t, N_0 is the quantity due to both new releases and residual up to the point just before time t, $t_{1/2}$ is the HL, and t is the time.

This resulted in a residual between 1 lb, with a relatively short HL, and 70 lb, a relatively long HL, for the TRI chemicals over the course of a 70-yr period. Since toxicity is based on annual exposure over a 70-yr life expectancy, this residual was then divided by 70 yrs, resulting in the chemical-specific PFs varying between 1/70 and 70/70, or 1, as presented in Table 8.1. TRI chemicals for which HL values were not available are not included in this table.

TABLE 8.1
Persistence Factors of TRI Chemicals

	Half-Lives (yrs)					Residual	
Chemical	Water	Soil	Sediment	Air	Average	(lb)	PF
1-(3-Chloroallyl)-3,5,7-triaza-1-azoniaadamantane chloride	0.49	0.99	4.38	0.00	1.47	3	0.04
1,1,1,2-Tetrachloro-2-fluoroethane	0.16	0.33	1.48	9.04	2.75	4	0.06
1,1,1,2-Tetrachloroethane	0.16	0.33	1.48	2.41	1.10	2	0.03
1,1,1-Trichloroethane	0.16	0.33	1.48	4.66	1.66	3	0.04
1,1,2,2-Tetrachloro-1-fluoroethane	0.16	0.33	1.48	3.56	1.38	3	0.04
1,1,2,2-Tetrachloroethane	0.16	0.33	1.48	0.17	0.54	1	0.02
1,1,2-Trichloroethane	0.10	0.21	0.93	0.23	0.37	1	0.02
1,1-Dichloro-1-fluoroethane	0.10	0.21	0.93	7.40	2.16	4	0.05
1,1-Dimethyl hydrazine	0.04	0.08	0.38	0.00	0.13	1	0.01
1,2,3-Trichloropropane	0.10	0.21	0.93	0.13	0.34	1	0.02
1,2,4-Trichlorobenzene	0.16	0.33	1.48	0.08	0.51	1	0.02
1,2,4-Trimethylbenzene	0.10	0.21	0.93	0.00	0.31	1	0.02
1,2-Butylene oxide	0.04	0.08	0.38	0.02	0.13	1	0.01
1,2-Dibromo-3-chloropropane	0.10	0.21	0.93	0.10	0.34	1	0.02
1,2-Dibromoethane	0.04	0.08	0.38	0.17	0.17	1	0.01
1,2-Dichloro-1,1,2-trifluoroethane	0.10	0.21	0.93	3.56	1.20	2	0.03
1,2-Dichloro-1,1-difluoroethane	0.10	0.21	0.93	2.74	1.00	2	0.03

TABLE 8.1 (continued)
Persistence Factors of TRI Chemicals

Chemical	Half-Lives (yrs)					Residual (lb)	PF
	Water	Soil	Sediment	Air	Average		
1,2-Dichlorobenzene	0.10	0.21	0.93	0.10	0.34	1	0.02
1,2-Dichloroethane	0.10	0.21	0.93	0.18	0.36	1	0.02
1,2-Dichloroethylene	0.10	0.21	0.93	0.02	0.31	1	0.02
1,2-Dichloropropane	0.10	0.21	0.93	0.10	0.33	1	0.02
1,2-Diphenylhydrazine	0.04	0.08	0.38	0.00	0.13	1	0.01
1,2-Phenylenediamine	0.10	0.21	0.93	0.00	0.31	1	0.02
1,3-Butadiene	0.04	0.08	0.38	0.00	0.13	1	0.01
1,3-Dichloro-1,1,2,2,3-pentafluoropropane	0.16	0.33	1.48	4.93	1.73	3	0.04
1,3-Dichlorobenzene	0.10	0.21	0.93	0.06	0.33	1	0.02
1,3-Dichloropropylene	0.10	0.21	0.93	0.01	0.31	1	0.02
1,3-Phenylenediamine	0.10	0.21	0.93	0.00	0.31	1	0.02
1,4-Dichloro-2-butene	0.10	0.21	0.93	0.00	0.31	1	0.02
1,4-Dichlorobenzene	0.10	0.21	0.93	0.14	0.34	1	0.02
1,4-Dioxane	0.04	0.08	0.38	0.00	0.13	1	0.01
1-Bromo-1-(bromomethyl)-1,3-propanedicarbonitrile	0.10	0.21	0.93	0.07	0.33	1	0.02
1-Chloro-1,1,2,2-tetrafluoroethane	0.10	0.21	0.93	82.19	20.86	28	0.39
1-Chloro-1,1-difluoroethane	0.10	0.21	0.93	13.70	3.73	6	0.08
2,2-Dichloro-1,1,1-trifluoroethane	0.16	0.33	1.48	1.15	0.78	2	0.02
2,3-Dichloropropene	0.10	0.21	0.93	0.00	0.31	1	0.02
2,4,5-Trichlorophenol	0.16	0.33	1.48	0.02	0.50	1	0.02
2,4,6-Trichlorophenol	0.16	0.33	1.48	0.07	0.51	1	0.02
2,4-D	0.10	0.21	0.93	0.01	0.31	1	0.02
2,4-D 2-Ethylhexyl ester	0.10	0.21	0.93	0.00	0.31	1	0.02
2,4-D Butoxyethyl ester	0.10	0.21	0.93	0.00	0.31	1	0.02
2,4-D Isopropyl ester	0.10	0.21	0.93	0.01	0.31	1	0.02
2,4-D Sodium salt	0.16	0.33	1.48	0.01	0.50	1	0.02
2,4-Db	0.10	0.21	0.93	0.00	0.31	1	0.02
2,4-Diaminotoluene	0.10	0.21	0.93	0.00	0.31	1	0.02
2,4-Dichlorophenol	0.10	0.21	0.93	0.04	0.32	1	0.02
2,4-Dimethylphenol	0.04	0.08	0.38	0.00	0.13	1	0.01
2,4-Dinitrophenol	0.10	0.21	0.93	0.07	0.33	1	0.02
2,4-Dinitrotoluene	0.10	0.21	0.93	0.21	0.36	1	0.02
2,6-Dinitrotoluene	0.10	0.21	0.93	0.21	0.36	1	0.02
2,6-Xylidine	0.10	0.21	0.93	0.00	0.31	1	0.02
2-Acetylaminofluorene	0.10	0.21	0.93	0.00	0.31	1	0.02
2-Chloro-1,1,1,2-tetrafluoroethane	0.16	0.33	1.48	4.66	1.66	3	0.04

continued

TABLE 8.1 (continued) Persistence Factors of TRI Chemicals

Chemical	Half-Lives (yrs)					Residual	
	Water	Soil	Sediment	Air	Average	(lb)	PF
2-Chloro-1,1,1-trifluoroethane	0.10	0.21	0.93	2.74	1.00	2	0.03
2-Ethoxyethanol	0.04	0.08	0.38	0.00	0.13	1	0.01
2-Mercaptobenzothiazole	0.04	0.08	0.38	0.00	0.13	1	0.01
2-Methoxyethanol	0.04	0.08	0.38	0.00	0.13	1	0.01
2-Methyllactonitrile	0.10	0.21	0.93	0.04	0.32	1	0.02
2-Methylpyridine	0.10	0.21	0.93	0.04	0.32	1	0.02
2-Nitrophenol	0.04	0.08	0.38	0.05	0.14	1	0.01
2-Nitropropane	0.04	0.08	0.38	0.17	0.17	1	0.01
2-Phenylphenol	0.04	0.08	0.38	0.00	0.13	1	0.01
3,3-Dichloro-1,1,1,2,2-pentafluoropropane	0.49	0.99	4.38	1.70	1.89	3	0.05
3,3′-Dichlorobenzidine dihydrochloride	0.16	0.33	1.48	0.08	0.51	1	0.02
3,3′-Dimethoxybenzidine	0.10	0.21	0.93	0.00	0.31	1	0.02
3,3′-Dimethoxybenzidine dihydrochloride	0.10	0.21	0.93	0.00	0.31	1	0.02
3,3′-Dimethylbenzidine	0.10	0.21	0.93	0.00	0.31	1	0.02
3-Chloro-2-methyl-1-propene	0.04	0.08	0.38	0.00	0.13	1	0.01
3-Iodo-2-propynyl butylcarbamate	0.04	0.08	0.38	0.00	0.13	1	0.01
4,4′-Diaminodiphenyl ether	0.10	0.21	0.93	0.00	0.31	1	0.02
4,4′-Isopropylidenediphenol	0.10	0.21	0.93	0.00	0.31	1	0.02
4,4′-Methylenebis(2-chloroaniline)	0.16	0.33	1.48	0.00	0.49	1	0.02
4,4′-Methylenedianiline	0.10	0.21	0.93	0.00	0.31	1	0.02
4,6-Dinitro-O-cresol	0.10	0.21	0.93	0.15	0.35	1	0.02
4-Aminoazobenzene	0.10	0.21	0.93	0.00	0.31	1	0.02
4-Aminobiphenyl	0.10	0.21	0.93	0.00	0.31	1	0.02
4-Dimethylaminoazobenzene	0.16	0.33	1.48	0.00	0.49	1	0.02
4-Nitrophenol	0.04	0.08	0.38	0.01	0.13	1	0.01
5-Nitro-O-toluidine	0.10	0.21	0.93	0.00	0.31	1	0.02
Abamectin	0.01	0.08	0.33	0.00	0.11	1	0.01
Acephate	0.10	0.21	0.93	0.00	0.31	1	0.02
Acetaldehyde	0.04	0.08	0.38	0.00	0.13	1	0.01
Acetamide	0.04	0.08	0.38	0.02	0.13	1	0.01
Acetonitrile	0.04	0.08	0.38	0.18	0.17	1	0.01
Acetophenone	0.04	0.08	0.38	0.02	0.13	1	0.01
Acifluorfen, sodium salt	0.49	0.99	4.38	0.11	1.49	3	0.04
Acrolein	0.04	0.08	0.38	0.00	0.13	1	0.01
Acrylamide	0.04	0.08	0.38	0.00	0.13	1	0.01
Acrylic acid	0.02	0.05	0.21	0.00	0.07	1	0.01
Acrylonitrile	0.04	0.08	0.38	0.01	0.13	1	0.01

TABLE 8.1 (continued)
Persistence Factors of TRI Chemicals

	Half-Lives (yrs)					Residual	
Chemical	Water	Soil	Sediment	Air	Average	(lb)	PF
Alachlor	0.16	0.33	1.48	0.00	0.49	1	0.02
Aldicarb	0.10	0.21	0.93	0.00	0.31	1	0.02
Aldrin	0.49	0.99	4.38	0.00	1.47	3	0.04
Allyl alcohol	0.04	0.08	0.38	0.00	0.13	1	0.01
Allyl chloride	0.04	0.08	0.38	0.00	0.13	1	0.01
Allylamine	0.04	0.08	0.38	0.00	0.13	1	0.01
alpha-Naphthylamine	0.10	0.21	0.93	0.00	0.31	1	0.02
Aluminum (fume or dust)	Long	Long	Long	Long	Long	70	1.00
Aluminum oxide (fibrous forms)	Long	Long	Long	Long	Long	70	1.00
Aluminum phosphide	Long	Long	Long	Long	Long	70	1.00
Ametryn	0.16	0.33	1.48	0.00	0.49	1	0.02
Amitraz	0.10	0.21	0.93	0.00	0.31	1	0.02
Amitrole	0.04	0.08	0.38	0.01	0.13	1	0.01
Ammonia	0.01	0.02	0.04	0.00	0.02	1	0.01
Aniline	0.05	0.03	27.40	0.00	6.87	10	0.15
Anthracene	0.16	0.33	1.48	0.00	0.49	1	0.02
Antimony	Long	Long	Long	Long	Long	70	1.00
Arsenic	Long	Long	Long	Long	Long	70	1.00
Asbestos (friable)	Long	Long	Long	Long	Long	70	1.00
Atrazine	0.16	0.33	1.48	0.00	0.49	1	0.02
Barium	Long	Long	Long	Long	Long	70	1.00
Benfluralin	0.49	0.99	4.38	0.00	1.47	3	0.04
Benzal chloride	0.10	0.21	0.93	0.02	0.32	1	0.02
Benzene	0.10	0.21	0.93	0.04	0.32	1	0.02
Benzidine	0.10	0.21	0.93	0.00	0.31	1	0.02
Benzo(G,H,I)perylene	0.16	0.33	1.48	0.00	0.49	1	0.02
Benzoic trichloride	0.16	0.33	1.48	0.13	0.52	1	0.02
Benzoyl chloride	0.04	0.08	0.38	0.03	0.13	1	0.01
Benzoyl peroxide	0.10	0.21	0.93	0.01	0.31	1	0.02
Benzyl chloride	0.04	0.08	0.38	0.01	0.13	1	0.01
Beryllium	Long	Long	Long	Long	Long	70	1.00
Bifenthrin	0.49	0.99	4.38	0.00	1.47	3	0.04
Biphenyl	0.04	0.08	0.38	0.01	0.13	1	0.01
Bis(2-chloro-1-methylethyl) ether	0.10	0.21	0.93	0.02	0.31	1	0.02
Bis(2-chloroethoxy)methane	0.10	0.21	0.93	0.01	0.31	1	0.02
Bis(2-chloroethyl) ether	0.10	0.21	0.93	0.01	0.31	1	0.02
Bis(chloromethyl) ether	0.10	0.21	0.93	0.06	0.33	1	0.02
Bis(tributyltin) oxide	0.03	0.31	8.22	0.00	2.14	4	0.05
Boron trichloride	0.04	0.08	0.38	0.49	0.25	1	0.02
Boron trifluoride	0.08	0.21	0.21	0.04	0.13	1	0.01

continued

TABLE 8.1 (continued)
Persistence Factors of TRI Chemicals

	Half-Lives (yrs)					Residual	
Chemical	Water	Soil	Sediment	Air	Average	(lb)	PF
Bromacil	0.10	0.21	0.93	0.00	0.31	1	0.02
Bromine	Long	Long	Long	Long	Long	70	1.00
Bromochlorodifluoromethane	0.10	0.21	0.93	0.49	0.43	1	0.02
Bromoform	0.10	0.21	0.93	1.04	0.57	1	0.02
Bromomethane	0.04	0.08	0.38	1.10	0.40	1	0.02
Bromotrifluoromethane	0.10	0.21	0.93	0.49	0.43	1	0.02
Bromoxynil	0.10	0.21	0.93	0.21	0.36	1	0.02
Bromoxynil octanoate	0.10	0.21	0.93	0.01	0.31	1	0.02
Butyl acrylate	0.02	0.05	0.21	0.00	0.07	1	0.01
Butyraldehyde	0.04	0.08	0.38	0.00	0.13	1	0.01
C.I. basic green 4	0.16	0.33	1.48	0.00	0.49	1	0.02
C.I. direct blue 218	0.49	0.99	4.38	0.00	1.47	3	0.04
C.I. solvent orange 7	0.16	0.33	1.48	0.00	0.49	1	0.02
Cadmium	Long	Long	Long	Long	Long	70	1.00
Calcium cyanamide	0.10	0.21	0.93	0.00	0.31	1	0.02
Captan	0.16	0.33	1.48	0.00	0.49	1	0.02
Carbaryl	0.10	0.21	0.93	0.00	0.31	1	0.02
Carbofuran	0.10	0.21	0.93	0.00	0.31	1	0.02
Carbon disulfide	0.04	0.08	0.38	0.49	0.25	1	0.02
Carbon tetrachloride	0.16	0.33	1.48	0.49	0.62	1	0.02
Carbonyl sulfide	0.04	0.08	0.38	0.49	0.25	1	0.02
Carboxin	0.10	0.21	0.93	0.00	0.31	1	0.02
Catechol	0.04	0.08	0.38	0.00	0.13	1	0.01
Certain glycol ethers	0.02	0.04	0.08	0.00	0.03	1	0.01
Chlordane	0.49	0.99	4.38	0.01	1.47	3	0.04
Chlorendic acid	0.49	0.99	4.38	0.01	1.47	3	0.04
Chlorimuron ethyl	0.16	0.33	1.48	0.00	0.49	1	0.02
Chlorine	0.04	0.08	0.38	0.49	0.25	1	0.02
Chlorine dioxide	0.04	0.08	0.38	0.49	0.25	1	0.02
Chloroacetic acid	0.04	0.08	0.38	0.05	0.14	1	0.01
Chlorobenzene	0.04	0.08	0.38	0.06	0.14	1	0.01
Chlorobenzilate	0.16	0.33	1.48	0.01	0.50	1	0.02
Chlorodifluoromethane	0.04	0.08	0.38	9.32	2.46	4	0.06
Chloroethane	0.04	0.08	0.38	0.11	0.15	1	0.01
Chloroform	0.10	0.21	0.93	0.41	0.41	1	0.02
Chloromethane	0.04	0.08	0.38	1.01	0.38	1	0.02
Chloromethyl methyl ether	0.04	0.08	0.38	0.02	0.13	1	0.01
Chlorophenols	0.01	0.01	0.02	0.00	0.01	1	0.01
Chloropicrin	0.16	0.33	1.48	0.33	0.58	1	0.02
Chloroprene	0.04	0.08	0.38	0.00	0.13	1	0.01
Chlorotetrafluoroethane	0.10	0.21	0.93	82.19	20.86	28	0.39
Chlorothalonil	0.49	0.99	4.38	7.12	3.25	5	0.07

TABLE 8.1 (continued)
Persistence Factors of TRI Chemicals

Chemical	Half-Lives (yrs)					Residual	PF
	Water	Soil	Sediment	Air	Average	(lb)	
Chlorotrifluoromethane	0.10	0.21	0.93	0.49	0.43	1	0.02
Chlorpyrifos methyl	0.16	0.33	1.48	0.00	0.49	1	0.02
Chlorsulfuron	0.16	0.33	1.48	0.02	0.50	1	0.02
Chromium	Long	Long	Long	Long	Long	70	1.00
Cobalt	Long	Long	Long	Long	Long	70	1.00
Cobalt compounds	Long	Long	Long	Long	Long	70	1.00
Copper	Long	Long	Long	Long	Long	70	1.00
Creosote	1.00	0.33	0.33	0.03	0.42	1	0.02
Cresol (mixed isomers)	0.04	0.08	0.38	0.00	0.13	1	0.01
Crotonaldehyde	0.04	0.08	0.38	0.00	0.13	1	0.01
Cumene	0.04	0.08	0.38	0.01	0.13	1	0.01
Cumene hydroperoxide	0.10	0.21	0.93	0.01	0.31	1	0.02
Cupferron	0.04	0.08	0.38	0.00	0.13	1	0.01
Cyanazine	0.49	0.99	4.38	0.00	1.47	3	0.04
Cyclohexane	0.04	0.08	0.38	0.01	0.13	1	0.01
Cyclohexanol	0.04	0.08	0.38	0.00	0.13	1	0.01
Cyfluthrin	0.49	0.99	4.38	0.00	1.47	3	0.04
Cyhalothrin	0.49	0.99	4.38	0.00	1.47	3	0.04
Dazomet	0.10	0.21	0.93	0.00	0.31	1	0.02
Dazomet, sodium salt	0.10	0.21	0.93	0.00	0.31	1	0.02
Decabromodiphenyl oxide	0.49	0.99	4.38	1.26	1.78	3	0.04
Desmedipham	0.10	0.21	0.93	0.00	0.31	1	0.02
Di(2-ethylhexyl) phthalate	0.04	0.08	0.38	0.00	0.13	1	0.01
Diallate	0.16	0.33	1.48	0.00	0.49	1	0.02
Diaminotoluene (mixed isomers)	0.10	0.21	0.93	0.00	0.31	1	0.02
Diazinon	0.10	0.21	0.93	0.00	0.31	1	0.02
Dibenzofuran	0.04	0.08	0.38	0.01	0.13	1	0.01
Dibromotetrafluoroethane	0.10	0.21	0.93	0.49	0.43	1	0.02
Dibutyl phthalate	0.02	0.05	0.21	0.00	0.07	1	0.01
Dicamba	0.10	0.21	0.93	0.01	0.31	1	0.02
Dichloran	0.16	0.33	1.48	0.33	0.58	1	0.02
Dichlorobenzene (mixed isomers)	0.10	0.21	0.93	0.10	0.34	1	0.02
Dichlorobromomethane	0.10	0.21	0.93	0.55	0.45	1	0.02
Dichlorodifluoromethane	0.10	0.21	0.93	0.49	0.43	1	0.02
Dichlorofluoromethane	0.10	0.21	0.93	1.48	0.68	2	0.02
Dichloromethane	0.10	0.21	0.93	0.30	0.39	1	0.02
Dichloropentafluoropropane	0.10	0.21	0.93	0.03	0.32	1	0.02
Dichlorotetrafluoroethane (Cfc-114)	0.16	0.33	1.48	0.49	0.62	1	0.02
Dichlorotrifluoroethane	0.00	0.00	0.00	0.00	0.00	1	0.01
Dichlorvos	0.10	0.21	0.93	0.00	0.31	1	0.02

continued

TABLE 8.1 (continued)
Persistence Factors of TRI Chemicals

	Half-Lives (yrs)					Residual	
Chemical	Water	Soil	Sediment	Air	Average	(lb)	PF
Dicyclopentadiene	0.04	0.08	0.38	0.00	0.13	1	0.01
Diethanolamine	0.02	0.05	0.21	0.00	0.07	1	0.01
Diethyl sulfate	0.04	0.08	0.38	0.02	0.13	1	0.01
Diflubenzuron	0.49	0.99	4.38	0.00	1.47	3	0.04
Diglycidyl resorcinol ether	0.10	0.21	0.93	0.00	0.31	1	0.02
Dihydrosafrole	0.10	0.21	0.93	0.00	0.31	1	0.02
Diisocyanates	27.40	27.40	27.40	27.40	27.40	33	0.47
Dimethipin	0.10	0.21	0.93	0.00	0.31	1	0.02
Dimethoate	0.04	0.08	0.38	0.00	0.13	1	0.01
Dimethyl chlorothiophosphate	0.04	0.08	0.38	0.00	0.13	1	0.01
Dimethyl phthalate	0.04	0.08	0.38	0.08	0.15	1	0.01
Dimethyl sulfate	0.04	0.08	0.38	0.23	0.18	1	0.01
Dimethylamine	0.04	0.08	0.38	0.00	0.13	1	0.01
Dimethylamine dicamba	0.16	0.33	1.48	0.01	0.49	1	0.02
Dimethylcarbamyl chloride	0.04	0.08	0.38	0.00	0.13	1	0.01
Dinitrobutyl phenol	0.10	0.21	0.93	0.01	0.31	1	0.02
Dinitrotoluene (mixed isomers)	0.10	0.21	0.93	0.23	0.37	1	0.02
Dioxin and dioxin-like compounds	0.01	2.74	12.00	0.00	3.69	6	0.08
Diphenylamine	0.10	0.21	0.93	0.00	0.31	1	0.02
Dipotassium endothall	0.10	0.21	0.93	0.00	0.31	1	0.02
Disodium cyanodithioimidocarbonate	0.04	0.08	0.38	0.00	0.13	1	0.01
Diuron	0.10	0.21	0.93	0.00	0.31	1	0.02
D-trans-Allethrin	0.04	0.08	0.38	0.00	0.13	1	0.01
Epichlorohydrin	0.04	0.08	0.38	0.10	0.15	1	0.01
Ethoprop	0.04	0.08	0.38	0.00	0.13	1	0.01
Ethyl acrylate	0.04	0.08	0.38	0.00	0.13	1	0.01
Ethyl chloroformate	0.04	0.08	0.38	0.03	0.13	1	0.01
Ethyl dipropylthiocarbamate	0.10	0.21	0.93	0.00	0.31	1	0.02
Ethylbenzene	0.04	0.08	0.38	0.01	0.13	1	0.01
Ethylene	0.04	0.08	0.38	0.00	0.13	1	0.01
Ethylene glycol	0.02	0.05	0.21	0.01	0.07	1	0.01
Ethylene oxide	0.04	0.08	0.38	0.58	0.27	1	0.02
Ethylene thiourea	0.04	0.08	0.38	0.00	0.13	1	0.01
Ethylenebisdithiocarbamic acid, salts, and esters	0.04	0.08	0.38	0.00	0.13	1	0.01
Ethyleneimine	0.04	0.08	0.38	0.01	0.13	1	0.01
Ethylidene dichloride	0.10	0.21	0.93	0.16	0.35	1	0.02
Fenarimol	0.16	0.33	1.48	0.01	0.50	1	0.02
Fenbutatin oxide	52.75	0.25	0.25	0.00	13.31	19	0.27
Fenoxycarb	0.10	0.21	0.93	0.00	0.31	1	0.02

TABLE 8.1 (continued)
Persistence Factors of TRI Chemicals

	Half-Lives (yrs)						
Chemical	Water	Soil	Sediment	Air	Average	Residual (lb)	PF
Fenpropathrin	0.16	0.33	1.48	0.00	0.49	1	0.02
Ferbam	0.00	0.00	0.00	0.00	0.00	1	0.01
Fluazifop butyl	0.16	0.33	1.48	0.00	0.49	1	0.02
Fluometuron	0.16	0.33	1.48	0.01	0.49	1	0.02
Fluorine	Long	Long	Long	Long	Long	70	1.00
Fluorouracil	0.04	0.08	0.38	0.01	0.13	1	0.01
Folpet	0.16	0.33	1.48	0.00	0.49	1	0.02
Fomesafen	0.49	0.99	4.38	0.03	1.47	3	0.04
Formaldehyde	0.04	0.08	0.38	0.00	0.13	1	0.01
Formic acid	0.02	0.05	0.21	0.10	0.10	1	0.01
Freon 113	0.16	0.33	1.48	0.49	0.62	1	0.02
Heptachlor	0.49	0.99	4.38	0.00	1.47	3	0.04
Hexachloro-1,3-butadiene	0.49	0.99	4.38	1.48	1.84	3	0.05
Hexachlorobenzene	0.49	0.99	4.38	0.16	1.51	3	0.04
Hexachlorocyclopentadiene	0.49	0.99	4.38	0.11	1.49	3	0.04
Hexachloroethane	0.49	0.99	4.38	0.49	1.59	3	0.04
Hexachlorophene	0.49	0.99	4.38	0.02	1.47	3	0.04
Hexazinone	0.10	0.21	0.93	0.00	0.31	1	0.02
Hydramethylnon	0.49	0.99	4.38	0.00	1.47	3	0.04
Hydrazine	0.02	0.00	0.00	0.00	0.01	1	0.01
Hydrochloric acid	0.00	0.00	0.00	0.00	0.00	1	0.01
Hydrogen cyanide	0.04	0.08	0.38	1.48	0.50	1	0.02
Hydrogen fluoride	0.27	0.27	0.27	0.27	0.27	1	0.02
Hydroquinone	0.04	0.08	0.38	0.00	0.13	1	0.01
Iron pentacarbonyl	27.40	27.40	27.40	27.40	27.40	33	0.47
Isobutyraldehyde	0.04	0.08	0.38	0.00	0.13	1	0.01
Isodrin	0.49	0.99	4.38	0.00	1.47	3	0.04
Isosafrole	0.10	0.21	0.93	0.00	0.31	1	0.02
Lactofen	0.49	0.99	4.38	0.01	1.47	3	0.04
Lead	Long	Long	Long	Long	Long	70	1.00
Lindane	0.49	0.99	4.38	0.23	1.52	3	0.04
Linuron	0.16	0.33	1.48	0.00	0.49	1	0.02
Lithium carbonate	0.04	0.08	0.38	0.49	0.25	1	0.02
Malathion	0.04	0.08	0.38	0.00	0.13	1	0.01
Maleic anhydride	0.04	0.08	0.38	0.01	0.13	1	0.01
Malononitrile	0.04	0.08	0.38	1.26	0.44	1	0.02
Maneb	0.10	0.21	0.93	0.00	0.31	1	0.02
Manganese	Long	Long	Long	Long	Long	70	1.00
M-Cresol	0.04	0.08	0.38	0.00	0.13	1	0.01
M-Dinitrobenzene	0.10	0.21	0.93	1.48	0.68	2	0.02
Mecoprop	0.10	0.21	0.93	0.00	0.31	1	0.02

continued

TABLE 8.1 (continued)
Persistence Factors of TRI Chemicals

	Half-Lives (yrs)						
Chemical	Water	Soil	Sediment	Air	Average	Residual (lb)	PF
Mercury	Long	Long	Long	Long	Long	70	1.00
Merphos	0.02	0.05	0.21	0.00	0.07	1	0.01
Methacrylonitrile	0.04	0.08	0.38	0.00	0.13	1	0.01
Metham sodium	0.04	0.08	0.38	0.00	0.13	1	0.01
Methanol	0.02	0.05	0.21	0.05	0.08	1	0.01
Methazole	0.16	0.33	1.48	0.01	0.50	1	0.02
Methiocarb	0.10	0.21	0.93	0.00	0.31	1	0.02
Methoxone	0.04	0.08	0.38	0.00	0.13	1	0.01
Methoxone sodium salt	0.10	0.21	0.93	0.00	0.31	1	0.02
Methoxychlor	0.49	0.99	4.38	0.00	1.47	3	0.04
Methyl acrylate	0.04	0.08	0.38	0.00	0.13	1	0.01
Methyl chlorocarbonate	0.04	0.08	0.38	0.21	0.18	1	0.01
Methyl hydrazine	0.04	0.08	0.38	0.00	0.13	1	0.01
Methyl iodide	0.04	0.08	0.38	0.60	0.28	1	0.02
Methyl isobutyl ketone	0.04	0.08	0.38	0.00	0.13	1	0.01
Methyl isocyanate	0.04	0.08	0.38	0.33	0.21	1	0.01
Methyl isothiocyanate	0.04	0.08	0.38	0.33	0.21	1	0.01
Methyl methacrylate	0.04	0.08	0.38	0.00	0.13	1	0.01
Methyl parathion	0.10	0.21	0.93	0.00	0.31	1	0.02
Methyl tert-butyl ether	0.04	0.08	0.38	0.01	0.13	1	0.01
Methylene bromide	0.04	0.08	0.38	0.38	0.22	1	0.01
Metribuzin	0.10	0.21	0.93	0.00	0.31	1	0.02
Molinate	0.10	0.21	0.93	0.00	0.31	1	0.02
Molybdenum trioxide	0.55	10.00	10.00	0.04	5.15	8	0.11
Monochloropentafluoroethane	0.16	0.33	1.48	0.49	0.62	1	0.02
M-Xylene	0.04	0.08	0.38	0.00	0.13	1	0.01
Myclobutanil	0.10	0.21	0.93	0.01	0.31	1	0.02
N,N-Dimethylaniline	0.10	0.21	0.93	0.00	0.31	1	0.02
N,N-Dimethylformamide	0.04	0.08	0.38	0.00	0.13	1	0.01
Nabam	0.10	0.21	0.93	0.00	0.31	1	0.02
Naled	0.16	0.33	1.48	0.01	0.49	1	0.02
Naphthalene	0.10	0.21	0.93	0.00	0.31	1	0.02
N-Butyl alcohol	0.02	0.05	0.21	0.01	0.07	1	0.01
N-Hexane	0.02	0.05	0.21	0.01	0.07	1	0.01
Nickel	Long	Long	Long	Long	Long	70	1.00
Nicotine and salts	0.01	0.00	0.00	0.00	0.00	1	0.01
Nitrapyrin	0.49	0.99	4.38	1.26	1.78	3	0.04
Nitrate compounds	0.27	0.27	0.27	0.27	0.27	1	0.02
Nitric acid	0.03	0.00	0.00	0.08	0.03	1	0.01
Nitrilotriacetic acid	0.02	0.05	0.21	0.00	0.07	1	0.01
Nitrobenzene	0.04	0.08	0.38	0.33	0.21	1	0.01
Nitroglycerin	0.10	0.21	0.93	0.04	0.32	1	0.02

TABLE 8.1 (continued)
Persistence Factors of TRI Chemicals

Chemical	Half-Lives (yrs) Water	Soil	Sediment	Air	Average	Residual (lb)	PF
N-Methyl-2-pyrrolidone	0.04	0.08	0.38	0.00	0.13	1	0.01
N-Methylolacrylamide	0.04	0.08	0.38	0.00	0.13	1	0.01
N-Nitrosodiethylamine	0.10	0.21	0.93	0.00	0.31	1	0.02
N-Nitrosodimethylamine	0.10	0.21	0.93	0.02	0.31	1	0.02
N-Nitrosodi-N-butylamine	0.04	0.08	0.38	0.00	0.13	1	0.01
N-Nitrosodi-N-propylamine	0.10	0.21	0.93	0.00	0.31	1	0.02
N-Nitrosodiphenylamine	0.10	0.21	0.93	0.00	0.31	1	0.02
N-Nitroso-N-ethylurea	0.04	0.08	0.38	0.01	0.13	1	0.01
N-Nitroso-N-methylurea	0.04	0.08	0.38	0.03	0.13	1	0.01
N-Nitrosopiperidine	0.10	0.21	0.93	0.00	0.31	1	0.02
Norflurazon	0.16	0.33	1.48	0.00	0.49	1	0.02
O-Anisidine	0.10	0.21	0.93	0.00	0.31	1	0.02
O-Cresol	0.04	0.08	0.38	0.00	0.13	1	0.01
Octachloronaphthalene	0.49	0.99	4.38	1.15	1.75	3	0.04
Octachlorostyrene	0.49	0.99	4.38	0.04	1.48	3	0.04
O-Dinitrobenzene	0.10	0.21	0.93	2.05	0.82	2	0.03
Oryzalin	0.16	0.33	1.48	0.00	0.49	1	0.02
O-Toluidine	0.04	0.08	0.38	0.00	0.13	1	0.01
O-Toluidine hydrochloride	0.04	0.08	0.38	0.00	0.13	1	0.01
Oxydemeton methyl	0.04	0.08	0.38	0.00	0.13	1	0.01
Oxydiazon	0.16	0.33	1.48	0.00	0.49	1	0.02
Oxyfluorfen	0.49	0.99	4.38	0.00	1.47	3	0.04
O-Xylene	0.04	0.08	0.38	0.00	0.13	1	0.01
Ozone	0.00	0.00	0.00	0.02	0.01	1	0.01
Paraldehyde	0.04	0.08	0.38	0.00	0.13	1	0.01
Paraquat dichloride	0.10	0.21	0.93	0.00	0.31	1	0.02
Parathion	0.10	0.21	0.93	0.00	0.31	1	0.02
P-Chloroaniline	0.10	0.21	0.93	0.00	0.31	1	0.02
P-Cresidine	0.10	0.21	0.93	0.00	0.31	1	0.02
P-Cresol	0.04	0.08	0.38	0.00	0.13	1	0.01
P-Dinitrobenzene	0.10	0.21	0.93	2.05	0.82	2	0.03
Pendimethalin	0.16	0.33	1.48	0.00	0.49	1	0.02
Pentachlorobenzene	0.49	0.99	4.38	0.77	1.66	3	0.04
Pentachloroethane	0.49	0.99	4.38	2.27	2.03	3	0.05
Pentachlorophenol	0.49	0.99	4.38	0.08	1.49	3	0.04
Pentobarbital sodium	0.10	0.21	0.93	0.00	0.31	1	0.02
Peracetic acid	0.04	0.08	0.38	0.01	0.13	1	0.01
Permethrin	0.16	0.33	1.48	0.00	0.49	1	0.02
Phenanthrene	0.16	0.33	1.48	0.00	0.49	1	0.02
Phenol	0.04	0.08	0.38	0.00	0.13	1	0.01
Phenothrin	0.10	0.21	0.93	0.00	0.31	1	0.02

continued

TABLE 8.1 (continued)
Persistence Factors of TRI Chemicals

	Half-Lives (yrs)						
Chemical	Water	Soil	Sediment	Air	Average	Residual (lb)	PF
Phenytoin	0.10	0.21	0.93	0.00	0.31	1	0.02
Phosgene	0.04	0.08	0.38	0.49	0.25	1	0.02
Phosphine	0.00	0.00	0.00	0.00	0.00	1	0.01
Phosphorus (yellow or white)	0.03	0.03	0.03	0.03	0.03	1	0.01
Phthalic anhydride	0.04	0.08	0.38	0.06	0.14	1	0.01
Picloram	0.16	0.33	1.48	0.05	0.51	1	0.02
Picric acid	0.16	0.33	1.48	0.30	0.57	1	0.02
Piperonyl butoxide	0.10	0.21	0.93	0.00	0.31	1	0.02
P-Nitroaniline	0.10	0.21	0.93	0.00	0.31	1	0.02
Polybrominated biphenyls	27.40	27.40	27.40	27.40	27.40	33	0.47
Polychlorinated alkanes	27.40	27.40	27.40	27.40	27.40	33	0.47
Polychlorinated biphenyls	0.49	0.99	4.38	0.04	1.48	3	0.04
Polycyclic aromatic compounds	27.40	27.40	27.40	27.40	27.40	33	0.47
Potassium bromate	Long	Long	Long	Long	Long	70	1.00
Potassium dimethyldithiocarbamate	0.04	0.08	0.38	0.00	0.13	1	0.01
Potassium N-methyldithiocarbamate	0.04	0.08	0.38	0.00	0.13	1	0.01
P-Phenylenediamine	0.10	0.21	0.93	0.00	0.31	1	0.02
Profenofos	0.16	0.33	1.48	0.00	0.49	1	0.02
Prometryn	0.16	0.33	1.48	0.00	0.49	1	0.02
Pronamide	0.16	0.33	1.48	0.00	0.49	1	0.02
Propachlor	0.10	0.21	0.93	0.00	0.31	1	0.02
Propane sultone	0.04	0.08	0.38	0.02	0.13	1	0.01
Propanil	0.16	0.33	1.48	0.01	0.50	1	0.02
Propargite	0.16	0.33	1.48	0.00	0.49	1	0.02
Propargyl alcohol	0.04	0.08	0.38	0.00	0.13	1	0.01
Propiconazole	0.16	0.33	1.48	0.00	0.49	1	0.02
Propionaldehyde	0.04	0.08	0.38	0.00	0.13	1	0.01
Propoxur	0.10	0.21	0.93	0.00	0.31	1	0.02
Propylene	0.04	0.08	0.38	0.00	0.13	1	0.01
Propylene oxide	0.04	0.08	0.38	0.08	0.15	1	0.01
Propyleneimine	0.04	0.08	0.38	0.01	0.13	1	0.01
P-Xylene	0.04	0.08	0.38	0.00	0.13	1	0.01
Pyridine	0.04	0.08	0.38	0.12	0.16	1	0.01
Quinoline	0.04	0.08	0.38	0.00	0.13	1	0.01
Quinone	0.04	0.08	0.38	0.00	0.13	1	0.01
Quintozene	0.49	0.99	4.38	6.03	2.97	5	0.07
Quizalofop-ethyl	0.16	0.33	1.48	0.00	0.49	1	0.02
Resmethrin	0.10	0.21	0.93	0.00	0.31	1	0.02
S,S,S-Tributyltrithiophosphate	0.02	0.05	0.21	0.00	0.07	1	0.01
Saccharin	0.04	0.08	0.38	0.01	0.13	1	0.01

TABLE 8.1 (continued)
Persistence Factors of TRI Chemicals

	Half-Lives (yrs)						
Chemical	Water	Soil	Sediment	Air	Average	Residual (lb)	PF
Safrole	0.10	0.21	0.93	0.00	0.31	1	0.02
sec-Butyl alcohol	0.04	0.08	0.38	0.00	0.13	1	0.01
Selenium	Long	Long	Long	Long	Long	70	1.00
Sethoxydim	0.10	0.21	0.93	0.00	0.31	1	0.02
Silver	Long	Long	Long	Long	Long	70	1.00
Silver compounds	Long	Long	Long	Long	Long	70	1.00
Simazine	0.16	0.33	1.48	0.00	0.49	1	0.02
Sodium azide	0.27	0.27	0.27	0.27	0.27	1	0.02
Sodium dicamba	0.16	0.33	1.48	0.01	0.50	1	0.02
Sodium dimethyldithiocarbamate	0.04	0.08	0.38	0.00	0.13	1	0.01
Sodium nitrite	0.03	0.03	0.03	0.03	0.03	1	0.01
Sodium O-phenylphenoxide	0.04	0.08	0.38	0.00	0.13	1	0.01
Strychnine and salts	27.40	27.40	27.40	27.40	27.40	33	0.47
Styrene	0.04	0.08	0.38	0.00	0.13	1	0.01
Styrene oxide	0.04	0.08	0.38	0.01	0.13	1	0.01
Sulfuric acid (1994 and after "acid aerosols" only)	0.03	0.03	0.03	0.03	0.03	1	0.01
Sulfuryl fluoride	27.40	27.40	27.40	27.40	27.40	33	0.47
Tebuthiuron	0.10	0.21	0.93	0.01	0.31	1	0.02
Temephos	0.10	0.21	0.93	0.00	0.31	1	0.02
tert-Butyl alcohol	0.04	0.08	0.38	0.04	0.14	1	0.01
Tetrabromobisphenol A	0.49	0.99	4.38	0.01	1.47	3	0.04
Tetrachloroethylene	0.16	0.33	1.48	0.26	0.56	1	0.02
Tetrachlorvinphos	0.16	0.33	1.48	0.00	0.49	1	0.02
Tetracycline hydrochloride	0.16	0.33	1.48	0.00	0.49	1	0.02
Tetramethrin	0.10	0.21	0.93	0.00	0.31	1	0.02
Thallium	Long	Long	Long	Long	Long	70	1.00
Thiabendazole	0.04	0.08	0.38	0.00	0.13	1	0.01
Thioacetamide	0.04	0.08	0.38	0.00	0.13	1	0.01
Thiobencarb	0.10	0.21	0.93	0.00	0.31	1	0.02
Thiodicarb	0.10	0.21	0.93	0.00	0.31	1	0.02
Thiophanate-methyl	0.16	0.33	1.48	0.00	0.49	1	0.02
Thiourea	0.04	0.08	0.38	0.00	0.13	1	0.01
Thiram	0.10	0.21	0.93	0.00	0.31	1	0.02
Thorium dioxide	Long	Long	Long	Long	Long	70	1.00
Titanium tetrachloride	Long	Long	Long	Long	Long	70	1.00
Toluene	0.04	0.08	0.38	0.01	0.13	1	0.01
Toluene diisocyanate (mixed isomers)	0.10	0.21	0.93	0.01	0.31	1	0.02
Toxaphene	0.49	0.99	4.38	0.02	1.47	3	0.04

continued

TABLE 8.1 (continued)
Persistence Factors of TRI Chemicals

	Half-Lives (yrs)					Residual	
Chemical	**Water**	**Soil**	**Sediment**	**Air**	**Average**	**(lb)**	**PF**
trans-1,3-Dichloropropene	0.10	0.21	0.93	0.01	0.31	1	0.02
trans-1,4-Dichloro-2-butene	0.10	0.21	0.93	0.00	0.31	1	0.02
Triadimefon	0.16	0.33	1.48	0.00	0.49	1	0.02
Triallate	0.16	0.33	1.48	0.00	0.49	1	0.02
Tribenuron methyl	0.16	0.33	1.48	0.01	0.50	1	0.02
Tributyltin methacrylate	27.40	27.40	27.40	27.40	27.40	33	0.47
Trichlorfon	0.16	0.33	1.48	0.01	0.49	1	0.02
Trichloroacetyl chloride	0.16	0.33	1.48	0.49	0.62	1	0.02
Trichloroethylene	0.10	0.21	0.93	0.02	0.31	1	0.02
Trichlorofluoromethane	0.16	0.33	1.48	0.49	0.62	1	0.02
Triclopyr triethylammonium salt	0.49	0.99	4.38	0.00	1.47	3	0.04
Triethylamine	0.10	0.21	0.93	0.00	0.31	1	0.02
Trifluralin	0.49	0.99	4.38	0.00	1.47	3	0.04
Triphenyltin hydroxide	27.40	27.40	27.40	27.40	27.40	33	0.47
Tris(2,3-dibromopropyl) phosphate	0.16	0.33	1.48	0.00	0.49	1	0.02
Trypan blue	0.49	0.99	4.38	0.00	1.47	3	0.04
Urethane	0.04	0.08	0.38	0.01	0.13	1	0.01
Vanadium	Long	Long	Long	Long	Long	70	1.00
Vinclozolin	0.16	0.33	1.48	0.00	0.49	1	0.02
Vinyl acetate	0.04	0.08	0.38	0.00	0.13	1	0.01
Vinyl chloride	0.04	0.08	0.38	0.01	0.13	1	0.01
Vinylidene chloride	0.10	0.21	0.93	0.00	0.31	1	0.02
Warfarin and salts	27.40	27.40	27.40	27.40	27.40	33	0.47
Xylene (mixed isomers)	0.04	0.08	0.38	0.00	0.13	1	0.01
Zinc (fume or dust)	Long	Long	Long	Long	Long	70	1.00

Source: U.S. Environmental Protection Agency, Persistent, Bioaccumulative, and Toxic (PBT) Profiler. 2006. http://www.pbtprofiler.net/before.asp.

BIBLIOGRAPHY

U.S. Environmental Protection Agency. 2006. Persistent, Bioaccumulative, and Toxic (PBT) Profiler. http://www.pbtprofiler.net/before.asp (accessed January 20, 2010).

9 Quantifying Bioconcentration

INTRODUCTION

Compounds that concentrate in the food chain can result in a much higher dose of toxic chemical than simply from the initial release of the chemical to the environment. Bioconcentration is the process by which an organism develops an internal concentration of a chemical that is higher than that of its environment. Bioconcentration is represented by published bioconcentration factors (BCFs), or the ratio of chemical concentration in the organism to that in surrounding water. Bioconcentration results when the organism takes in and absorbs the chemical in its tissue at a rate faster than it is excreted or metabolized. As the organism is consumed by other organisms, the chemical can concentrate further up the food chain. For example, although mercury is only present in small amounts in seawater, it is absorbed by algae, generally as methyl mercury. It is efficiently absorbed but only slowly excreted by organisms (Croteau et al. 2005). Mercury builds up in the adipose tissue of successive levels in the food chain. At each step, mercury that is eaten accumulates until that organism is eaten. The higher the level in the food chain, the higher the concentration of mercury will be in a particular organism. This process explains why predatory fish such as swordfish and sharks or birds like osprey and eagles have higher concentrations of mercury in their tissue than could be accounted for by direct environmental exposure alone. For example, herring contains mercury at approximately 0.01 part per million (ppm), and shark contains mercury at greater than 1 ppm.

BIOCONCENTRATION ADJUSTMENT FACTOR

Bioconcentration provides a path for increasing exposure to a chemical through consumption of seafood. Therefore, the associated analysis of the impact on human health is dependent on the relative ingestion of a chemical through consumption of seafood as compared to the direct amount of chemical consumed through breathing (air) and through consumption of water or indirect consumption of soil. Bioconcentration is specific in that it refers to uptake and accumulation of a substance from water alone, whereas bioaccumulation refers to uptake from all sources combined (e.g., water, food, air). For the purpose of this analysis, bioconcentration associated with uptake through the consumption of seafood in water was used as a surrogate due to the availability of reliable BCF data. Bioaccumulation across other environmental media (i.e., the chemical uptake in plants via the soil and the air and the plants are then eaten; by cattle in the soil and air, which then affects meat and milk that humans may consume) is taken into account in general terms by accounting for media in addition to water, as detailed further in this chapter.

For this examination, we have compared the amount of a chemical ingested due to fish consumption in water in addition to our direct consumption of the water itself. In the U.S. Environmental Protection Agency (USEPA) regulatory human health risk assessments, the standard assumption is that humans consume 2 L or kg of water per day, or 730 kg/yr, using the standard conversion factor of 365 days per year. The average fish consumption in the United States is approximately 20 kg/yr. Japan averages 69 kg/yr, and the Maldives averages 169 kg/yr. Those who fish in Japan have been found to consume 90 kg/yr, and Eskimos have been found to consume 150 kg/yr. Therefore, the water-to-seafood consumption ratio for humans varies between a factor of 4.3 for the most sensitive population (highest consumers of fish) to 36.5 for average U.S. population. The tendency is to base protection of human health on impacts to the most sensitive population, which we also followed for our analysis.

Along the same vein as the concept noted regarding bioaccumulation across other environmental media (in addition to water), one other consideration is that this analysis is based on the water pathway, and our analysis of overall toxicity impacts of Toxics Release Inventory (TRI) chemicals is based on release and transfer of chemicals not only to water but also to the air and land. Therefore, we based the effect of bioconcentration on equal distribution to these three pathways, as detailed further detail here. More specifically, the impact of water bioconcentration, as represented by published BCFs (USEPA PBT Profiler; USEPA 2006), were adjusted to develop chemical-specific bioconcentration adjustment factors (BAFs). The chemical-specific BAFs represent the likely toxic impact of bioconcentration to the sensitive (highest consumers of fish) population of humans and were quantified by taking the published chemical-specific BCF for each chemical and dividing it by a factor of 3 (representing bioaccumulation or uptake across three different media, e.g., water, air, food from soil, etc.) multiplied by 4.3 (conservative water-to-food consumption ratio for humans) as follows:

$$\text{Bioconcentration Adjustment Factor (BAF)} = 1 + [(\text{BCF} - 1)/(3 \times 4.3)]$$

If the BCF is 1 (i.e., no bioconcentration), then there is no adjustment to the BCF; that is, the BAF will equal the BCF of 1.

Table 9.1 presents the published BCFs and the associated developed chemical-specific BAFs for TRI chemicals.

TABLE 9.1
Bioconcentration Factors (BCFs) and Associated Developed Bioconcentration Adjustment Factors (BAFs) of TRI Chemicals

Chemical Name	BCF	BAF
(1,1′-Biphenyl)-4,4′-diamine, 3,3′-dimethyl-	35	4
(1,1′-Biphenyl)-4,4′-diamine, 3,3′-dimethyl-, dihydrochloride (9ci)	120	10
1-(3-Chloroallyl)-3,5,7-triaza-1-azoniaadamantane chloride	3.2	1
1,1,1,2-Tetrachloro-2-fluoroethane (Hcfc-121a)	79	7
1,1,1,2-Tetrachloroethane	99	9
1,1,1-Trichloroethane	8.9	2
1,1,1-Trifluoro-2,2-dichloroethane	26	3
1,1,2,2-Tetrachloro-1-fluoroethane	79	7
1,1,2,2-Tetrachloroethane	8	2
1,1,2-Trichloroethane	10	2
1,1′-Bi(ethylene oxide)	3.2	1
1,1-Dichloro-1,2,2,3,3-pentafluoropropane(Hcfc-225cc)	140	12
1,1-Dichloro-1,2,2-trifluoroethane	26	3
1,1-Dichloro-1,2,3,3,3-pentafluoropropane(Hcfc-225eb)	140	12
1,1-Dichloro-1-fluoroethane	37	4
1,1-Dichloroethane	14	2
1,1-Dichloroethylene	24	3
1,1-Dimethyl hydrazine	3.2	1
1,1′-Methylenebis(4-isocyanatobenzene)	—	1
1,2,3-Trichloropropane	31	3
1,2,4-Trichlorobenzene	720	57
1,2,4-Trimethylbenzene	340	27
1,2-Butylene Oxide	3.2	1
1,2-Dibromo-3-Chloropropane (Dbcp)	100	9
1,2-Dibromoethane	10	2
1,2-Dichloro-1,1,2,3,3-pentafluoropropane(Hcfc-225bb)	140	12
1,2-Dichloro-1,1,2-trifluoroethane	26	3
1,2-Dichloro-1,1,3,3,3-pentafluoropropane(Hcfc-225da)	140	12
1,2-Dichloro-1,1-difluoroethane	34	4
1,2-Dichlorobenzene	150	13
1,2-Dichloroethane	2	1
1,2-Dichloroethylene	15	2
1,2-Dichloropropane	10	2
1,2-Diphenylhydrazine	100	9
1,3-Butadiene	19	2
1,3-Dichloro-1,1,2,2,3-pentafluoropropane	140	12
1,3-Dichloro-1,1,2,3,3-pentafluoropropane(Hcfc-225ea)	140	12
1,3-Dichlorobenzene	575	45
1,3-Dichloropropene (mixed isomers)	32	3
1,4-Dichloro-2-butene	56	5
1,4-Dichlorobenzene	150	13

continued

TABLE 9.1 (continued)
Bioconcentration Factors (BCFs) and Associated Developed Bioconcentration Adjustment Factors (BAFs) of TRI Chemicals

Chemical Name	BCF	BAF
1,4-Dioxane	3.2	1
1,4-Phenylenediamine dihydrochloride	3.2	1
1-Amino-2-methylanthraquinone	730	58
1-Bromo-1-(bromomethyl)-1,3-propanedicarbonitrile	10	2
1-Chloro-1,1,2,2-tetrafluoroethane	15	2
1-Chloro-1,1-difluoroethane	21	3
2,2′,6,6′-Tetrabromo-4,4′-isopropylidenediphenol	—	1
2,2-Dichloro-1,1,1,3,3-pentafluoropropane(Hcfc-225aa)	140	12
2,3,5-Trimethylphenyl methylcarbamate	52	5
2,3-Dichloro-1,1,1,2,3-pentafluoropropane(Hcfc-225ba)	140	12
2,3-Dichloropropene	41	4
2,4,5-Trichlorophenol	1910	149
2,4,6-Trichlorophenol	309	25
2,4,6-Trinitrophenol	1	1
2,4-D	10	2
2,4-D 2-Ethyl-4-methylpentyl ester	9700	753
2,4-D 2-Ethylhexyl ester	34,000	2,637
2,4-D Butoxyethyl ester	770	61
2,4-D Butyl ester	1,300	102
2,4-D Chlorocrotyl ester	1,300	102
2,4-D Propylene glycol butyl ether ester	1,200	94
2,4-D Sodium salt	3.2	1
2,4-D, Isopropyl ester	460	37
2,4-Db	71	6
2,4-Diaminoanisole	2.4	1
2,4-Diaminoanisole sulfate	3.2	1
2,4-Diaminotoluene	3.2	1
2,4-Dichlorophenol	62	6
2,4-Dimethylphenol	48	5
2,4-Dinitrophenol	11	2
2,4-Dinitrotoluene	204	17
2,6-Dinitrotoluene	23	3
2,6-Xylidine	15	2
2-Acetylaminofluorene	140	12
2-Aminoanthraquinone	190	16
2-Aminonaphthalene	32	3
2-Chlor-1,3-butadiene	49	5
2-Chloro-1,1,1,2-tetrafluoroethane	15	2
2-Chloro-1,1,1-trifluoroethane	19	2
2-Chloroacetophenone	17	2
2-Mercaptobenzothiazole	41	4
2-Methyllactonitrile	—	1

TABLE 9.1 (continued)
Bioconcentration Factors (BCFs) and Associated Developed Bioconcentration Adjustment Factors (BAFs) of TRI Chemicals

Chemical Name	BCF	BAF
2-Methylpyridine	4.1	1
2-Nitrophenol	14	2
2-Nitropropane	10	2
2-Phenylphenol	130	11
3,3-Dichloro-1,1,1,2,2-pentafluoropropane	140	12
3,3′-Dichlorobenzidine	329	26
3,3′-Dichlorobenzidine dihydrochloride	270	22
3,3′-Dichlorobenzidine sulfate	270	22
3,3′-Dimethoxybenzidine	14	2
3,3′-Dimethoxybenzidine dihydrochloride	4	1
3,3′-Dimethoxybenzidine hydrochloride(O-dianisidine hydrochloride)	—	1
3,3′-Dimethylbenzidine dihydrofluoride	35	4
3-Chloro-1,1,1-trifluoropropane	45	4
3-Chloro-2-methyl-1-propene	45	4
3-Chloropropionitrile	1.4	1
3-Iodo-2-propynyl butylcarbamate	43	4
4,4′-Diaminodiphenyl ether	6.4	1
4,4′-Diaminodiphenyl sulfide	27	3
4,4′-Isopropylidenediphenol	100	9
4,4′-Methylenebis(2-chloroaniline)	550	44
4,4′-Methylenebis(N,N-dimethyl)benzenamine	1,200	94
4,4′-Methylenedianiline	9.5	2
4,6-Dinitro-O-cresol	24	3
4-Aminoazobenzene	230	19
4-Aminobiphenyl	88	8
4-Dimethylaminoazobenzene	1,800	140
4-Nitrobiphenyl	470	37
4-Nitrophenol	110	9
5-Nitro-O-anisidine	7.7	2
5-Nitro-O-toluidine	16	2
Abamectin	3,600	280
Acephate	3.2	1
Acetaldehyde	3.2	1
Acetamide	3.2	1
Acetonitrile	3.2	1
Acetophenone	9.3	2
Acifluorfen, sodium salt	3.2	1
Acrolein	350	28
Acrylamide	1	1
Acrylic acid	3.2	1

continued

TABLE 9.1 (continued)
Bioconcentration Factors (BCFs) and Associated Developed Bioconcentration Adjustment Factors (BAFs) of TRI Chemicals

Chemical Name	BCF	BAF
Acrylonitrile	48	5
Alachlor	280	23
Aldicarb	42	4
Aldrin	3,715	289
Allyl alcohol	3.2	1
Allyl amine	3.2	1
Allyl chloride	17	2
alpha-Lindane	1,950	152
alpha-Naphthylamine	30	3
Aluminum	231	19
Aluminum oxide (fibrous forms)	3.2	1
Aluminum phosphide	3.2	1
Ametryn	110	9
Amitraz	8,900	691
Amitrole	3.2	1
Ammonia	3.2	1
Ammonium nitrate (solution)	3.2	1
Ammonium sulfate (solution)	3.2	1
Anilazine	520	41
Aniline	9.3	2
Anthracene	1,900	148
Antimony	1	1
Arsenic	44	4
Asbestos (friable)	—	1
Atrazine	8.8	2
Auramine	110	9
Barium	3.2	1
Bendiocarb	12	2
Benfluralin	6,200	482
Benomyl	24	3
Benzal chloride	—	1
Benzamide	2.9	1
Benzene	5	1
Benzidine	93	8
Benzo(Ghi)perylene	—	1
Benzoic trichloride	—	1
Benzoyl chloride	—	1
Benzoyl peroxide	—	1
Benzyl chloride	33	3
Beryllium	19	2
beta-Propiolactone	3.2	1
Bifenthrin	21,000	1,629

TABLE 9.1 (continued)
Bioconcentration Factors (BCFs) and Associated Developed Bioconcentration Adjustment Factors (BAFs) of TRI Chemicals

Chemical Name	BCF	BAF
Biphenyl	377	30
Bis(2-chloro-1-methylethyl) ether	45	4
Bis(2-chloroethoxy)methane	5.7	1
Bis(2-chloroethyl) ether	11	2
Bis(2-ethylhexyl)phthalate	210	17
Bis(chloromethyl) ether	—	1
Bis(tributyltin) oxide	700	55
Boron trichloride	1.6	1
Boron trifluoride	3.2	1
Bromacil	24	3
Bromacil lithium salt (2,4(H,3h)-pyrimidinedione, ethyl-3 (1-methylpropyl), lithium salt)	16	2
Bromine	1.2	1
Bromochlorodifluoromethane	16	2
Bromotrifluoromethane	15	2
Bromoxynil	79	7
Bromoxynil octanoate	25,000	1,939
Brucine	3.3	1
Butyl acrylate	37	4
Butyraldehyde	3.2	1
C.I. acid green 3	3.2	1
C.I. acid red 114	550,000	42,637
C.I. basic green 4	1.6	1
C.I. basic red 1	18,000	1,396
C.I. direct blue 218	3.2	1
C.I. direct brown 95	3.2	1
C.I. disperse yellow 3	620	49
C.I. food red 15	18	2
C.I. food red 5	47	5
C.I. solvent orange 7	61,000	4,730
C.I. solvent yellow 14	9,100	706
C.I. solvent yellow 3	1,100	86
C.I. vat yellow 4	35,000	2,714
Cadmium	64	6
Calcium cyanamide	3.2	1
Camphechlor	34,050	2,640
Captan	10	2
Carbaryl	300	24
Carbofuran	34	4
Carbon disulfide	18	2
Carbon tetrachloride	23	3

continued

TABLE 9.1 (continued)
Bioconcentration Factors (BCFs) and Associated Developed Bioconcentration Adjustment Factors (BAFs) of TRI Chemicals

Chemical Name	BCF	BAF
Carbonyl sulfide	11	2
Carboxin	25	3
Catechol	3.2	1
Chinomethionat (6-methyl-1,3-dithiolo[4,5-B]quinox	440	35
Chloramben	16	2
Chlordane	11,050	858
Chlorendic acid	140	12
Chlorimuron ethyl	47	5
Chlorine	3.2	1
Chlorine dioxide	3.2	1
Chloroacetic acid	3.2	1
Chlorobenzene	79	7
Chlorobenzilate	2,400	187
Chlorodifluoromethane	3.9	1
Chloroethane	7.2	1
Chloroform	4.8	1
Chloromethane	2.9	1
Chloromethyl methyl ether	—	1
Chlorophenols	46	4
Chloropicrin	23	3
Chlorotetrafluoroethane	15	2
Chlorothalonil	120	10
Chlorotrifluoromethane	11	2
Chlorpyrifos methyl	1,100	86
Chlorsulfuron	19	2
Chromium	16	2
Cobalt	4,430	344
Copper	36	4
Creosotes	—	1
Cresol (mixed isomers)	18	2
Crotonaldehyde	3.2	1
Cumene	35	4
Cumene hydroperoxide	26	3
Cupferron	3.2	1
Cyanazine	29	3
Cyanide compounds	3.2	1
Cycloate	520	41
Cyclohexane	240	20
Cyclohexanol	5.1	1
Cyfluthrin	20,000	1,551
Cyhalothrin	87,000	6,745
Dazomet	6.8	1

TABLE 9.1 (continued)
Bioconcentration Factors (BCFs) and Associated Developed Bioconcentration Adjustment Factors (BAFs) of TRI Chemicals

Chemical Name	BCF	BAF
Dazomet, sodium salt	6.8	1
Decabromodiphenyl oxide	4,900	381
Desmedipham	220	18
Diallate	1,500	117
Diaminotoluene (mixed isomers)	3.2	1
Diazinon	460	37
Diazomethane	19	2
Dibenzofuran	1,350	106
Dibromotetrafluoroethane (Halon 2402)	100	9
Dibutyl phthalate	866	68
Dicamba	28	3
Dichloran	79	7
Dichloro-1,1,2-trifluoroethane	26	3
Dichlorobenzene (mixed isomers)	180	15
Dichlorobromomethane	19	2
Dichlorodifluoromethane	26	3
Dichlorofluoromethane	8.9	2
Dichloromethane	5.2	1
Dichlorophene	1,000	78
Dichlorotetrafluoroethane (Cfc-114)	82	7
Dichlorotrifluoroethane	26	3
Dichlorpentafluoro-propane	140	12
Dichlorvos	7.7	2
Diclofop methyl	1,900	148
Dicofol	13,900	1,078
Dicyclopentadiene	150	13
Diethanolamine	3.2	1
Diethatyl ethyl	320	26
Diethyl sulfate	4.3	1
Diflubenzuron	520	41
Diglycidyl resorcinol ether (Dgre)	5.1	1
Dihydrosafrole	310	25
Diisocyanates	58	5
Dimethipin	5.3	1
Dimethoate	2	1
Dimethyl chlorothiophosphate	6.7	1
Dimethyl phthalate	58	5
Dimethyl sulfate	1.9	1
Dimethylamine	3.2	1
Dimethylamine dicamba	4.3	1
Dimethylcarbamoyl chloride	—	1

continued

TABLE 9.1 (continued)
Bioconcentration Factors (BCFs) and Associated Developed Bioconcentration Adjustment Factors (BAFs) of TRI Chemicals

Chemical Name	BCF	BAF
Dinitrobutyl phenol	30	3
Dinitrotoluene (mixed isomers)	27	3
Dinocap	21,000	1,629
Di-N-propylnitrosamine	6.4	1
Dioxin and dioxin-like compounds	—	1
Diphenamid	26	3
Diphenylamine	30	3
Dipotassium endothall	17	2
Dipropyl isocinchomeronate	300	24
Direct black 38	3,100	241
Direct blue 6	56	5
Disodium cyanodithioimidocarbonate	3.2	1
Dithiobiuret	7.2	1
Diuron	64	6
Dodine	16	2
Dodine	240	20
D-trans-Allethrin	2,500	195
Epichlorohydrin	3.2	1
Ethoprop	320	26
Ethyl acrylate	5.9	1
Ethyl chloroformate	1.8	1
Ethyl dipropylthiocarbamate	160	13
Ethylbenzene	8.4	2
Ethylene	4.3	1
Ethylene glycol	10	2
Ethylene glycol monoethyl ether	—	1
Ethylene glycol monomethyl ether	—	1
Ethylene oxide	3.2	1
Ethylene thiourea	2.3	1
Ethylenebisdithiocarbamic acid, salts, and esters	2.8	1
Ethyleneimine	3.2	1
Famphur	29	3
Fenarimol	320	26
Fenbutatin oxide	5,300	412
Fenoxaprop ethyl(2-(4-((6-chloro-2-benzoxazolyen)oxy)penoxy)propanic acid, ethyl ester)	3,400	264
Fenoxycarb	1,100	86
Fenpropathrin	13,000	1,009
Fenthion	760	60
Fenvalerate	30,000	2,327
Ferbam	40	4
Fluazifop-butyl	1,500	117

TABLE 9.1 (continued)
Bioconcentration Factors (BCFs) and Associated Developed Bioconcentration Adjustment Factors (BAFs) of TRI Chemicals

Chemical Name	BCF	BAF
Fluometuron	41	4
Fluorine	3.2	1
Fluoroacetic acid, sodium salt	3.2	1
Fluorouracil	3.2	1
Fluvalinate	88,000	6,823
Folpet	86	8
Fomesafen	94	8
Formaldehyde	0	1
Formic acid	3.2	1
Freon 113	150	13
gamma-Lindane	1,300	102
Glycol ethers	—	1
Heptachlor	19,953	1,548
Hexachloro-1,3-butadiene	11,400	885
Hexachlorobenzene	66,000	5,117
Hexachlorocyclopentadiene	120	10
Hexachloroethane	440	35
Hexachloronaphthalene	24,000	1,861
Hexachlorophene (Hcp)	320,000	24,807
Hexamethylphosphoramide	3.2	1
Hexazinone	15	2
Hydramethylnon	34	4
Hydrazine	—	1
Hydrazine sulfate	1.8	1
Hydrochloric acid	3.2	1
Hydrofluoric acid	3.2	1
Hydrogen cyanide	—	1
Hydroquinone	40	4
Imazalil	470	37
Iron pentacarbonyl	—	1
Isobutyraldehyde	3.2	1
Isodrin	20,180	1,565
Isofenphos	800	63
Isopropyl alcohol	3.2	1
Isosafrole	210	17
Lactofen	2,700	210
Lead	42	4
Linuron	160	13
Lithium carbonate	3	1
Malathion	36	4
Maleic anhydride	—	1

continued

TABLE 9.1 (continued)
Bioconcentration Factors (BCFs) and Associated Developed Bioconcentration Adjustment Factors (BAFs) of TRI Chemicals

Chemical Name	BCF	BAF
Malononitrile	3.2	1
Maneb	220	18
Manganese	3.2	1
M-Cresol	20	2
M-Dinitrobenzene	74	7
Mechlorethamine	2.9	1
Mecoprop	140	12
Mercaptodimethur	98	9
Mercury	36,000	2,792
Merphos	15,000	1,164
Methacrylonitrile	2	1
Metham sodium	3.2	1
Methanamine, N-methyl-N-nitroso	3.2	1
Methanol	3	1
Methazole	160	13
Methoxone	170	14
Methoxone sodium salt ((4-chloro-2-methylpgenoxy) acetate sodium salt)	3.2	1
Methoxychlor	8,128	631
Methyl acrylate	1.4	1
Methyl bromide	4.7	1
Methyl chlorocarbonate	—	1
Methyl ethyl ketone	3.2	1
Methyl hydrazine	3.2	1
Methyl iodide	8.3	2
Methyl isobutyl ketone	5.8	1
Methyl isocyanate	—	1
Methyl isothiocyanate	3.1	1
Methyl methacrylate	6.6	1
Methyl parathion	40	4
Methyl tert-butyl ether	1.5	1
Methylene bromide	12	2
Metiram	2.8	1
Metribuzin	12	2
Mevinphos	3.2	1
Michler's ketone	510	40
Molinate	160	13
Molybdenum trioxide	10	2
Monochloropentafluoroethane	44	4
Monuron	18	2
M-Phenylenediamine	3.2	1
Mustard gas	40	4
M-Xylene	13	2

TABLE 9.1 (continued)
Bioconcentration Factors (BCFs) and Associated Developed Bioconcentration Adjustment Factors (BAFs) of TRI Chemicals

Chemical Name	BCF	BAF
Myclobutanil	100	9
N,N-Dimethylaniline	10	2
N,N-Dimethylformamide	3.2	1
Nabam	3.2	1
Naled	6.6	1
Naphthalene	60	6
N-Butyl alcohol	2.7	1
N-Ethyl-N-nitrosourea	3	1
N-Hexane	540	43
Nickel	47	5
Nicotine and salts	4.6	1
Nitrapyrin	230	19
Nitrate compounds (water dissociable)	—	1
Nitric acid	3.2	1
Nitrilotriacetic acid	3.2	1
Nitrobenzene	13	2
Nitrofen	1,550	121
Nitroglycerin	10	2
N-Methyl-2-pyrrolidone	3.2	1
N-Methylolacrylamide	3.2	1
N-Nitrosodiethylamine	3.2	1
N-Nitrosodi-N-butylamine	59	5
N-Nitrosodiphenylamine	219	18
N-Nitrosomethylvinylamine	1.8	1
N-Nitrosomorpholine	3.2	1
N-Nitroso-N-methylurea	2.8	1
N-Nitrosonornicotine	3.2	1
N-Nitrosopiperidine	3.2	1
Norflurazon	33	3
O-Anisidine	4.6	1
O-Anisidine hydrochloride	4.6	1
O-Cresol	18	2
Octachloronaphthalene	100,000	7,753
Octachlorostyrene	—	1
O-Dinitrobenzene	11	2
O-Phenylenediamine	1.6	1
O-Phenylphenate, sodium	48	5
Oryzalin	70	6
Osmium oxide Oso4 (T-4)	10	2
O-Toluidine	5.9	1
O-Toluidine hydrochloride	10	2

continued

TABLE 9.1 (continued)
Bioconcentration Factors (BCFs) and Associated Developed Bioconcentration Adjustment Factors (BAFs) of TRI Chemicals

Chemical Name	BCF	BAF
Oxydemeton methyl	3.2	1
Oxydiazon	2,600	202
Oxyfluorfen	2,300	179
O-Xylene	11	2
Ozone	3.2	1
P-Anisidine	3.1	1
Paraldehyde	3.2	1
Paraquat	3.2	1
Parathion	480	38
P-Chloroaniline	14	2
P-Chloro-O-toluidine	31	3
P-Chlorophenyl isocyanate	170	14
P-Cresidine	10	2
P-Cresol	18	2
P-Dinitrobenzene	7.6	2
Pebulate	480	38
Pendimethalin	1,944	152
Pentachlorobenzene	7,500	582
Pentachloroethane	67	6
Pentachlorophenol	110	9
Pentobarbital sodium	23	3
Peracetic acid	3.2	1
Perchloromethyl mercaptan	260	21
Permethrin	51,000	3,954
Phenanthrene	2,160	168
Phenol	45	4
Phenothrin	8,400	652
Phenytoin	44	4
Phosgene	—	1
Phosphine	—	1
Phosphorus (yellow or white)	3.2	1
Phthalic anhydride	—	1
Picloram	20	2
Piperonyl butoxide	2,400	187
Pirimiphos methyl	920	72
P-Nitroaniline	6.7	1
P-Nitrosodiphenylamine	150	13
Polybrominated biphenyls (PBBs)	18,200	1,412
Polychlorinated alkanes	—	1
Polychlorinated biphenyls	47,000	3,644
Polycyclic aromatic compounds	912	72
Potassium bromate	3.2	1

TABLE 9.1 (continued)
Bioconcentration Factors (BCFs) and Associated Developed Bioconcentration Adjustment Factors (BAFs) of TRI Chemicals

Chemical Name	BCF	BAF
Potassium dimethyldithiocarbamate	3.9	1
Potassium N-methyldithiocarbamate	3.2	1
P-Phenylenediamine	1.6	1
Profenofos	2,100	164
Prometryn	270	22
Pronamide	240	20
Propachlor	27	3
Propane sultone	3.2	1
Propanil	1.6	1
Propargite	3,700	288
Propargyl alcohol	3.2	1
Propetamphos	470	37
Propiconazole	400	32
Propionaldehyde	3.2	1
Propoxur	8.4	2
Propylene	13	2
Propylene oxide	3.2	1
Propyleneimine	3.2	1
P-Xylene	19	2
Pyridine	3.2	1
Quinoline	8	2
Quinone	3.2	1
Quintozene	912	72
Quizalofop-ethyl	1,100	86
Resmethrin	3,900	303
S,S,S-Tributyltrithiophosphate	13,000	1,009
Saccharin	2.9	1
Safrole	250	20
sec-Butyl alcohol	3.2	1
Selenium	4.8	1
Sethoxydim	1,300	102
Silver	0.5	1
Simazine	27	3
Sodium azide	1.5	1
Sodium dicamba	28	3
Sodium dimethyldithiocarbamate	3.2	1
Sodium nitrite	3.2	1
Sodium pentachlorophenate	21	3
Strychnine and salts	—	1
Styrene	13	2
Styrene oxide	9.9	2

continued

TABLE 9.1 (continued)
Bioconcentration Factors (BCFs) and Associated Developed Bioconcentration Adjustment Factors (BAFs) of TRI Chemicals

Chemical Name	BCF	BAF
Sulfuric acid	3.2	1
Sulfuryl fluoride	15	2
Sulprofos	8,600	668
Tebuthiuron	14	2
Temephos	20,000	1,551
Terbacil	16	2
tert-Butyl alcohol	3.2	1
Tetrachloroethylene	23	3
Tetrachlorvinphos	280	23
Tetracycline hydrochloride	3.2	1
Tetramethrin	2,300	179
Thallium	116	10
Thiabendazole	44	4
Thioacetamide	3.2	1
Thiobencarb	230	19
Thiodicarb	12	2
Thiophanate ethyl	22	3
Thiophanate-methyl	6.8	1
Thiosemicarbazide	3.4	1
Thiourea	3.2	1
Thiram	100	9
Thorium dioxide	10	2
Titanium tetrachloride	2.7	1
Toluene	7.4	1
Toluene diisocyanate (mixed isomers)	—	1
Toluene-2,4-diisocyanate	—	1
Toluene-2,6-diisocyanate	—	1
trans-1,3-Dichloropropene	21	3
trans-1,4-Dichloro-2-butene	56	5
Triadimefon	75	7
Triallate	1,800	140
Triaziquone	3.2	1
Tribenuron methyl	68	6
Tribromomethane	3.2	1
Tributyltin fluoride	1,200	94
Tributyltin methacrylate	770	61
Trichlorfon	3.2	1
Trichloroacetyl chloride	—	1
Trichloroethylene	17	2
Trichlorofluoromethane	49	5
Triclopyr triethylammonium salt	8.1	2
Triethylamine	7.4	1

TABLE 9.1 (continued)
Bioconcentration Factors (BCFs) and Associated Developed Bioconcentration Adjustment Factors (BAFs) of TRI Chemicals

Chemical Name	BCF	BAF
Trifluralin	5,674	441
Triforine	28	3
Triphenyltin chloride	900	71
Triphenyltin hydroxide	280	23
Tris(2,3-dibromopropyl) phosphate	2.8	1
Trypan blue	3.2	1
Urethane	3.2	1
Vanadium	3.2	1
Vinclozolin	130	11
Vinyl acetate	2.3	1
Vinyl bromide	9.2	2
Vinyl chloride	10	2
Warfarin and salts	56	5
Xylene (mixed isomers)	150	13
Zinc	47	5
Zineb	5.7	1

Source: U.S. Environmental Protection Agency, Persistent, Bioaccumulative, and Toxic (PBT) Profiler. 2006. http://www.pbtprofiler.net/before.asp.

BIBLIOGRAPHY

Croteau, M., S. N. Luoma, and A. R Stewart. 2005. Trophic Transfer of Metals along Freshwater Food Webs: Evidence of Cadmium Biomagnification in Nature. *Limnology and Oceanography,* 50(5): 1511–1519.

Josupeit, H. 1996. *Global Overview on Fish Consumption, CIHEAM-IAMZ,* 9–23.

Kromhout, D. 1993. Epidemiological aspects of fish in the diet. *Proceedings of the Nutrition Society* 52, 431–439.

NOAA Fisheries: Office of Science and Technology. 2007. Fisheries of the United States—2005. http://www.st.nmfs.noaa.gov/st1/fus/fus05/ (accessed January 20, 2010).

U.S. Environmental Protection Agency. 2002. *Estimated Per Capita Fish Consumption in the United States*. Washington, DC: EPA-821-C- 02-003.

U.S. Environmental Protection Agency. 2006. Persistent, Bioaccumulative, and Toxic (PBT) Profiler. http://www.pbtprofiler.net/before.asp (accessed January 20, 2010).

10 Developing Effective Toxicity Factors

INTRODUCTION

In the previous chapters, we derived factors for toxicity (Chapter 6), mobility (Chapter 7), persistence (Chapter 8), and bioconcentration (Chapter 9) for the chemicals reported in the Toxics Release Inventory (TRI). This chapter develops an effective toxicity factor (ETF) in units of doses/capita-pound that can then be used to evaluate the relative overall toxic impacts of the various TRI chemical releases.

This is an initial proposed approach; we welcome suggestions on how to improve our analysis or use the same or additional toxicity data to relatively rank chemicals differently. The important point to keep in mind is that the objective of this approach was not to come up with an absolute number that we claim to represent the actual and fixed toxic impact of a chemical. Rather, the proposed methodology was developed solely for the purposes of coming up with a logical means of applying published and routinely used toxicological data to then develop chemical-specific "toxicity" factors to conduct an apples-to-apples comparison across chemicals. This would relatively rank the potential effect among various toxic chemicals such that reduction in the use of toxic chemicals can be targeted and managed accordingly.

The purpose of this is to build on the emphasis of the U.S. Environmental Protection Agency (USEPA) on persistent, bioaccumulative, and toxic (PBT) chemicals and develop an overall relative toxicity index to guide policy in reducing these releases. In this chapter, the quantified ETF (doses/capita-pound) is then multiplied by the 2007 TRI releases (pounds), resulting in a chemical-specific toxicity unit (TU) in doses per capita, or a measure of the toxic impact of the release of each chemical. Ranking the list of 2007 TRI releases by the respective TUs, in other words, by highest overall relative toxicity, shows the trend of the top 10 chemicals on the TRI list representing 99.98 percent of the toxic impact of the releases; yet, these same 10 chemicals represent only 17 percent of the volume of releases reported. This drives home the point that volume alone cannot be used to guide policy in reducing toxic chemical use.

EFFECTIVE TOXICITY FACTOR

Table 10.1 presents the proposed ETFs for TRI chemicals. The ETFs were derived by multiplying together each of the following factors developed in previous chapters:

- TF, toxicity factor (doses/capita-pound) (Chapter 6)
- MF, mobility factor (Chapter 7)
- PF, persistence factor (Chapter 8)
- BAF, bioconcentration adjustment factor (Chapter 9)

For those chemicals that did not have the data needed for quantifying the MF, PF, and BAF factors, the TF was multiplied by the median value for the MF, PF, and BAF across all the TRI chemicals, or 0.002.

TOXICITY IMPACT

Table 10.2 shows the 2007 TRI release data ranked by pounds released and includes the calculated TUs. TU was calculated by multiplying the ETF by the amount released in pounds and represents the estimated toxicity impact per capita of the release of that chemical. A TU of 1 means that the pounds of a chemical released in a given year coupled with the quantified ETF of the chemical has reached the toxicity threshold if distributed evenly over the population of the United States. As an example, chromium was estimated to have a TU of 70 million doses/capita based on the quantified ETF for chromium coupled with the 2007 TRI chromium release data. If the U.S. population were equally exposed to this amount, each person would receive 70 million times the acceptable dose of this compound.

Table 10.3 rearranges the release data and ranks them by TU. One can see that the top 10 chemicals on this list represent 99.98 percent of the TUs released; yet, they represent only 17 percent of the volume of releases reported.

One distortion in this analysis is the apparent relative impact of chromium. Undoubtedly, chromium releases are important, but the toxicity is based on a USEPA assumption that, in general, a sixth of the chromium releases are hexavalent chromium. All types of chromium releases are reported in the TRI together, yet hexavalent chromium is highly toxic, trivalent chromium is only moderately toxic, and metallic chromium is benign. As an example of a beneficial TRI modification, it would be useful to revise the TRI such that hexavalent and trivalent chromium were reported separately and metallic chromium eliminated from reporting. The same logic applies for cobalt and cobalt compounds. In addition, the commercial production of polychlorinated biphenyls (PCBs) has been banned in the United States since 1976; the only PCB releases are due to the presence of PCBs in products and materials produced before the ban took place and due to taking PCB-containing equipment out of service/remediation waste.

Figures 10.1 and 10.2 present the top 2007 TRI release data ranked by releases in pounds as compared to the top 2007 TRI chemicals ranked by TU, respectively. The toxicity data in Figure 10.2 are shown on a logarithmic scale as the variation in toxicity ranges widely.

The quantified chemical-specific ETFs were then applied to the 1988 through 2007 yearly chemical-specific TRI releases (pounds) and the TUs summed by year to come up with a TU for each individual year from 1988 through 2007, as shown in Figure 10.3 (along with the total TRI releases in pounds for these same years).

As noted in Chapter 5 and shown on Figure 10.3, after the first TRI reporting year (1988), the releases of TRI chemicals (pounds) declined steadily through 1996, despite the addition of 286 new chemicals and the addition of federal facilities in 1994. The 1998 report included seven new industry sectors to the reports, which caused the reported releases to more than double. Following this expansion, the release volumes again declined. As far as the total TUs are concerned, although the total TUs were consistently lower than the total release volumes over the years, the overall TU trend/curve followed the same pattern as that for total release volumes.

TABLE 10.1
Combined Impact of Toxicity, Mobility, Persistence, and Bioaccumulation (Effective Toxicity Factor, ETF)

Chemical Name	TF (Doses/ Capita-lb)	MF	PF	BAF	ETF (Doses/ Capita-lb)
(1,1′-Biphenyl)-4,4′-diamine, 3,3′-dimethyl-	6.39E–01	0.01804	5.51E–04	5.51E–04	5.51E–04
1-(3-Chloroallyl)-3,5,7-triaza-1-azoniaadamantane chloride	—	0.50002	—	—	—
1,1,1,2-Tetrachloro-2-fluoroethane (Hcfc-121a)	—	0.00724	—	—	—
1,1,1,2-Tetrachloroethane	3.78E–03	0.07942	7.64E–05	7.64E–05	7.64E–05
1,1,1-Trichloroethane	9.04E–08	0.20068	1.21E–09	1.21E–09	1.21E–09
1,1,1-Trifluoro-2,2-dichloroethane	—	0.50124	—	—	—
1,1,2,2-Tetrachloro-1-fluoroethane	—	0.00724	—	—	—
1,1,2,2-Tetrachloroethane	2.94E–02	0.06776	5.52E–05	5.52E–05	5.52E–05
1,1,2-Trichloroethane	8.23E–03	0.11230	2.26E–05	2.26E–05	2.26E–05
1,1-Dichloro-1-fluoroethane	—	0.39382	—	—	—
1,1-Dichloroethane	8.22E–04	0.31005	7.17E–06	7.17E–06	7.17E–06
1,1-Dichloroethylene	2.70E–06	0.46467	4.59E–08	4.59E–08	4.59E–08
1,1-Dimethyl hydrazine	3.54E–01	0.72727	2.51E–03	2.51E–03	2.51E–03
1,1′-Methylenebis(4-isocyanatobenzene)	5.11E–04	0.002	1.02E–06	—	—
1,2,3-Trichloropropane	4.06E–01	0.04657	8.71E–04	8.71E–04	8.71E–04
1,2,4-Trichlorobenzene	2.92E–04	0.01411	4.08E–06	4.08E–06	4.08E–06
1,2,4-Trimethylbenzene	4.38E–05	0.02221	3.49E–07	3.49E–07	3.49E–07
1,2-Butylene oxide	1.53E–05	0.39474	6.05E–08	6.05E–08	6.05E–08
1,2-Dibromo-3-chloropropane (Dbcp)	1.89E+00	0.03376	7.58E–03	7.58E–03	7.58E–03
1,2-Dibromoethane	3.00E–01	0.09237	4.58E–04	4.58E–04	4.58E–04

continued

TABLE 10.1 (continued)
Combined Impact of Toxicity, Mobility, Persistence, and Bioaccumulation (Effective Toxicity Factor, ETF)

Chemical Name	TF (Doses/ Capita-lb)	MF	PF	BAF	ETF (Doses/ Capita-lb)
1,2-Dichloro-1,1,2-trifluoroethane	—	0.00683	—	—	—
1,2-Dichloro-1,1-difluoroethane	—	0.01582	—	—	—
1,2-Dichlorobenzene	2.18E–06	0.02771	1.04E–08	1.04E–08	1.04E–08
1,2-Dichloroethane	1.33E–02	0.18782	3.80E–05	3.80E–05	3.80E–05
1,2-Dichloroethylene	6.45E–06	0.35607	6.34E–08	6.34E–08	6.34E–08
1,2-Dichloropropane	5.24E–03	0.14401	1.75E–05	1.75E–05	1.75E–05
1,2-Diphenylhydrazine	1.14E–01	0.00753	6.20E–05	6.20E–05	6.20E–05
1,3-Butadiene	2.07E–01	0.51356	2.12E–03	2.12E–03	2.12E–03
1,3-Dichloro-1,1,2,2,3-pentafluoropropane	—	0.00201	—	—	—
1,3-Dichlorobenzene	—	0.03310	—	—	—
1,3-Dichloropropene (mixed isomers)	7.05E–03	0.11715	3.70E–05	3.70E–05	3.70E–05
1,4-Dichloro-2-butene	1.29E+00	0.04563	4.07E–03	4.07E–03	4.07E–03
1,4-Dichlorobenzene	3.69E–03	0.01856	1.19E–05	1.19E–05	1.19E–05
1,4-Dioxane	3.00E–03	0.61048	1.79E–05	1.79E–05	1.79E–05
1-Chloro-1,1,2,2-tetrafluoroethane	6.97E–02	0.51005	2.92E–02	2.92E–02	2.92E–02
1-Chloro-1,1-difluoroethane	6.14E–09	0.51871	6.82E–10	6.82E–10	6.82E–10
2,2′,6,6′-Tetrabromo-4,4′-isopropylidenediphenol	5.81E–02	0.00003	6.32E–08	6.32E–08	6.32E–08
2,3-Dichloropropene	2.90E–06	0.15523	2.43E–08	2.43E–08	2.43E–08
2,4,5-Trichlorophenol	5.81E–07	0.02001	2.97E–08	2.97E–08	2.97E–08
2,4,6-Trichlorophenol	1.65E–03	0.01580	1.13E–05	1.13E–05	1.13E–05
2,4,6-Trinitrophenol	—	0.05636	—	—	—
2,4-D	5.81E–06	0.01425	1.85E–09	1.85E–09	1.85E–09
2,4-D 2-Ethylhexyl ester	2.85E–02	0.00009	9.24E–05	9.24E–05	9.24E–05
2,4-D Butoxyethyl ester	2.67E–02	0.00177	3.77E–05	3.77E–05	3.77E–05
2,4-D Butyl ester	5.81E–06	0.002	1.16E–08	—	—
2,4-D Sodium salt	1.15E–02	0.28941	6.67E–05	6.67E–05	6.67E–05
2,4-Db	7.26E–06	0.002	1.45E–08	—	—
2,4-Diaminotoluene	5.58E–01	0.13690	1.18E–03	1.18E–03	1.18E–03
2,4-Dichlorophenol	1.94E–05	0.03895	5.77E–08	5.77E–08	5.77E–08
2,4-Dimethylphenol	2.90E–06	0.05003	5.62E–09	5.62E–09	5.62E–09
2,4-Dinitrophenol	2.90E–05	0.02677	1.86E–08	1.86E–08	1.86E–08
2,4-Dinitrotoluene	4.53E–02	0.00841	9.09E–05	9.09E–05	9.09E–05
2,6-Dinitrotoluene	5.81E–05	0.00708	9.59E–09	9.59E–09	9.59E–09
2,6-Xylidine	9.87E–03	0.04540	1.23E–05	1.23E–05	1.23E–05
2-Acetylaminofluorene	6.19E–01	0.00119	1.14E–04	1.14E–04	1.14E–04
2-Aminonaphthalene	1.05E–01	0.002	2.09E–04	—	—
2-Chlor-1,3-butadiene	4.67E–05	0.25405	4.68E–07	4.68E–07	4.68E–07

TABLE 10.1 (continued)
Combined Impact of Toxicity, Mobility, Persistence, and Bioaccumulation (Effective Toxicity Factor, ETF)

Chemical Name	TF (Doses/ Capita-lb)	MF	PF	BAF	ETF (Doses/ Capita-lb)
2-Chloro-1,1,1,2-tetrafluoroethane	9.87E–03	0.51005	4.33E–04	4.33E–04	4.33E–04
2-Chloro-1,1,1-trifluoroethane	0.00E+00	0.04798	0.00E+00	0.00E+00	0.00E+00
2-Chloroacetophenone	1.02E–02	0.002	2.05E–05	—	—
2-Mercaptobenzothiazole	8.13E–03	0.00549	1.53E–06	1.53E–06	1.53E–06
2-Methyllactonitrile	2.45E–05	0.51622	1.56E–07	1.56E–07	1.56E–07
2-Methylpyridine	8.13E–03	0.55130	7.43E–05	7.43E–05	7.43E–05
2-Nitrophenol	1.28E–02	0.03205	7.25E–06	7.25E–06	7.25E–06
2-Nitropropane	8.28E–01	0.13059	1.79E–03	1.79E–03	1.79E–03
2-Phenylphenol	1.10E–04	0.01404	1.42E–07	1.42E–07	1.42E–07
3,3-Dichloro-1,1,1,2,2-pentafluoropropane	—	0.00201	—	—	—
3,3′-Dichlorobenzidine	1.30E–01	0.02997	1.49E–03	1.49E–03	1.49E–03
3,3′-Dichlorobenzidine dihydrochloride	4.18E–03	0.02997	4.78E–05	4.78E–05	4.78E–05
3,3′-Dichlorobenzidine sulfate	1.30E–01	0.02997	1.49E–03	1.49E–03	1.49E–03
3,3′-Dimethoxybenzidine	8.13E–04	0.00389	8.34E–08	8.34E–08	8.34E–08
3,3′-Dimethoxybenzidine dihydrochloride	4.06E–03	0.01347	8.88E–07	8.88E–07	8.88E–07
3-Chloro-2-methyl-1-propene	2.47E–03	0.20163	1.83E–05	1.83E–05	1.83E–05
3-Iodo-2-propynyl butylcarbamate	5.81E–06	0.00565	1.17E–09	1.17E–09	1.17E–09
4,4′-Diaminodiphenyl ether	1.16E–03	0.01184	2.56E–07	2.56E–07	2.56E–07
4,4′-Isopropylidenediphenol	1.16E–06	0.00549	7.27E–10	7.27E–10	7.27E–10
4,4′-Methylenebis(2-chloroaniline)	1.38E–01	0.00188	1.92E–04	1.92E–04	1.92E–04
4,4′-Methylenebis(N,N-dimethyl)benzenamine	6.66E–03	0.002	1.33E–05	—	—
4,4′-Methylenedianiline	2.34E–01	0.04707	2.40E–04	2.40E–04	2.40E–04
4,6-Dinitro-O-cresol	5.81E–04	0.00720	1.62E–07	1.62E–07	1.62E–07
4-Aminoazobenzene	1.09E–03	0.00284	7.65E–07	7.65E–07	7.65E–07
4-Aminobiphenyl	3.06E+00	0.00762	2.37E–03	2.37E–03	2.37E–03
4-Dimethylaminoazobenzene	6.66E–01	0.00026	4.08E–04	4.08E–04	4.08E–04
4-Nitrophenol	2.90E–05	0.05397	1.25E–07	1.25E–07	1.25E–07
5-Nitro-O-anisidine	7.14E–03	0.002	1.43E–05	—	—
5-Nitro-O-toluidine	1.92E–03	0.03127	1.71E–06	1.71E–06	1.71E–06
Abamectin	1.16E–03	0.00113	2.75E–06	2.75E–06	2.75E–06
Acephate	5.20E–04	0.45224	3.62E–06	3.62E–06	3.62E–06
Acetaldehyde	7.09E–04	0.99342	6.89E–06	6.89E–06	6.89E–06
Acetamide	6.14E–05	0.50602	3.10E–07	3.10E–07	3.10E–07

continued

TABLE 10.1 (continued)
Combined Impact of Toxicity, Mobility, Persistence, and Bioaccumulation (Effective Toxicity Factor, ETF)

Chemical Name	TF (Doses/ Capita-lb)	MF	PF	BAF	ETF (Doses/ Capita-lb)
Acetone	7.44E–08	0.002	1.49E–10	—	—
Acetonitrile	5.11E–06	0.65603	3.86E–08	3.86E–08	3.86E–08
Acetophenone	5.81E–07	0.05058	4.09E–10	4.09E–10	4.09E–10
Acifluorfen, sodium salt	5.81E–04	0.002	1.16E–06	—	—
Acrolein	1.55E–02	0.49925	1.81E–03	1.81E–03	1.81E–03
Acrylamide	6.60E–01	0.40152	2.22E–03	2.22E–03	2.22E–03
Acrylic acid	3.07E–04	0.53245	1.04E–06	1.04E–06	1.04E–06
Acrylonitrile	5.24E–02	0.30172	6.18E–04	6.18E–04	6.18E–04
Alachlor	3.26E–03	0.00785	9.87E–06	9.87E–06	9.87E–06
Aldicarb	5.81E–05	0.03893	1.24E–07	1.24E–07	1.24E–07
Aldrin	2.49E+00	0.00022	5.96E–03	5.96E–03	5.96E–03
Allyl alcohol	1.03E–03	0.58112	5.86E–06	5.86E–06	5.86E–06
Allyl amine	1.53E–05	0.50002	7.48E–08	7.48E–08	7.48E–08
Allyl chloride	3.37E–03	0.36348	2.29E–05	2.29E–05	2.29E–05
alpha-Naphthylamine	1.05E–01	0.002	2.09E–04	—	—
Aluminum	6.14E–05	0.01972	2.28E–05	2.28E–05	2.28E–05
Aluminum oxide (fibrous forms)	—	0.00003	—	—	—
Aluminum phosphide	1.45E–04	0.21910	3.72E–05	3.72E–05	3.72E–05
Ametryn	6.45E–06	0.00725	7.54E–09	7.54E–09	7.54E–09
Amitraz	2.32E–05	0.002	4.65E–08	—	—
Amitrole	7.67E–06	0.26459	2.00E–08	2.00E–08	2.00E–08
Ammonia	3.07E–06	0.84713	8.87E–10	8.87E–10	8.87E–10
Aniline	1.14E–03	0.10480	2.91E–05	2.91E–05	2.91E–05
Anthracene	1.94E–07	0.00013	6.55E–11	6.55E–11	6.55E–11
Antimony and antimony compounds	1.45E–04	0.00052	7.49E–08	7.49E–08	7.49E–08
Arsenic and arsenic compounds	1.43E+00	0.00052	3.19E–03	3.19E–03	3.19E–03
Asbestos (friable)	3.83E–06	0.00011	3.80E–10	3.80E–10	3.80E–10
Atrazine	1.34E–02	0.00296	1.08E–06	1.08E–06	1.08E–06
Auramine	—	—	—	—	—
Barium and barium compounds	6.14E–04	0.00052	3.71E–07	3.71E–07	3.71E–07
Bendiocarb	—	—	—	—	—
Benfluralin	1.94E–07	0.00031	1.06E–09	1.06E–09	1.06E–09
Benomyl	1.16E–06	0.002	2.32E–09	—	—
Benzal chloride	6.14E–07	0.01784	1.34E–10	1.34E–10	1.34E–10
Benzene	5.61E–03	0.17928	1.76E–05	1.76E–05	1.76E–05
Benzidine	3.39E+01	0.00903	3.27E–02	3.27E–02	3.27E–02
Benzo(Ghi)perylene	6.14E–07	0.00002	2.30E–13	2.30E–13	2.30E–13
Benzoic trichloride	7.55E–01	0.01067	1.32E–04	1.32E–04	1.32E–04
Benzoyl chloride	3.07E–07	0.04662	1.13E–10	1.13E–10	1.13E–10
Benzoyl peroxide	3.07E–07	0.00152	5.70E–12	5.70E–12	5.70E–12

TABLE 10.1 (continued)
Combined Impact of Toxicity, Mobility, Persistence, and Bioaccumulation (Effective Toxicity Factor, ETF)

Chemical Name	TF (Doses/ Capita-lb)	MF	PF	BAF	ETF (Doses/ Capita-lb)
Benzyl chloride	2.52E–02	0.02959	2.20E–05	2.20E–05	2.20E–05
Beryllium and beryllium compounds	7.52E–01	0.00007	1.18E–04	1.18E–04	1.18E–04
Bifenthrin	5.81E–06	0.00017	6.13E–08	6.13E–08	6.13E–08
Biphenyl	1.16E–06	0.00176	5.15E–10	5.15E–10	5.15E–10
Bis(2-chloro-1-methylethyl) ether	7.13E–03	0.03734	1.55E–05	1.55E–05	1.55E–05
Bis(2-chloroethoxy)methane	1.94E–05	0.04437	1.54E–08	1.54E–08	1.54E–08
Bis(2-chloroethyl) ether	1.65E–01	0.08086	3.13E–04	3.13E–04	3.13E–04
Bis(2-ethylhexyl) adipate	6.98E–05	0.002	1.40E–07	—	—
Bis(2-ethylhexyl)phthalate	1.55E–03	0.00028	6.14E–08	6.14E–08	6.14E–08
Bis(chloromethyl) ether	3.18E+01	0.002	6.36E–02	—	—
Bis(tributyltin) oxide	1.94E–04	0.00502	2.74E–06	2.74E–06	2.74E–06
Boron trichloride	—	0.37000	—	—	—
Boron trifluoride	4.38E–04	1.00000	4.40E–06	4.40E–06	4.40E–06
Bromacil	3.07E–07	0.01429	1.60E–10	1.60E–10	1.60E–10
Bromine	—	0.33142	—	—	—
Bromochlorodifluoromethane	3.07E–07	0.00833	8.73E–11	8.73E–11	8.73E–11
Bromotrifluoromethane	2.05E–07	0.00896	6.04E–11	6.04E–11	6.04E–11
Bromoxynil	2.90E–06	0.00574	1.67E–09	1.67E–09	1.67E–09
Bromoxynil octanoate	2.90E–06	0.00018	1.34E–08	1.34E–08	1.34E–08
Brucine	—	—	—	—	—
Butyl acrylate	1.70E–07	0.05864	2.06E–10	2.06E–10	2.06E–10
Butyl benzyl phthalate	1.11E–04	0.002	2.21E–07		
Butyraldehyde	1.53E–08	0.13324	2.00E–11	2.00E–11	2.00E–11
C.I. direct blue 218	2.05E–09	0.00002	2.39E–14	2.39E–14	2.39E–14
Cadmium and cadmium compounds	1.32E+00	0.00007	5.11E–04	5.11E–04	5.11E–04
Calcium cyanamide	—	0.44932	—	—	—
Camphechlor	1.62E–01	0.00041	6.58E–03	6.58E–03	6.58E–03
Captan	3.36E–04	0.00118	1.15E–08	1.15E–08	1.15E–08
Carbaryl	5.81E–07	0.00527	9.72E–10	9.72E–10	9.72E–10
Carbofuran	1.16E–05	0.00898	4.88E–09	4.88E–09	4.88E–09
Carbon disulfide	1.02E–06	0.30965	8.61E–09	8.61E–09	8.61E–09
Carbon tetrachloride	1.22E–02	0.18616	1.21E–04	1.21E–04	1.21E–04
Carbonyl sulfide	7.36E–10	0.51746	7.96E–12	7.96E–12	7.96E–12
Carboxin	5.81E–07	0.00707	1.54E–10	1.54E–10	1.54E–10
Catechol	7.36E–10	0.34263	2.46E–12	2.46E–12	2.46E–12
Chloramben	3.87E–06	0.002	7.74E–09	—	—
Chlordane	2.90E–09	0.00018	1.63E–11	1.63E–11	1.63E–11
Chlorendic acid	7.36E–10	0.02960	9.56E–12	9.56E–12	9.56E–12

continued

TABLE 10.1 (continued)
Combined Impact of Toxicity, Mobility, Persistence, and Bioaccumulation (Effective Toxicity Factor, ETF)

Chemical Name	TF (Doses/ Capita-lb)	MF	PF	BAF	ETF (Doses/ Capita-lb)
Chlorimuron ethyl	8.30E–05	0.01734	1.12E–07	1.12E–07	1.12E–07
Chlorine	2.05E–03	0.53969	1.52E–05	1.52E–05	1.52E–05
Chlorine dioxide	1.54E–03	1.00000	2.12E–05	2.12E–05	2.12E–05
Chloroacetic acid	2.90E–05	0.47017	1.41E–07	1.41E–07	1.41E–07
Chlorobenzene	9.04E–06	0.07362	4.15E–08	4.15E–08	4.15E–08
Chlorobenzilate	1.59E–02	0.00183	9.30E–05	9.30E–05	9.30E–05
Chlorodifluoromethane	6.14E–09	0.52632	2.28E–10	2.28E–10	2.28E–10
Chloroethane	3.07E–08	0.53768	2.26E–10	2.26E–10	2.26E–10
Chloroform	8.87E–03	0.27402	4.83E–05	4.83E–05	4.83E–05
Chloromethane	3.41E–06	0.53647	3.07E–08	3.07E–08	3.07E–08
Chloromethyl methyl ether	3.51E–01	0.39772	1.10E–03	1.10E–03	1.10E–03
Chlorophenols	8.38E–03	0.02451	8.22E–06	8.22E–06	8.22E–06
Chloropicrin	7.67E–04	0.09469	3.69E–06	3.69E–06	3.69E–06
Chlorotetrafluoroethane	4.38E–03	0.50005	1.80E–03	1.80E–03	1.80E–03
Chlorothalonil	4.57E–04	0.00040	1.39E–07	1.39E–07	1.39E–07
Chlorotrifluoromethane	7.36E–10	0.50474	1.04E–11	1.04E–11	1.04E–11
Chlorpyrifos methyl	5.81E–06	0.00121	1.03E–08	1.03E–08	1.03E–08
Chlorsulfuron	4.47E–06	0.08370	1.54E–08	1.54E–08	1.54E–08
Chromium and chromium compounds	3.68E+00	0.14726	1.17E+00	1.17E+00	1.17E+00
Cobalt and cobalt compounds	2.81E+00	0.00007	6.37E–02	6.37E–02	6.37E–02
Copper and copper compounds	1.45E–06	0.00007	3.55E–10	3.55E–10	3.55E–10
Creosotes	7.36E–10	0.00003	3.33E–16	3.33E–16	3.33E–16
Cresol (mixed isomers)	1.09E–06	0.05950	1.26E–09	1.26E–09	1.26E–09
Crotonaldehyde	1.10E–01	0.29178	3.14E–04	3.14E–04	3.14E–04
Cumene	1.35E–06	0.03636	1.50E–09	1.50E–09	1.50E–09
Cumene hydroperoxide	1.35E–06	0.06085	3.17E–09	3.17E–09	3.17E–09
Cyanazine	4.88E–02	0.00654	3.77E–05	3.77E–05	3.77E–05
Cyanide compounds	1.05E–04	0.60976	1.28E–06	1.28E–06	1.28E–06
Cyclohexane	5.11E–08	0.16287	1.36E–09	1.36E–09	1.36E–09
Cyclohexanol	2.90E–07	0.11818	3.78E–10	3.78E–10	3.78E–10
Cyfluthrin	8.72E–03	0.00013	6.63E–05	6.63E–05	6.63E–05
Cyhalothrin	1.94E–04	0.002	3.87E–07	—	—
Dazomet	7.36E–10	0.02740	3.84E–13	3.84E–13	3.84E–13
Dazomet, sodium salt	7.36E–10	0.02740	3.84E–13	3.84E–13	3.84E–13
Decabromodiphenyl oxide	4.89E–05	0.00009	7.74E–08	7.74E–08	7.74E–08
Desmedipham	7.36E–10	0.00134	2.33E–13	2.33E–13	2.33E–13
Diallate	3.54E–03	0.00209	1.48E–05	1.48E–05	1.48E–05
Diaminotoluene (mixed isomers)	0.00E+00	0.13753	—	—	—
Diazinon	8.30E–05	0.00338	1.35E–07	1.35E–07	1.35E–07
Dibenzofuran	—	0.00178	—	—	—

TABLE 10.1 (continued)
Combined Impact of Toxicity, Mobility, Persistence, and Bioaccumulation (Effective Toxicity Factor, ETF)

Chemical Name	TF (Doses/ Capita-lb)	MF	PF	BAF	ETF (Doses/ Capita-lb)
Dibutyl phthalate	5.81E–07	0.00179	3.87E–10	3.87E–10	3.87E–10
Dicamba	1.94E–06	0.04569	3.62E–09	3.62E–09	3.62E–09
Dichlorobenzene (mixed isomers)	2.18E–06	0.02669	1.19E–08	1.19E–08	1.19E–08
Dichlorobromomethane	1.50E–02	0.15579	8.98E–05	8.98E–05	8.98E–05
Dichlorodifluoromethane	1.82E–06	0.50837	4.30E–08	4.30E–08	4.30E–08
Dichlorofluoromethane	7.67E–09	0.56856	1.47E–10	1.47E–10	1.47E–10
Dichloromethane	5.81E–04	0.39586	4.50E–06	4.50E–06	4.50E–06
Dichlorotetrafluoroethane (Cfc-114)	1.82E–06	0.50570	1.32E–07	1.32E–07	1.32E–07
Dichlorotrifluoroethane	1.82E–06	0.00003	4.68E–12	4.68E–12	4.68E–12
Dichlorpentafluoro-propane	1.82E–06	0.00003	9.03E–12	9.03E–12	9.03E–12
Dichlorvos	4.30E–02	0.04671	4.02E–05	4.02E–05	4.02E–05
Dicyclopentadiene	5.11E–05	0.02403	1.28E–07	1.28E–07	1.28E–07
Diethanolamine	7.36E–10	0.50030	2.32E–12	2.32E–12	2.32E–12
Diethyl phthalate	7.26E–08	0.002	1.45E–10	—	—
Diethyl sulfate	7.36E–10	0.05284	4.18E–13	4.18E–13	4.18E–13
Diflubenzuron	2.90E–06	0.002	5.81E–09	—	—
Diglycidyl resorcinol ether (Dgre)	7.36E–10	0.14292	1.82E–12	1.82E–12	1.82E–12
Dihydrosafrole	—	0.00806	—	—	—
Diisocyanates	5.11E–04	0.50002	5.11E–06	5.11E–06	5.11E–06
Dimethipin	2.90E–06	0.03393	1.73E–09	1.73E–09	1.73E–09
Dimethoate	2.90E–04	0.07911	2.06E–07	2.06E–07	2.06E–07
Dimethyl chlorothiophosphate	—	0.03318	—	—	—
Dimethyl phthalate	7.26E–08	0.03236	1.15E–10	1.15E–10	1.15E–10
Dimethyl sulfate	—	0.09634	—	—	—
Dimethylamine	4.38E–05	1.00000	4.28E–07	4.28E–07	4.28E–07
Dimethylamine dicamba	1.94E–06	0.42428	1.76E–08	1.76E–08	1.76E–08
Dimethylcarbamoyl chloride	—	0.36389	—	—	—
Dinitrobutyl phenol	5.81E–05	0.00857	2.13E–08	2.13E–08	2.13E–08
Dinitrotoluene (mixed isomers)	3.95E–02	0.01065	1.82E–05	1.82E–05	1.82E–05
Dinocap	—	—	—	—	—
Di-N-propylnitrosamine	1.02E+00	0.06233	1.19E–03	1.19E–03	1.19E–03
Dioxin and dioxin-like compounds	1.93E+04	0.00004	5.80E–02	5.80E–02	5.80E–02
Diphenylamine	2.32E–06	0.00403	3.99E–10	3.99E–10	3.99E–10
Dipotassium endothall	2.90E–06	0.00526	4.50E–10	4.50E–10	4.50E–10
Direct black 38	1.07E+00	0.002	2.15E–03	—	—
Diuron	2.90E–05	0.00326	7.32E–09	7.32E–09	7.32E–09
Dodine	1.45E–05	0.002	2.90E–08	—	—
Epichlorohydrin	1.26E–03	0.19119	2.60E–06	2.60E–06	2.60E–06

continued

TABLE 10.1 (continued)
Combined Impact of Toxicity, Mobility, Persistence, and Bioaccumulation (Effective Toxicity Factor, ETF)

Chemical Name	TF (Doses/ Capita-lb)	MF	PF	BAF	ETF (Doses/ Capita-lb)
Ethoprop	—	0.01403	—	—	—
Ethyl acrylate	2.79E–03	0.15892	5.10E–06	5.10E–06	5.10E–06
Ethyl chloroformate	—	0.08954	—	—	—
Ethyl dipropylthiocarbamate	2.32E–06	0.01308	5.32E–09	5.32E–09	5.32E–09
Ethylbenzene	1.41E–03	0.05449	1.01E–06	1.01E–06	1.01E–06
Ethylene	—	0.50572	—	—	—
Ethylene glycol	7.96E–07	0.50406	3.74E–09	3.74E–09	3.74E–09
Ethylene glycol monoethyl ether	1.68E–06	0.54188	7.01E–09	7.01E–09	7.01E–09
Ethylene glycol monomethyl ether	3.47E–05	0.54516	1.46E–07	1.46E–07	1.46E–07
Ethylene oxide	4.50E–02	1.00000	6.45E–04	6.45E–04	6.45E–04
Ethylene thiourea	7.33E–03	0.07093	4.77E–06	4.77E–06	4.77E–06
Ethylenebisdithiocarbamic acid, salts, and esters	—	0.50035	—	—	—
Ethyleneimine	—	0.72943	—	—	—
Fenarimol	4.39E+00	0.002	8.78E–03	—	—
Fenbutatin oxide	0.00E+00	0.002	0.00E+00	—	—
Fenoxycarb	5.23E–04	0.00124	7.34E–07	7.34E–07	7.34E–07
Fenpropathrin	2.32E–06	0.002	4.65E–09	—	—
Fenvalerate	2.32E–06	0.002	—	—	—
Fluometuron	4.47E–06	0.00526	1.65E–09	1.65E–09	1.65E–09
Fluorine	9.68E–07	0.50065	5.67E–07	5.67E–07	5.67E–07
Fluoroacetic acid, sodium salt	2.90E–03	—	5.81E–06	5.81E–06	5.81E–06
Fluorouracil	—	0.05269	—	—	—
Fluvalinate	2.32E–06	0.002	4.65E–09	—	—
Folpet	2.04E–04	0.00056	1.47E–08	1.47E–08	1.47E–08
Fomesafen	1.92E+00	0.00355	2.09E–03	2.09E–03	2.09E–03
Formaldehyde	4.02E–03	0.81623	2.53E–05	2.53E–05	2.53E–05
Formic acid	1.02E–04	0.60731	5.04E–07	5.04E–07	5.04E–07
Freon 113	1.22E–08	0.31273	9.37E–10	9.37E–10	9.37E–10
gamma-Lindane	1.59E–01	0.00141	8.75E–04	8.75E–04	8.75E–04
Glycol ethers	1.40E–07	0.06414	1.79E–10	1.79E–10	1.79E–10
Heptachlor	6.60E–01	0.00049	1.86E–02	1.86E–02	1.86E–02
Hexachloro-1,3-butadiene	1.13E–02	0.00792	3.57E–03	3.57E–03	3.57E–03
Hexachlorobenzene	2.34E–01	0.00010	4.53E–03	4.53E–03	4.53E–03
Hexachlorocyclopentadiene	1.54E–03	0.00580	3.46E–06	3.46E–06	3.46E–06
Hexachloroethane	2.10E–03	0.01185	3.47E–05	3.47E–05	3.47E–05
Hexachlorophene (Hcp)	1.94E–04	0.00593	1.06E–03	1.06E–03	1.06E–03
Hexazinone	1.76E–06	0.09085	4.38E–09	4.38E–09	4.38E–09
Hydramethylnon	9.68E–07	0.00005	7.00E–12	7.00E–12	7.00E–12
Hydrazine	1.68E+00	0.55736	7.19E–03	7.19E–03	7.19E–03
Hydrazine sulfate	1.68E+00	0.55736	7.19E–03	7.19E–03	7.19E–03

TABLE 10.1 (continued)
Combined Impact of Toxicity, Mobility, Persistence, and Bioaccumulation (Effective Toxicity Factor, ETF)

Chemical Name	TF (Doses/ Capita-lb)	MF	PF	BAF	ETF (Doses/ Capita-lb)
Hydrochloric acid	1.53E–05	0.95360	1.23E–07	1.23E–07	1.23E–07
Hydrofluoric acid	2.34E–05	0.17331	5.84E–08	5.84E–08	5.84E–08
Hydrogen cyanide	1.05E–04	0.95164	1.58E–06	1.58E–06	1.58E–06
Hydroquinone	3.25E–03	0.13463	1.47E–05	1.47E–05	1.47E–05
Iron pentacarbonyl	—	0.00002	—	—	—
Isobutyraldehyde	—	0.24896	—	—	—
Isodrin	—	0.00010	—	—	—
Isopropyl alcohol	4.38E–08	0.002	8.77E–11	—	—
Isosafrole	—	0.00885	—	—	—
Lactofen	4.47E–06	0.00017	6.10E–09	6.10E–09	6.10E–09
Lead and lead compounds	5.81E–01	0.00011	2.60E–04	2.60E–04	2.60E–04
Linuron	2.90E–05	0.00440	2.91E–08	2.91E–08	2.91E–08
Lithium carbonate	1.16E–05	0.05658	8.94E–09	8.94E–09	8.94E–09
Malathion	2.90E–06	0.00603	5.42E–10	5.42E–10	5.42E–10
Maleic anhydride	4.39E–04	0.04411	1.51E–07	1.51E–07	1.51E–07
Malononitrile	5.81E–04	0.18457	2.00E–06	2.00E–06	2.00E–06
Maneb	1.16E–05	0.002	2.32E–08	—	—
Manganese and manganese compounds	6.14E–03	0.00160	1.15E–05	1.15E–05	1.15E–05
M-Cresol	1.16E–06	0.08219	1.97E–09	1.97E–09	1.97E–09
M-Dinitrobenzene	5.81E–04	0.00557	4.81E–07	4.81E–07	4.81E–07
Mecoprop	5.81E–05	0.01247	1.12E–07	1.12E–07	1.12E–07
Mercury and mercury compounds	1.06E–03	0.00086	2.55E–03	2.55E–03	2.55E–03
Merphos	1.94E–03	0.00005	5.52E–07	5.52E–07	5.52E–07
Methacrylonitrile	1.02E–03	0.23274	2.14E–06	2.14E–06	2.14E–06
Metham sodium	—	0.42487	—	—	—
Methanamine, N-methyl-N-nitroso	7.26E+00	0.67398	3.49E–02	3.49E–02	3.49E–02
Methanol	1.93E–07	0.00063	2.76E–11	2.76E–11	2.76E–11
Methoxone	1.16E–04	0.01257	1.72E–07	1.72E–07	1.72E–07
Methoxychlor	1.16E–05	0.00021	5.67E–08	5.67E–08	5.67E–08
Methyl acrylate	1.94E–06	0.26289	4.39E–09	4.39E–09	4.39E–09
Methyl bromide	1.03E–04	0.56164	1.12E–06	1.12E–06	1.12E–06
Methyl chlorocarbonate	—	0.36735	—	—	—
Methyl ethyl ketone	1.58E–07	0.002	3.16E–10	—	—
Methyl hydrazine	1.68E+00	0.002	3.36E–03	—	—
Methyl iodide	—	0.42150	—	—	—
Methyl isobutyl ketone	8.28E–07	0.11335	1.08E–09	1.08E–09	1.08E–09
Methyl isocyanate	1.00E–02	0.42381	4.23E–05	4.23E–05	4.23E–05

continued

TABLE 10.1 (continued)
Combined Impact of Toxicity, Mobility, Persistence, and Bioaccumulation (Effective Toxicity Factor, ETF)

Chemical Name	TF (Doses/ Capita-lb)	MF	PF	BAF	ETF (Doses/ Capita-lb)
Methyl isothiocyanate	5.81E–07	—	1.16E–09	1.16E–09	1.16E–09
Methyl methacrylate	4.80E–07	0.17364	9.96E–10	9.96E–10	9.96E–10
Methyl parathion	2.32E–04	0.00309	3.79E–08	3.79E–08	3.79E–08
Methyl tert-butyl ether	1.84E–04	0.39913	6.47E–07	6.47E–07	6.47E–07
Methylene bromide	5.81E–06	0.17743	2.12E–08	2.12E–08	2.12E–08
Metribuzin	2.32E–06	0.01622	9.18E–10	9.18E–10	9.18E–10
Molinate	2.90E–05	0.002	5.81E–08	—	—
Molybdenum trioxide	—	0.01108	—	—	—
Monochloropentafluoroethane	—	0.50381	—	—	—
M-Phenylenediamine	9.68E–06	0.24476	3.64E–08	3.64E–08	3.64E–08
M-Xylene	4.67E–07	0.05077	3.82E–10	3.82E–10	3.82E–10
Myclobutanil	2.90E–07	0.00598	1.98E–10	1.98E–10	1.98E–10
N,N-Dimethylaniline	2.90E–05	0.03187	2.06E–08	2.06E–08	2.06E–08
N,N-Dimethylformamide	1.08E–05	0.53489	5.65E–08	5.65E–08	5.65E–08
Nabam	—	0.22362	—	—	—
Naled	2.90E–05	0.00087	6.18E–10	6.18E–10	6.18E–10
Naphthalene	1.05E–02	0.00798	6.16E–06	6.16E–06	6.16E–06
N-Butyl alcohol	5.81E–07	0.17194	6.19E–10	6.19E–10	6.19E–10
N-Ethyl-N-nitrosourea	3.93E+00	0.05944	2.27E–03	2.27E–03	2.27E–03
N-Hexane	1.41E–06	0.20024	6.66E–08	6.66E–08	6.66E–08
Nickel and nickel compounds	8.32E–02	0.00007	2.50E–05	2.50E–05	2.50E–05
Nicotine and salts	—	0.50354	—	—	—
Nitrapyrin	—	0.00520	—	—	—
Nitrate compounds	4.52E–08	0.50002	1.13E–09	1.13E–09	1.13E–09
Nitric acid	—	0.29479	0.00E+00	0.00E+00	0.00E+00
Nitrilotriacetic acid	—	0.12161	—	—	—
Nitrobenzene	1.23E–02	0.02988	7.68E–06	7.68E–06	7.68E–06
Nitroglycerin	1.57E–03	0.01886	6.71E–07	6.71E–07	6.71E–07
N-Methyl-2-pyrrolidone	0.00E+00	0.50002	—	—	—
N-Methylolacrylamide	6.60E–01	0.50002	3.22E–03	3.22E–03	3.22E–03
N-Nitrosodiethylamine	2.19E+01	0.17961	6.06E–02	6.06E–02	6.06E–02
N-Nitrosodi-N-butylamine	8.04E–01	0.02096	7.73E–04	7.73E–04	7.73E–04
N-Nitrosodiphenylamine	1.08E–03	0.00869	2.21E–06	2.21E–06	2.21E–06
N-Nitroso-N-methylurea	1.74E+01	0.16498	2.81E–02	2.81E–02	2.81E–02
N-Nitrosopiperidine	1.37E+00	0.14379	3.04E–03	3.04E–03	3.04E–03
Norflurazon	1.45E–06	0.00292	2.52E–10	2.52E–10	2.52E–10
O-Anisidine	7.14E–03	0.05121	6.15E–06	6.15E–06	6.15E–06
O-Cresol	1.16E–06	0.08935	2.01E–09	2.01E–09	2.01E–09
O-Dinitrobenzene	5.81E–04	0.00588	1.45E–07	1.45E–07	1.45E–07
O-Phenylenediamine	2.73E–03	0.10229	3.84E–06	3.84E–06	3.84E–06
Oryzalin	1.16E–06	0.002	2.32E–09	—	—

TABLE 10.1 (continued)
Combined Impact of Toxicity, Mobility, Persistence, and Bioaccumulation (Effective Toxicity Factor, ETF)

Chemical Name	TF (Doses/ Capita-lb)	MF	PF	BAF	ETF (Doses/ Capita-lb)
O-Toluidine	2.61E–02	0.07016	2.10E–05	2.10E–05	2.10E–05
O-Toluidine hydrochloride	1.89E–02	0.04616	1.23E–05	1.23E–05	1.23E–05
Oxydiazon	1.16E–05	0.00043	1.74E–08	1.74E–08	1.74E–08
Oxyfluorfen	1.94E–05	0.00019	2.41E–08	2.41E–08	2.41E–08
O-Xylene	4.67E–07	0.04723	5.15E–10	5.15E–10	5.15E–10
Ozone	—	0.93131	—	—	—
P-Anisidine	7.14E–03	0.002	1.43E–05	—	—
Paraldehyde	—	0.25857	—	—	—
Paraquat	1.29E–05	0.41835	8.31E–08	8.31E–08	8.31E–08
Parathion	9.68E–06	0.00177	8.58E–09	8.58E–09	8.58E–09
P-Chloroaniline	1.16E–02	0.03345	1.03E–05	1.03E–05	1.03E–05
P-Cresidine	—	0.02834	—	—	—
P-Cresol	1.16E–05	0.07933	1.78E–08	1.78E–08	1.78E–08
P-Dinitrobenzene	5.81E–04	0.00443	9.31E–08	9.31E–08	9.31E–08
Pebulate	1.16E–06	0.002	2.32E–09	—	—
Pendimethalin	1.45E–06	0.00036	1.36E–09	1.36E–09	1.36E–09
Pentachlorobenzene	7.26E–05	0.00391	6.81E–06	6.81E–06	6.81E–06
Pentachloroethane	5.23E–03	0.04440	6.95E–05	6.95E–05	6.95E–05
Pentachlorophenol	8.38E–03	0.00206	6.15E–06	6.15E–06	6.15E–06
Peracetic acid	—	0.50002	—	—	—
Permethrin	1.16E–06	0.00005	4.27E–09	4.27E–09	4.27E–09
Phenanthrene	—	0.00101	—	—	—
Phenol	1.73E–06	0.15199	9.66E–09	9.66E–09	9.66E–09
Phenothrin	—	0.00007	—	—	—
Phenytoin	—	0.00284	—	—	—
Phosgene	1.02E–03	0.84464	9.39E–06	9.39E–06	9.39E–06
Phosphine	1.22E–03	0.72627	2.25E–05	2.25E–05	2.25E–05
Phosphorus (yellow or white)	2.90E–03	0.23399	9.05E–07	9.05E–07	9.05E–07
Phospohoric acid	3.07E–05	0.002	6.14E–08	—	—
Phthalic anhydride	1.54E–05	0.03963	4.97E–09	4.97E–09	4.97E–09
Picloram	8.30E–07	0.01038	3.69E–10	3.69E–10	3.69E–10
Piperonyl butoxide	—	0.00191	—	—	—
Pirimiphos methyl	5.81E–06	0.002	1.16E–08	—	—
P-Nitroaniline	1.23E–03	0.01419	3.30E–07	3.30E–07	3.30E–07
P-Nitrosodiphenylamine	1.08E–03	0.002	2.16E–06	—	—
Polychlorinated alkanes (C10-C13)	7.26E–06	0.00003	9.18E–12	9.18E–12	9.18E–12
Polychlorinated Biphenyls	2.91E–01	0.00058	2.30E–02	2.30E–02	2.30E–02
Polycyclic aromatic compounds	7.61E–01	0.00003	9.63E–07	9.63E–07	9.63E–07
Potassium bromate	—	0.13136	—	—	—

continued

TABLE 10.1 (continued)
Combined Impact of Toxicity, Mobility, Persistence, and Bioaccumulation (Effective Toxicity Factor, ETF)

Chemical Name	TF (Doses/ Capita-lb)	MF	PF	BAF	ETF (Doses/ Capita-lb)
Potassium dimethyldithiocarbamate	—	0.08516	—	—	—
Potassium N-methyldithiocarbamate	—	0.33168	—	—	—
P-Phenylenediamine	3.06E–07	0.09746	4.10E–10	4.10E–10	4.10E–10
Prometryn	1.45E–05	0.00289	1.56E–08	1.56E–08	1.56E–08
Pronamide	7.74E–07	0.00210	5.43E–10	5.43E–10	5.43E–10
Propachlor	4.47E–06	0.01350	2.39E–09	2.39E–09	2.39E–09
Propane sultone	—	0.20704	—	—	—
Propanil	1.16E–05	0.00618	1.29E–09	1.29E–09	1.29E–09
Propargite	2.90E–06	0.03177	4.53E–07	4.53E–07	4.53E–07
Propargyl alcohol	2.90E–05	0.55608	1.58E–07	1.58E–07	1.58E–07
Propiconazole	4.47E–06	0.00526	1.28E–08	1.28E–08	1.28E–08
Propionaldehyde	3.83E–05	0.55464	2.08E–07	2.08E–07	2.08E–07
Propoxur	1.45E–05	0.02160	6.49E–09	6.49E–09	6.49E–09
Propylene	—	0.50707	—	—	—
Propylene oxide	1.51E–02	0.76669	1.23E–04	1.23E–04	1.23E–04
Propyleneimine	—	0.69196	—	—	—
P-Xylene	4.38E–07	0.05261	4.61E–10	4.61E–10	4.61E–10
Pyridine	5.81E–05	0.58112	3.68E–07	3.68E–07	3.68E–07
Quinoline	1.74E–01	0.04470	1.00E–04	1.00E–04	1.00E–04
Quinone	—	0.05812	—	—	—
Quintozene	1.51E–02	0.00046	3.40E–05	3.40E–05	3.40E–05
Quizalofop-ethyl	2.32E–07	0.002	4.65E–10	—	—
Resmethrin	1.94E–06	0.002	3.87E–09	—	—
S,S,S-Tributyltrithiophosphate	1.94E–03	0.00148	1.56E–05	1.56E–05	1.56E–05
Saccharin	—	0.03164	—	—	—
Safrole	3.21E–02	0.00994	8.52E–05	8.52E–05	8.52E–05
sec-Butyl alcohol	3.93E–08	0.27008	1.04E–10	1.04E–10	1.04E–10
Selenium and selenium compounds	2.70E–05	0.02321	8.10E–07	8.10E–07	8.10E–07
Sethoxydim	1.10E–02	0.002	2.21E–05	—	—
Silver and silver compounds	1.16E–05	0.00007	7.35E–10	7.35E–10	7.35E–10
Simazine	6.98E–03	0.00126	4.53E–07	4.53E–07	4.53E–07
Sodium azide	1.45E–05	0.09580	1.78E–08	1.78E–08	1.78E–08
Sodium dicamba	1.94E–06	0.30002	3.07E–08	3.07E–08	3.07E–08
Sodium dimethyldithiocarbamate	—	0.50002	—	—	—
Sodium nitrite	—	0.50002	—	—	—
Strychnine	1.94E–04	0.00599	2.32E–08	2.32E–08	2.32E–08
Styrene	5.97E–07	0.04936	4.74E–10	4.74E–10	4.74E–10
Styrene oxide	—	0.03732	—	—	—

TABLE 10.1 (continued)
Combined Impact of Toxicity, Mobility, Persistence, and Bioaccumulation (Effective Toxicity Factor, ETF)

Chemical Name	TF (Doses/ Capita-lb)	MF	PF	BAF	ETF (Doses/ Capita-lb)
Sulfuric acid	8.13E–09	0.50014	5.42E–12	5.42E–12	5.42E–12
Sulfuryl fluoride	—	0.01371	—	—	—
Tebuthiuron	8.30E–07	0.02503	5.51E–10	5.51E–10	5.51E–10
Temephos	2.90E–06	0.002	5.81E–09	—	—
Terbacil	4.47E–06	0.002	8.93E–09	—	—
Terephthalic acid	5.81E–08	0.002	1.16E–10	—	—
tert-Butyl alcohol	3.93E–08	0.61671	2.46E–10	2.46E–10	2.46E–10
Tetrachloroethylene	3.32E–02	0.07494	1.24E–04	1.24E–04	1.24E–04
Tetrachlorvinphos	1.40E–03	0.00167	9.02E–07	9.02E–07	9.02E–07
Tetracycline hydrochloride	—	0.24952	—	—	—
Tetramethrin	—	0.00069	—	—	—
Thallium and thallium compounds	8.93E–04	0.00007	5.83E–07	5.83E–07	5.83E–07
Thiabendazole	—	0.00355	—	—	—
Thioacetamide	—	0.20188	—	—	—
Thiobencarb	5.81E–06	0.00306	4.33E–10	4.33E–10	4.33E–10
Thiodicarb	1.94E–05	0.002	3.87E–08	—	—
Thiophanate-methyl	7.26E–07	0.01049	1.88E–10	1.88E–10	1.88E–10
Thiourea	7.33E–03	0.18843	1.35E–05	1.35E–05	1.35E–05
Thiram	1.16E–05	0.00279	9.21E–10	9.21E–10	9.21E–10
Titanium tetrachloride	3.07E–03	0.08353	2.90E–04	2.90E–04	2.90E–04
Toluene	7.87E–07	0.09654	9.55E–10	9.55E–10	9.55E–10
Toluene diisocyanate (mixed isomers)	1.00E–02	0.00574	6.99E–07	6.99E–07	6.99E–07
Toluene-2,4-diisocyanate	1.00E–02	0.00488	2.00E–05	2.00E–05	2.00E–05
Toluene-2,6-diisocyanate	1.00E–02	0.00563	2.08E–06	2.08E–06	2.08E–06
Trans-1,3-dichloropropene	7.05E–03	0.11302	2.68E–05	2.68E–05	2.68E–05
Trans-1,4-dichloro-2-butene	1.29E+00	0.04817	4.29E–03	4.29E–03	4.29E–03
Triadimefon	1.94E–06	0.002	3.87E–09	—	—
Triallate	4.47E–06	0.00120	1.28E–08	1.28E–08	1.28E–08
Tribenuron methyl	—	0.00355	—	—	—
Tribromomethane	7.99E–04	0.06999	1.22E–06	1.22E–06	1.22E–06
Tributyltin methacrylate	1.94E–04	0.002	3.87E–07	—	—
Trichlorfon	—	0.17326	—	—	—
Trichloroacetyl chloride	—	0.04873	—	—	—
Trichloroethylene	1.37E–03	0.15708	6.38E–06	6.38E–06	6.38E–06
Trichlorofluoromethane	6.32E–07	0.49200	2.88E–08	2.88E–08	2.88E–08
Triclopyr triethylammonium salt	2.90E–06	0.00866	1.45E–09	1.45E–09	1.45E–09
Triethylamine	4.38E–05	0.26397	2.27E–07	2.27E–07	2.27E–07
Trifluralin	4.55E–04	0.00040	3.02E–06	3.02E–06	3.02E–06
Tris(2,3-dibromopropyl) phosphate	—	0.00166	—	—	—

continued

TABLE 10.1 (continued)
Combined Impact of Toxicity, Mobility, Persistence, and Bioaccumulation (Effective Toxicity Factor, ETF)

Chemical Name	TF (Doses/ Capita-lb)	MF	PF	BAF	ETF (Doses/ Capita-lb)
Trypan blue	—	0.00016	—	—	—
Urethane	—	0.35566	—	—	—
Vanadium and vanadium compounds	8.30E–06	0.00007	6.39E–10	6.39E–10	6.39E–10
Vinclozolin	2.32E–06	0.002	4.65E–09	—	—
Vinyl acetate	1.59E–06	0.23794	3.48E–09	3.48E–09	3.48E–09
Vinyl bromide	9.92E–03	0.002	1.98E–05	—	—
Vinyl chloride	4.32E–02	0.54690	3.36E–04	3.36E–04	3.36E–04
Warfarin and salts	1.94E–04	0.00208	8.04E–09	8.04E–09	8.04E–09
Xylene (mixed isomers)	3.36E–06	0.05642	1.99E–08	1.99E–08	1.99E–08
Zinc and zinc compounds	1.94E–07	0.01972	1.74E–08	1.74E–08	1.74E–08
Zineb	1.16E–06	0.002	2.32E–09	—	—

TABLE 10.2
2007 TRI Releases Ranked by Volume (Pounds)

Chemical	ETF (Doses/Capita-lb)	2007 TRI Release (lb)	TU (Doses/Capita)
Zinc and zinc compounds	1.74E–08	800,318,423	13.9
Hydrochloric acid	1.23E–07	502,532,256	61.6
Lead and lead compounds	2.60E–04	495,875,564	129,166.8
Nitrate compounds	1.13E–09	270,689,909	0.3
Manganese and manganese compounds	1.15E–05	245,353,348	2,814.3
Barium and barium compounds	3.71E–07	244,837,073	90.7
Copper and copper compounds	3.55E–10	176,593,590	0.1
Ammonia	8.87E–10	153,483,048	0.1
Methanol	2.76E–11	151,136,956	0.0
Sulfuric acid	5.42E–12	138,083,554	0.0
Arsenic and arsenic compounds	3.19E–03	97,581,160	311,237.7
Hydrofluoric acid	5.84E–08	72,907,435	4.3
Chromium and chromium compounds	1.17E+00	60,402,489	70,822,503.1
Vanadium and vanadium compounds	6.39E–10	46,394,709	0.0
Toluene	9.55E–10	41,716,531	0.0
Styrene	4.74E–10	40,748,666	0.0
Aluminum	2.28E–05	39,901,864	909.9
Nickel and nickel compounds	2.50E–05	37,902,077	947.3
N-Hexane	6.66E–08	34,978,189	2.3
Xylene (mixed isomers)	1.99E–08	25,595,670	0.5
Formaldehyde	2.53E–05	21,933,684	555.8
Carbonyl sulfide	7.96E–12	19,902,093	0.0
Ethylene	—	18,577,924	0.0
Glycol ethers	1.79E–10	18,476,420	0.0
Nitric acid	0.00E+00	17,994,379	0.0
Acetonitrile	3.86E–08	17,941,970	0.7
Formic acid	5.04E–07	13,934,367	7.0
N-Butyl alcohol	6.19E–10	13,897,074	0.0
Antimony and antimony compounds	7.49E–08	12,326,053	0.9
Acetaldehyde	6.89E–06	11,309,426	77.9
Propylene	—	11,255,935	0.0
Asbestos (friable)	3.80E–10	10,430,381	0.0
Carbon disulfide	8.61E–09	8,935,912	0.1
Benzene	1.76E–05	8,465,367	148.8
1,2,4-Trimethylbenzene	3.49E–07	7,242,424	2.5
Acrylonitrile	6.18E–04	7,059,836	4,364.8
Cobalt and cobalt compounds	6.37E–02	6,970,293	444,242.0
Mercury and mercury compounds	2.55E–03	6,935,622	17,656.4
Cyanide compounds	1.28E–06	6,871,538	8.8
Chlorodifluoromethane	2.28E–10	6,682,534	0.0
Phenol	9.66E–09	6,677,767	0.1

continued

TABLE 10.2 (continued)
2007 TRI Releases Ranked by Volume (Pounds)

Chemical	ETF (Doses/Capita-lb)	2007 TRI Release (lb)	TU (Doses/Capita)
Ethylene glycol	3.74E–09	6,614,766	0.0
Acrylamide	2.22E–03	6,161,247	13,656.2
1-Chloro-1,1-difluoroethane	6.82E–10	6,045,042	0.0
Dichloromethane	4.50E–06	5,903,242	26.6
Chlorine	1.52E–05	5,643,223	85.9
Cyclohexane	1.36E–09	5,283,345	0.0
Ethylbenzene	1.01E–06	4,843,102	4.9
Methyl isobutyl ketone	1.08E–09	4,818,216	0.0
Biphenyl	5.15E–10	4,800,703	0.0
N-Methyl-2-pyrrolidone	—	4,713,092	0.0
Aluminum oxide (fibrous forms)	—	4,686,360	0.0
Acrylic acid	1.04E–06	4,625,008	4.8
Trichloroethylene	6.38E–06	4,485,202	28.6
Sodium nitrite	—	4,320,241	0.0
Selenium and selenium compounds	8.10E–07	4,286,943	3.5
Cadmium and cadmium compounds	5.11E–04	3,907,391	1,996.1
Naphthalene	6.16E–06	2,850,878	17.6
Methyl methacrylate	9.96E–10	2,733,371	0.0
Cyclohexanol	3.78E–10	2,616,534	0.0
N,N-Dimethylformamide	5.65E–08	2,431,907	0.1
Hydrogen cyanide	1.58E–06	2,309,862	3.7
Tetrachloroethylene	1.24E–04	2,237,864	277.5
Thallium and thallium compounds	5.83E–07	2,125,769	1.2
Vinyl acetate	3.48E–09	2,108,154	0.0
Polychlorinated biphenyls	2.30E–02	2,090,371	47,983.4
Chloromethane	3.07E–08	2,014,817	0.1
Creosotes	3.33E–16	1,950,641	0.0
1,3-Butadiene	2.12E–03	1,788,084	3,790.0
Triethylamine	2.27E–07	1,744,585	0.4
Acrolein	1.81E–03	1,696,876	3,066.4
Diethanolamine	2.32E–12	1,563,591	0.0
Diisocyanates	5.11E–06	1,472,453	7.5
Molybdenum trioxide	—	1,417,709	0.0
Polycyclic aromatic compounds	9.63E–07	1,365,384	1.3
P-Xylene	4.61E–10	1,311,540	0.0
Cresol (mixed isomers)	1.26E–09	1,240,288	0.0
4,4′-Isopropylidenediphenol	7.27E–10	1,131,912	0.0
Bis(2-ethylhexyl)phthalate	6.14E–08	1,126,569	0.1
Cumene	1.50E–09	1,106,755	0.0
Acetamide	3.10E–07	1,097,494	0.3
Aniline	2.91E–05	920,606	26.8
Silver and silver compounds	7.35E–10	911,796	0.0

TABLE 10.2 (continued)
2007 TRI Releases Ranked by Volume (Pounds)

Chemical	ETF (Doses/Capita-lb)	2007 TRI Release (lb)	TU (Doses/Capita)
Beryllium and beryllium compounds	1.18E–04	867,078	102.7
Nicotine and salts	—	844,069	0.0
tert-Butyl alcohol	2.46E–10	789,200	0.0
sec-Butyl alcohol	1.04E–10	777,850	0.0
2-Chlor-1,3-butadiene	4.68E–07	741,853	0.3
Acetophenone	4.09E–10	726,934	0.0
Chloroethane	2.26E–10	715,803	0.0
Maleic anhydride	1.51E–07	707,065	0.1
Chloroform	4.83E–05	706,555	34.1
Ozone	—	704,712	0.0
Pyridine	3.68E–07	702,367	0.3
Decabromodiphenyl oxide	7.74E–08	695,861	0.1
Nitrobenzene	7.68E–06	601,120	4.6
Methyl tert-butyl ether	6.47E–07	585,465	0.4
Freon 113	9.37E–10	565,482	0.0
2-Chloro-1,1,1,2-tetrafluoroethane	4.33E–04	551,358	238.6
Chlorobenzene	4.15E–08	545,708	0.0
Chlorine dioxide	2.12E–05	545,291	11.5
Allyl alcohol	5.86E–06	526,216	3.1
Atrazine	1.08E–06	515,072	0.6
Butyraldehyde	2.00E–11	510,567	0.0
1,2-Dichloroethane	3.80E–05	449,853	17.1
Hydroquinone	1.47E–05	430,989	6.3
Bromine	—	410,482	0.0
Phenanthrene	—	378,556	0.0
Vinyl chloride	3.36E–04	372,635	125.3
Methyl bromide	1.12E–06	363,571	0.4
Dimethylamine	4.28E–07	362,524	0.2
2,2′,6,6′-Tetrabromo-4,4′-isopropylidenediphenol	6.32E–08	343,273	0.0
Propylene oxide	1.23E–04	338,598	41.6
Dibutyl phthalate	3.87E–10	335,487	0.0
Propionaldehyde	2.08E–07	334,438	0.1
1,1-Dichloro-1-fluoroethane	—	327,321	0.0
O-Xylene	5.15E–10	325,496	0.0
Trichlorofluoromethane	2.88E–08	318,052	0.0
M-Xylene	3.82E–10	317,740	0.0
Dichlorotetrafluoroethane (Cfc-114)	1.32E–07	312,871	0.0
Carbon tetrachloride	1.21E–04	308,357	37.3
Ethylene oxide	6.45E–04	305,961	197.5
Benzoyl peroxide	5.70E–12	298,409	0.0

continued

TABLE 10.2 (continued)
2007 TRI Releases Ranked by Volume (Pounds)

Chemical	ETF (Doses/Capita-lb)	2007 TRI Release (lb)	TU (Doses/Capita)
Dimethyl phthalate	1.15E–10	287,316	0.0
Phosphorus (yellow or white)	9.05E–07	276,993	0.3
Phthalic anhydride	4.97E–09	268,997	0.0
Dichlorodifluoromethane	4.30E–08	268,445	0.0
3-Iodo-2-propynyl butylcarbamate	1.17E–09	261,184	0.0
M-Cresol	1.97E–09	192,773	0.0
Methyl acrylate	4.39E–09	186,646	0.0
1,4-Dioxane	1.79E–05	185,132	3.3
Dicyclopentadiene	1.28E–07	174,818	0.0
Anthracene	6.55E–11	171,896	0.0
Sodium dimethyldithiocarbamate	—	171,471	0.0
Fluorine	5.67E–07	168,844	0.1
Thiram	9.21E–10	166,819	0.0
Ethylene glycol monomethyl ether	1.46E–07	156,833	0.0
Epichlorohydrin	2.60E–06	155,813	0.4
Chlorothalonil	1.39E–07	154,189	0.0
1,1,1-Trifluoro-2,2-dichloroethane	—	146,385	0.0
Isobutyraldehyde	—	141,932	0.0
Lithium carbonate	8.94E–09	137,578	0.0
1,2-Dichloro-1,1,2-trifluoroethane	—	137,476	0.0
Butyl benzyl phthalate	2.21E–07	137,168	0.0
Nitroglycerin	6.71E–07	130,438	0.1
2-Methyllactonitrile	1.56E–07	126,169	0.0
Titanium tetrachloride	2.90E–04	123,546	35.8
2-Mercaptobenzothiazole	1.53E–06	121,620	0.2
1,2-Dichloropropane	1.75E–05	115,710	2.0
M-Phenylenediamine	3.64E–08	115,161	0.0
Sulfuryl fluoride	—	112,245	0.0
P-Cresol	1.78E–08	104,857	0.0
Benzo(Ghi)perylene	2.30E–13	98,286	0.0
Urethane	—	94,810	0.0
Polychlorinated alkanes (C10–C13)	9.18E–12	94,418	0.0
Methacrylonitrile	2.14E–06	92,267	0.2
1,2-Dichlorobenzene	1.04E–08	90,119	0.0
Cumene hydroperoxide	3.17E–09	87,693	0.0
Pendimethalin	1.36E–09	86,336	0.0
Ethylene glycol monoethyl ether	7.01E–09	82,419	0.0
1,1,1-Trichloroethane	1.21E–09	81,479	0.0
1,4-Dichlorobenzene	1.19E–05	79,266	0.9
Ethyl acrylate	5.10E–06	78,135	0.4
Fomesafen	2.09E–03	69,115	144.8
4,4′-Methylenedianiline	2.40E–04	67,423	16.2

TABLE 10.2 (continued)
2007 TRI Releases Ranked by Volume (Pounds)

Chemical	ETF (Doses/Capita-lb)	2007 TRI Release (lb)	TU (Doses/Capita)
Nitrilotriacetic acid	—	63,971	0.0
Propiconazole	1.28E–08	59,899	0.0
1,1-Dichloroethylene	4.59E–08	57,013	0.0
Catechol	2.46E–12	52,005	0.0
Diphenylamine	3.99E–10	51,677	0.0
Toluene diisocyanate (mixed isomers)	6.99E–07	50,026	0.0
Chlorophenols	8.22E–06	49,196	0.4
2-Chloro-1,1,1-trifluoroethane	0.00E+00	48,681	0.0
Ethoprop	—	48,311	0.0
2,4,6-Trinitrophenol	—	45,015	0.0
Hexachlorobenzene	4.53E–03	43,018	194.7
P-Phenylenediamine	4.10E–10	41,300	0.0
Dicamba	3.62E–09	39,276	0.0
2-Methylpyridine	7.43E–05	39,138	2.9
2,4-Dimethylphenol	5.62E–09	38,433	0.0
Allyl chloride	2.29E–05	35,188	0.8
Nabam	—	35,054	0.0
Metham sodium	—	35,042	0.0
2-Nitrophenol	7.25E–06	33,232	0.2
Ethylenebisdithiocarbamic acid, salts, and esters	—	31,277	0.0
Iron Pentacarbonyl	—	30,204	0.0
Thiabendazole	—	30,026	0.0
2-Nitropropane	1.79E–03	28,571	51.0
Pentachloroethane	6.95E–05	27,513	1.9
Methyl iodide	—	26,745	0.0
1,2,4-Trichlorobenzene	4.08E–06	26,339	0.1
Propargyl alcohol	1.58E–07	25,936	0.0
Diaminotoluene (mixed isomers)	—	25,554	0.0
Peracetic acid	—	23,025	0.0
1,1,2-Trichloroethane	2.26E–05	22,367	0.5
Monochloropentafluoroethane	—	21,429	0.0
2,4-D	1.85E–09	21,006	0.0
O-Cresol	2.01E–09	20,118	0.0
Toluene-2,4-diisocyanate	2.00E–05	18,955	0.4
2,4-Diaminotoluene	1.18E–03	18,220	21.4
Dibenzofuran	—	17,440	0.0
Diuron	7.32E–09	17,369	0.0
Hydrazine	7.19E–03	16,759	120.5
Metribuzin	9.18E–10	16,718	0.0
3,3-Dichloro-1,1,1,2,2-pentafluoropropane	—	16,497	0.0

continued

TABLE 10.2 (continued)
2007 TRI Releases Ranked by Volume (Pounds)

Chemical	ETF (Doses/Capita-lb)	2007 TRI Release (lb)	TU (Doses/Capita)
O-Toluidine	2.10E–05	16,348	0.3
Sodium azide	1.78E–08	16,100	0.0
Quinoline	1.00E–04	15,825	1.6
Aluminum phosphide	3.72E–05	15,468	0.6
Phosgene	9.39E–06	15,290	0.1
M-Dinitrobenzene	4.81E–07	14,660	0.0
1,2-Dichloro-1,1-difluoroethane	—	14,632	0.0
Tetramethrin	—	14,616	0.0
2,4-Dinitrotoluene	9.09E–05	13,594	1.2
2,4-Dichlorophenol	5.77E–08	13,541	0.0
Boron trifluoride	4.40E–06	13,391	0.1
Benzyl chloride	2.20E–05	13,323	0.3
O-Phenylenediamine	3.84E–06	12,849	0.0
Bromotrifluoromethane	6.04E–11	11,682	0.0
Dimethoate	2.06E–07	11,522	0.0
Dichlorobenzene (mixed isomers)	1.19E–08	9,955	0.0
Sodium dicamba	3.07E–08	9,541	0.0
N-Methylolacrylamide	3.22E–03	9,276	29.9
Potassium dimethyldithiocarbamate	—	8,956	0.0
1-Chloro-1,1,2,2-tetrafluoroethane	2.92E–02	8,626	252.1
Malathion	5.42E–10	8,557	0.0
Chloroacetic acid	1.41E–07	8,358	0.0
Dichlorpentafluoro-propane	9.03E–12	8,180	0.0
Dinitrobutyl phenol	2.13E–08	8,132	0.0
Trifluralin	3.02E–06	7,295	0.0
1,1-Dichloroethane	7.17E–06	7,253	0.1
3-Chloro-2-methyl-1-propene	1.83E–05	6,536	0.1
Chlordane	1.63E–11	6,353	0.0
4,4′-Methylenebis(2-chloroaniline)	1.92E–04	6,233	1.2
Allyl amine	7.48E–08	6,094	0.0
Hexachlorocyclopentadiene	3.46E–06	5,990	0.0
1,4-Dichloro-2-butene	4.07E–03	5,727	23.3
2-Phenylphenol	1.42E–07	5,715	0.0
1,3-Dichloropropene (mixed isomers)	3.70E–05	5,695	0.2
Diethyl sulfate	4.18E–13	5,346	0.0
Dichlorofluoromethane	1.47E–10	5,042	0.0
Methylene bromide	2.12E–08	4,998	0.0
1,2-Dichloroethylene	6.34E–08	4,399	0.0
Benzoyl chloride	1.13E–10	4,317	0.0
1,2-Dibromoethane	4.58E–04	4,236	1.9
Dinitrotoluene (mixed isomers)	1.82E–05	4,103	0.1
Crotonaldehyde	3.14E–04	4,008	1.3

TABLE 10.2 (continued)
2007 TRI Releases Ranked by Volume (Pounds)

Chemical	ETF (Doses/Capita-lb)	2007 TRI Release (lb)	TU (Doses/Capita)
2,3-Dichloropropene	2.43E–08	3,812	0.0
Toluene-2,6-diisocyanate	2.08E–06	3,795	0.0
Oxydiazon	1.74E–08	3,679	0.0
Chloromethyl methyl ether	1.10E–03	3,600	3.9
Bromochlorodifluoromethane	8.73E–11	3,491	0.0
Quintozene	3.40E–05	3,115	0.1
Chloropicrin	3.69E–06	3,081	0.0
Folpet	1.47E–08	3,047	0.0
Methyl parathion	3.79E–08	2,765	0.0
Pentachlorophenol	6.15E–06	2,740	0.0
4,4′-Diaminodiphenyl ether	2.56E–07	2,708	0.0
Dimethyl sulfate	—	2,626	0.0
Boron trichloride	—	2,492	0.0
1,3-Dichloro-1,1,2,2,3-pentafluoropropane	—	2,300	0.0
1,1,1,2-Tetrachloroethane	7.64E–05	2,249	0.2
Norflurazon	2.52E–10	2,206	0.0
Diazinon	1.35E–07	2,194	0.0
Bis(tributyltin) oxide	2.74E–06	2,001	0.0
Ethylene thiourea	4.77E–06	1,945	0.0
1-(3-Chloroallyl)-3,5,7-triaza-1-azoniaadamantane chloride	—	1,932	0.0
1,1,2,2-Tetrachloroethane	5.52E–05	1,861	0.1
1,2-Butylene oxide	6.05E–08	1,828	0.0
1,3-Dichlorobenzene	—	1,827	0.0
Dimethylamine dicamba	1.76E–08	1,783	0.0
Hexachloroethane	3.47E–05	1,751	0.1
Acephate	3.62E–06	1,736	0.0
Captan	1.15E–08	1,722	0.0
Dichlorvos	4.02E–05	1,715	0.1
C.I. direct blue 218	2.39E–14	1,659	0.0
3,3′-Dichlorobenzidine dihydrochloride	4.78E–05	1,565	0.1
gamma-Lindane	8.75E–04	1,555	1.4
N,N-Dimethylaniline	2.06E–08	1,546	0.0
Propyleneimine	—	1,482	0.0
1,2,3-Trichloropropane	8.71E–04	1,474	1.3
Pentachlorobenzene	6.81E–06	1,464	0.0
2,6-Dinitrotoluene	9.59E–09	1,343	0.0
Thiourea	1.35E–05	1,333	0.0
O-Dinitrobenzene	1.45E–07	1,272	0.0
P-Dinitrobenzene	9.31E–08	1,272	0.0
Propanil	1.29E–09	1,261	0.0

continued

TABLE 10.2 (continued)
2007 TRI Releases Ranked by Volume (Pounds)

Chemical	ETF (Doses/Capita-lb)	2007 TRI Release (lb)	TU (Doses/Capita)
Methyl isocyanate	4.23E–05	1,259	0.1
Camphechlor	6.58E–03	1,212	8.0
Dazomet	3.84E–13	1,198	0.0
2,4-D 2-Ethylhexyl ester	9.24E–05	1,158	0.1
Heptachlor	1.86E–02	1,133	21.1
Aldrin	5.96E–03	1,128	6.7
Methoxychlor	5.67E–08	1,050	0.0
Safrole	8.52E–05	1,000	0.1
Hexachloro-1,3-butadiene	3.57E–03	934	3.3
Thiodicarb	3.87E–08	890	0.0
Carbaryl	9.72E–10	847	0.0
Bis(2-chloro-1-methylethyl) ether	1.55E–05	788	0.0
P-Chloroaniline	1.03E–05	761	0.0
2,4-Dinitrophenol	1.86E–08	757	0.0
Di-N-propylnitrosamine	1.19E–03	751	0.9
2-Acetylaminofluorene	1.14E–04	750	0.1
Benzal chloride	1.34E–10	708	0.0
Warfarin and salts	8.04E–09	702	0.0
Hexachlorophene (Hcp)	1.06E–03	690	0.7
Hexazinone	4.38E–09	652	0.0
Benzoic trichloride	1.32E–04	646	0.1
Permethrin	4.27E–09	644	0.0
O-Anisidine	6.15E–06	638	0.0
Tetracycline hydrochloride	—	631	0.0
Tribenuron methyl	—	626	0.0
Saccharin	—	601	0.0
Pronamide	5.43E–10	598	0.0
Thiophanate-methyl	1.88E–10	528	0.0
2,4,6-Trichlorophenol	1.13E–05	513	0.0
4-Nitrophenol	1.25E–07	512	0.0
Dihydrosafrole	—	510	0.0
Bis(2-chloroethoxy)methane	1.54E–08	500	0.0
2,4,5-Trichlorophenol	2.97E–08	500	0.0
N-Nitrosopiperidine	3.04E–03	500	1.5
Dipotassium endothall	4.50E–10	500	0.0
Fluorouracil	—	500	0.0
N-Nitrosodiethylamine	6.06E–02	500	30.3
N-Nitrosodi-N-butylamine	7.73E–04	500	0.4
Simazine	4.53E–07	491	0.0
Styrene oxide	—	466	0.0
Chlorotrifluoromethane	1.04E–11	415	0.0
Carbofuran	4.88E–09	391	0.0

TABLE 10.2 (continued)
2007 TRI Releases Ranked by Volume (Pounds)

Chemical	ETF (Doses/Capita-lb)	2007 TRI Release (lb)	TU (Doses/Capita)
Trans-1,3-dichloropropene	2.68E–05	389	0.0
Picloram	3.69E–10	379	0.0
Alachlor	9.87E–06	373	0.0
Aldicarb	1.24E–07	372	0.0
Potassium N-methyldithiocarbamate	—	358	0.0
Tetrachlorvinphos	9.02E–07	355	0.0
Bis(2-chloroethyl) ether	3.13E–04	347	0.1
4-Aminoazobenzene	7.65E–07	335	0.0
2,4-D Butoxyethyl ester	3.77E–05	327	0.0
Ethyl chloroformate	—	325	0.0
Paraldehyde	—	323	0.0
P-Nitroaniline	3.30E–07	321	0.0
Dioxin and dioxin-like compounds	5.80E–02	319	18.5
Ametryn	7.54E–09	313	0.0
Methyl chlorocarbonate	—	301	0.0
Dichlorobromomethane	8.98E–05	296	0.0
Hydramethylnon	7.00E–12	274	0.0
Mecoprop	1.12E–07	267	0.0
Dimethylcarbamoyl chloride	—	260	0.0
Tris(2,3-dibromopropyl) phosphate	—	260	0.0
P-Cresidine	—	260	0.0
Propane sultone	—	260	0.0
Tebuthiuron	5.51E–10	260	0.0
4-Dimethylaminoazobenzene	4.08E–04	256	0.1
1,2-Dibromo-3-chloropropane (Dbcp)	7.58E–03	255	1.9
Isosafrole	—	255	0.0
Diallate	1.48E–05	255	0.0
5-Nitro-O-toluidine	1.71E–06	255	0.0
3,3′-Dimethoxybenzidine dihydrochloride	8.88E–07	255	0.0
Prometryn	1.56E–08	255	0.0
Potassium bromate	—	250	0.0
Trichlorfon	—	250	0.0
Oxyfluorfen	2.41E–08	224	0.0
Myclobutanil	1.98E–10	212	0.0
Paraquat	8.31E–08	202	0.0
Tribromomethane	1.22E–06	191	0.0
Cyanazine	3.77E–05	189	0.0
Benfluralin	1.06E–09	175	0.0
Linuron	2.91E–08	142	0.0
Dimethipin	1.73E–09	139	0.0
Propoxur	6.49E–09	129	0.0

continued

TABLE 10.2 (continued)
2007 TRI Releases Ranked by Volume (Pounds)

Chemical	ETF (Doses/Capita-lb)	2007 TRI Release (lb)	TU (Doses/Capita)
Calcium cyanamide	—	127	0.0
Amitrole	2.00E–08	123	0.0
Methoxone	1.72E–07	122	0.0
4,6-Dinitro-O-cresol	1.62E–07	104	0.0
Chlorendic acid	9.56E–12	96	0.0
Parathion	8.58E–09	89	0.0
Desmedipham	2.33E–13	51	0.0
Ethyl dipropylthiocarbamate	5.32E–09	50	0.0
Propargite	4.53E–07	48	0.0
Chlorsulfuron	1.54E–08	42	0.0
Phenytoin	—	40	0.0
Triallate	1.28E–08	38	0.0
Cyfluthrin	6.63E–05	34	0.0
Chlorobenzilate	9.30E–05	32	0.0
Bromoxynil octanoate	1.34E–08	27	0.0
Isodrin	—	22	0.0
Phenothrin	—	20	0.0
Chlorimuron ethyl	1.12E–07	19	0.0
Benzidine	3.27E–02	16	0.5
Chlorotetrafluoroethane	1.80E–03	15	0.0
1,1-Dimethyl hydrazine	2.51E–03	15	0.0
Carboxin	1.54E–10	14	0.0
Bromoxynil	1.67E–09	13	0.0
Chlorpyrifos methyl	1.03E–08	12	0.0
4-Aminobiphenyl	2.37E–03	11	0.0
Abamectin	2.75E–06	10	0.0
O-Toluidine hydrochloride	1.23E–05	10	0.0
(1,1′-Biphenyl)-4,4′-diamine, 3,3′-dimethyl-	5.51E–04	10	0.0
1,2-Diphenylhydrazine	6.20E–05	10	0.0
3,3′-Dimethoxybenzidine	8.34E–08	10	0.0
Malononitrile	2.00E–06	10	0.0
Naled	6.18E–10	10	0.0
N-Ethyl-N-nitrosourea	2.27E–03	10	0.0
N-Nitrosodiphenylamine	2.21E–06	10	0.0
N-Nitroso-N-methylurea	2.81E–02	10	0.3
Quinone	—	10	0.0
Thioacetamide	—	10	0.0
Triclopyr triethylammonium salt	1.45E–09	10	0.0
Trypan blue	—	10	0.0
2,4-D Sodium salt	6.67E–05	9	0.0
Bromacil	1.60E–10	8	0.0
Trichloroacetyl chloride	—	6	0.0

TABLE 10.2 (continued)
2007 TRI Releases Ranked by Volume (Pounds)

Chemical	ETF (Doses/Capita-lb)	2007 TRI Release (lb)	TU (Doses/Capita)
1,1,1,2-Tetrachloro-2-fluoroethane (Hcfc-121a)	—	6	0.0
Phosphine	2.25E–05	5	0.0
1,1,2,2-Tetrachloro-1-fluoroethane	—	5	0.0
Lactofen	6.10E–09	5	0.0
S,S,S-Tributyltrithiophosphate	1.56E–05	5	0.0
Bifenthrin	6.13E–08	4	0.0
Ethyleneimine	—	4	0.0
Merphos	5.52E–07	3	0.0
Nitrapyrin	—	2	0.0
2,6-Xylidine	1.23E–05	1	0.0
Diglycidyl resorcinol ether (Dgre)	1.82E–12	1	0.0
Fenoxycarb	7.34E–07	1	0.0
Fluometuron	1.65E–09	1	0.0
Piperonyl butoxide	—	1	0.0
Propachlor	2.39E–09	1	0.0
Strychnine	2.32E–08	1	0.0
trans-1,4-Dichloro-2-butene	4.29E–03	1	0.0

TABLE 10.3
2007 TRI Releases Ranked by Toxicity Unit (TU)

Chemical	ETF (Doses/Capita-lb)	2007 TRI Release (lb)	TU (Doses/Capita)
Chromium and chromium compounds	1.17E+00	60,402,489	70,822,503.1
Cobalt and cobalt compounds	6.37E–02	6,970,293	444,242.0
Arsenic and arsenic compounds	3.19E–03	97,581,160	311,237.7
Lead and lead compounds	2.60E–04	495,875,564	129,166.8
Polychlorinated biphenyls	2.30E–02	2,090,371	47,983.4
Mercury and mercury compounds	2.55E–03	6,935,622	17,656.4
Acrylamide	2.22E–03	6,161,247	13,656.2
Acrylonitrile	6.18E–04	7,059,836	4,364.8
1,3-Butadiene	2.12E–03	1,788,084	3,790.0
Acrolein	1.81E–03	1,696,876	3,066.4
Manganese and manganese compounds	1.15E–05	245,353,348	2,814.3
Cadmium and cadmium compounds	5.11E–04	3,907,391	1,996.1
Nickel and nickel compounds	2.50E–05	37,902,077	947.3
Aluminum	2.28E–05	39,901,864	909.9
Formaldehyde	2.53E–05	21,933,684	555.8
Tetrachloroethylene	1.24E–04	2,237,864	277.5
1-Chloro-1,1,2,2-tetrafluoroethane	2.92E–02	8,626	252.1
2-Chloro-1,1,1,2-tetrafluoroethane	4.33E–04	551,358	238.6
Ethylene oxide	6.45E–04	305,961	197.5
Hexachlorobenzene	4.53E–03	43,018	194.7
Benzene	1.76E–05	8,465,367	148.8
Fomesafen	2.09E–03	69,115	144.8
Vinyl Chloride	3.36E–04	372,635	125.3
Hydrazine	7.19E–03	16,759	120.5
Beryllium and beryllium compounds	1.18E–04	867,078	102.7
Barium and barium compounds	3.71E–07	244,837,073	90.7
Chlorine	1.52E–05	5,643,223	85.9
Acetaldehyde	6.89E–06	11,309,426	77.9
Hydrochloric acid	1.23E–07	502,532,256	61.6
2-Nitropropane	1.79E–03	28,571	51.0
Propylene oxide	1.23E–04	338,598	41.6
Carbon tetrachloride	1.21E–04	308,357	37.3
Titanium tetrachloride	2.90E–04	123,546	35.8
Chloroform	4.83E–05	706,555	34.1
N-Nitrosodiethylamine	6.06E–02	500	30.3
N-Methylolacrylamide	3.22E–03	9,276	29.9
Trichloroethylene	6.38E–06	4,485,202	28.6
Aniline	2.91E–05	920,606	26.8
Dichloromethane	4.50E–06	5,903,242	26.6
1,4-Dichloro-2-butene	4.07E–03	5,727	23.3
2,4-Diaminotoluene	1.18E–03	18,220	21.4
Heptachlor	1.86E–02	1,133	21.1

TABLE 10.3 (continued)
2007 TRI Releases Ranked by Toxicity Unit (TU)

Chemical	ETF (Doses/Capita-lb)	2007 TRI Release (lb)	TU (Doses/Capita)
Dioxin and dioxin-like compounds	5.80E–02	319	18.5
Naphthalene	6.16E–06	2,850,878	17.6
1,2-Dichloroethane	3.80E–05	449,853	17.1
4,4′-Methylenedianiline	2.40E–04	67,423	16.2
Zinc and zinc compounds	1.74E–08	800,318,423	13.9
Chlorine dioxide	2.12E–05	545,291	11.5
Cyanide compounds	1.28E–06	6,871,538	8.8
Camphechlor	6.58E–03	1,212	8.0
Diisocyanates	5.11E–06	1,472,453	7.5
Formic acid	5.04E–07	13,934,367	7.0
Aldrin	5.96E–03	1,128	6.7
Hydroquinone	1.47E–05	430,989	6.3
Ethylbenzene	1.01E–06	4,843,102	4.9
Acrylic acid	1.04E–06	4,625,008	4.8
Nitrobenzene	7.68E–06	601,120	4.6
Hydrofluoric acid	5.84E–08	72,907,435	4.3
Chloromethyl methyl ether	1.10E–03	3,600	3.9
Hydrogen cyanide	1.58E–06	2,309,862	3.7
Selenium and selenium compounds	8.10E–07	4,286,943	3.5
Hexachloro-1,3-butadiene	3.57E–03	934	3.3
1,4-Dioxane	1.79E–05	185,132	3.3
Allyl alcohol	5.86E–06	526,216	3.1
2-Methylpyridine	7.43E–05	39,138	2.9
1,2,4-Trimethylbenzene	3.49E–07	7,242,424	2.5
N-Hexane	6.66E–08	34,978,189	2.3
1,2-Dichloropropane	1.75E–05	115,710	2.0
1,2-Dibromoethane	4.58E–04	4,236	1.9
1,2-Dibromo-3-chloropropane (Dbcp)	7.58E–03	255	1.9
Pentachloroethane	6.95E–05	27,513	1.9
Quinoline	1.00E–04	15,825	1.6
N-Nitrosopiperidine	3.04E–03	500	1.5
gamma-Lindane	8.75E–04	1,555	1.4
Polycyclic aromatic compounds	9.63E–07	1,365,384	1.3
1,2,3-Trichloropropane	8.71E–04	1,474	1.3
Crotonaldehyde	3.14E–04	4,008	1.3
Thallium and thallium compounds	5.83E–07	2,125,769	1.2
2,4-Dinitrotoluene	9.09E–05	13,594	1.2
4,4′-Methylenebis(2-chloroaniline)	1.92E–04	6,233	1.2
1,4-Dichlorobenzene	1.19E–05	79,266	0.9
Antimony and antimony compounds	7.49E–08	12,326,053	0.9
Di-N-propylnitrosamine	1.19E–03	751	0.9

continued

TABLE 10.3 (continued)
2007 TRI Releases Ranked by Toxicity Unit (TU)

Chemical	ETF (Doses/Capita-lb)	2007 TRI Release (lb)	TU (Doses/Capita)
Allyl chloride	2.29E–05	35,188	0.8
Hexachlorophene (Hcp)	1.06E–03	690	0.7
Acetonitrile	3.86E–08	17,941,970	0.7
Aluminum phosphide	3.72E–05	15,468	0.6
Atrazine	1.08E–06	515,072	0.6
Benzidine	3.27E–02	16	0.5
Xylene (mixed isomers)	1.99E–08	25,595,670	0.5
1,1,2-Trichloroethane	2.26E–05	22,367	0.5
Methyl bromide	1.12E–06	363,571	0.4
Epichlorohydrin	2.60E–06	155,813	0.4
Chlorophenols	8.22E–06	49,196	0.4
Ethyl acrylate	5.10E–06	78,135	0.4
Triethylamine	2.27E–07	1,744,585	0.4
N-Nitrosodi-N-butylamine	7.73E–04	500	0.4
Toluene-2,4-diisocyanate	2.00E–05	18,955	0.4
Methyl tert-butyl ether	6.47E–07	585,465	0.4
2-Chlor-1,3-butadiene	4.68E–07	741,853	0.3
O-Toluidine	2.10E–05	16,348	0.3
Acetamide	3.10E–07	1,097,494	0.3
Nitrate compounds	1.13E–09	270,689,909	0.3
Benzyl chloride	2.20E–05	13,323	0.3
N-Nitroso-N-methylurea	2.81E–02	10	0.3
Pyridine	3.68E–07	702,367	0.3
Phosphorus (yellow or white)	9.05E–07	276,993	0.3
2-Nitrophenol	7.25E–06	33,232	0.2
1,3-Dichloropropene (mixed isomers)	3.70E–05	5,695	0.2
Methacrylonitrile	2.14E–06	92,267	0.2
2-Mercaptobenzothiazole	1.53E–06	121,620	0.2
1,1,1,2-Tetrachloroethane	7.64E–05	2,249	0.2
Dimethylamine	4.28E–07	362,524	0.2
Phosgene	9.39E–06	15,290	0.1
N,N-Dimethylformamide	5.65E–08	2,431,907	0.1
Ammonia	8.87E–10	153,483,048	0.1
3-Chloro-2-methyl-1-propene	1.83E–05	6,536	0.1
Bis(2-chloroethyl) ether	3.13E–04	347	0.1
1,2,4-Trichlorobenzene	4.08E–06	26,339	0.1
2,4-D 2-Ethylhexyl ester	9.24E–05	1,158	0.1
Maleic anhydride	1.51E–07	707,065	0.1
Quintozene	3.40E–05	3,115	0.1
4-Dimethylaminoazobenzene	4.08E–04	256	0.1
1,1,2,2-Tetrachloroethane	5.52E–05	1,861	0.1
Fluorine	5.67E–07	168,844	0.1

TABLE 10.3 (continued)
2007 TRI Releases Ranked by Toxicity Unit (TU)

Chemical	ETF (Doses/Capita-lb)	2007 TRI Release (lb)	TU (Doses/Capita)
Nitroglycerin	6.71E–07	130,438	0.1
2-Acetylaminofluorene	1.14E–04	750	0.1
Safrole	8.52E–05	1,000	0.1
Benzoic trichloride	1.32E–04	646	0.1
Carbon disulfide	8.61E–09	8,935,912	0.1
3,3′-Dichlorobenzidine dihydrochloride	4.78E–05	1,565	0.1
Dinitrotoluene (mixed isomers)	1.82E–05	4,103	0.1
Propionaldehyde	2.08E–07	334,438	0.1
Bis(2-ethylhexyl)phthalate	6.14E–08	1,126,569	0.1
Dichlorvos	4.02E–05	1,715	0.1
Phenol	9.66E–09	6,677,767	0.1
Copper and copper compounds	3.55E–10	176,593,590	0.1
Chloromethane	3.07E–08	2,014,817	0.1
Hexachloroethane	3.47E–05	1,751	0.1
Boron trifluoride	4.40E–06	13,391	0.1
Decabromodiphenyl oxide	7.74E–08	695,861	0.1
Methyl isocyanate	4.23E–05	1,259	0.1
1,1-Dichloroethane	7.17E–06	7,253	0.1

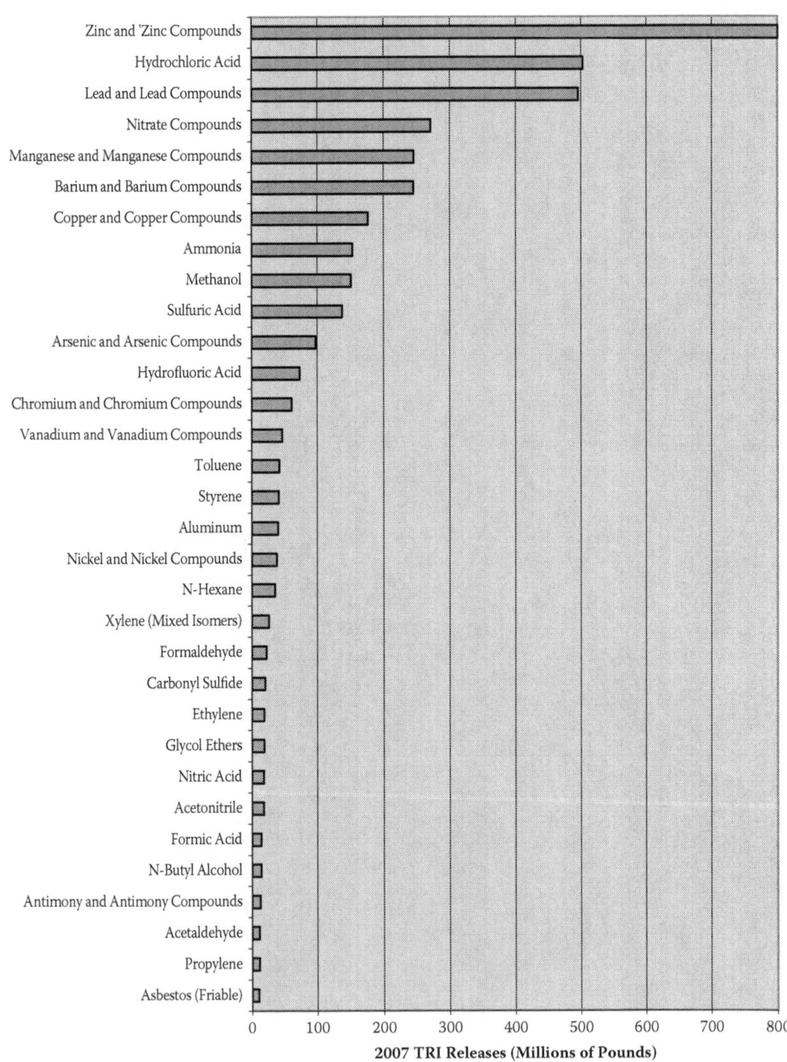

FIGURE 10.1 Top 2007 TRI release data ranked by total releases in pounds.

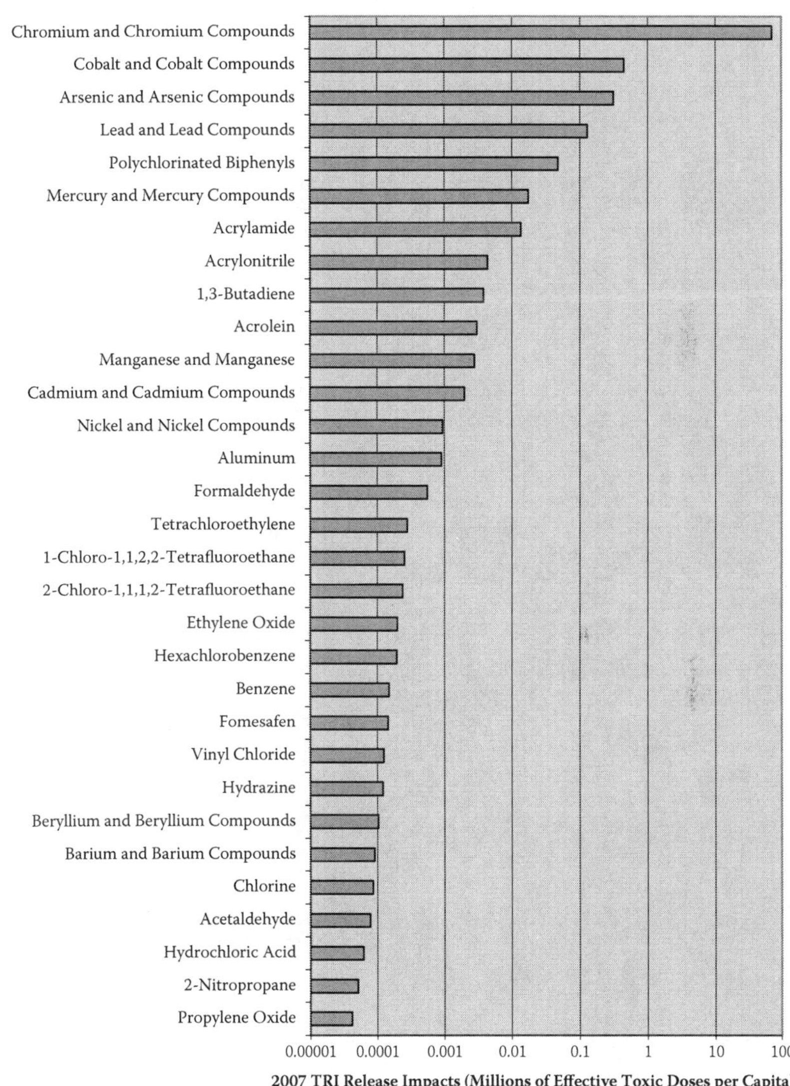

FIGURE 10.2 Top 2007 TRI release data ranked by toxicity units (TUs) in doses/capita.

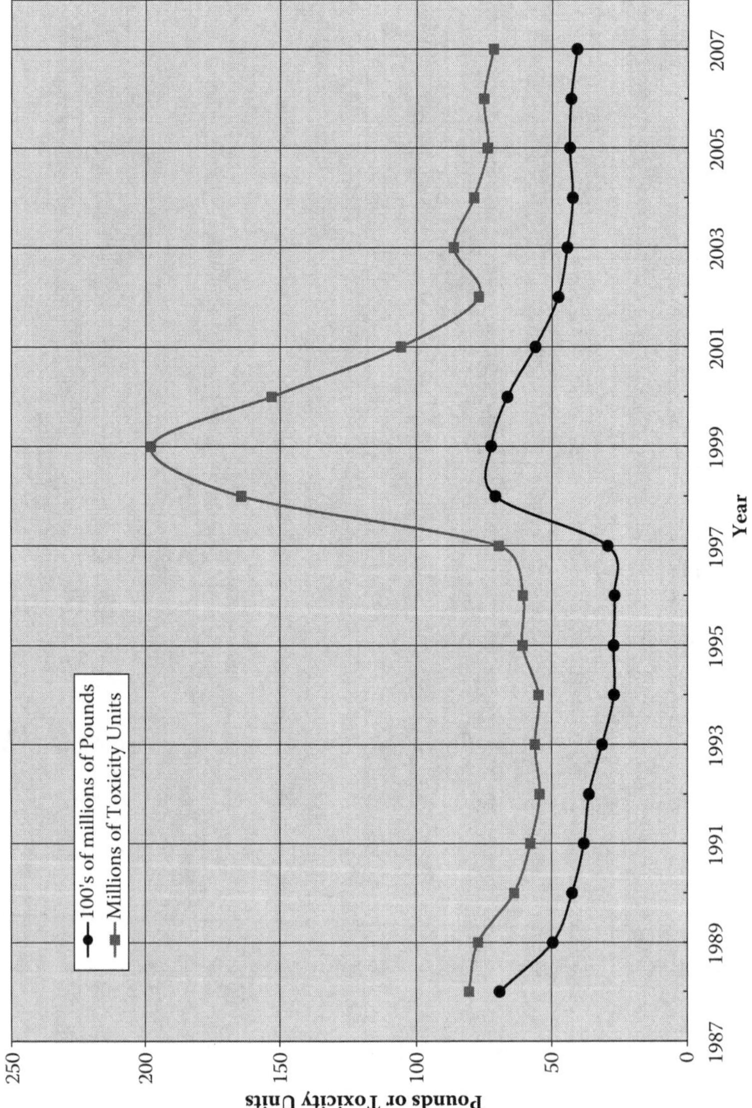

FIGURE 10.3 Yearly TRI releases (pounds) and toxicity units from 1988 through 2007.

11 Focusing on Impact Chemicals

INTRODUCTION

Successful toxic chemical use reduction programs target those chemicals that have the greatest potential toxic impact. Basing programs on total poundage of chemicals, versus a basis on toxicity, could result in reduction efforts focused on high-volume, low-toxicity releases. Under this scenario, one could conceivably reduce the total volume of a chemical by replacing the chemical, which may have a low toxicity, with a smaller volume of a much more toxic compound.

This chapter presents existing toxic chemical use reduction programs that have focused on impact chemicals, including the basis for the program, program requirements, the role of stakeholders, financial impacts, and the effectiveness of the program.

U.S. ENVIRONMENTAL PROTECTION AGENCY 33/50 PROGRAM

The U.S. Environmental Protection Agency (USEPA) 33/50 Program (USEPA 1999) takes its name from its goal, which was to reduce the releases and off-site transfers of 17 high toxicity targeted chemicals (Table 11.1) by 50 percent in 1995 (with an interim goal of 33 percent by 1992) using 1988 Toxics Release Inventory (TRI) reporting data as a baseline.

The program was a voluntary pollution prevention (P2) initiative implemented in 1991 by a number of large chemical companies in cooperation with the EPA.

TABLE 11.1
Seventeen Targeted Chemicals: U.S. EPA 33/50 Program

Benzene	Tetrachloroethylene	Cadmium and cadmium compounds
Carbon tetrachloride	Toluene	Chromium and chromium compounds
Chloroform	1,1,1-Trichloroethane	Cyanide compounds
Dichloromethane	Trichloroethylene	Lead and lead compounds
Methyl ethyl ketone	Xylenes	Mercury and mercury compounds
Methyl isobutyl ketone		Nickel and nickel compounds

Source: U.S. Environmental Protection Agency. *33/50 Program: The Final Record.* Office of Pollution Prevention and Toxics. Washington, DC: EPA-745-R-99-004, 1999.

Because the 33/50 Program was voluntary, the EPA sent invitations to companies to solicit their participation. In the spring and summer of 1991, the EPA sent invitations to 5,000 companies to participate in the program, focusing on facilities reporting to TRI on any of the targeted chemicals from 1988 to 1994. Another 2,500 more invitations were sent over the next three years. The EPA concentrated the majority of their outreach on the top 600 companies with the largest releases and transfers. A total of 1,294 facilities (16 percent of those invited) accepted the invitation to participate in the program.

Participants in the program examined production processes and identified cost-effective P2 practices. The companies were then to write a simple letter to the EPA stating their reduction goals and plans. The EPA asked that companies focus their reduction and release strategies on their waste management hierarchy, evaluating source reduction first and then looking for opportunities to recycle, treat, or dispose of wastes. Any steps taken to reduce the targeted chemicals were not enforceable unless the activities were otherwise required by law or regulation. The success of the 33/50 Program was measured according to whether reductions had been achieved nationwide, with reductions looked at as an aggregate of total releases of all 17 targeted chemicals.

Basis for Program

The public release of TRI data in 1988 was the basis for the 33/50 Program. Soon after the first release of TRI data to the public in 1989, public citizen groups placed full-page ads in the *New York Times* highlighting the top 10 corporate air, water, and land polluters. Some of these companies approached the EPA and pledged their commitment to improve their environmental performance, which in effect started the 33/50 Program (Arora and Cason 1996).

The EPA then conducted high-level meetings with chief executives of major corporations, industry trade associations, and environmental groups, looking for ways to use the TRI data to reduce chemical releases. These meetings led to the 33/50 Program.

The purpose of the 33/50 Program was to demonstrate whether voluntary regulation by the industries could supplement the traditional command-and-control approach of the EPA by bringing about targeted reductions more quickly than regulations alone would. This program is an example of voluntary environmental regulation that was national in scope, involved multimedia (integrating all media to reduce releases to air, water, and land), and was prevention oriented. The program also sought to foster a P2 ethic, encouraging companies to consider and apply P2 approaches to reduce their environmental releases rather than traditional end-of-the-pipe methods for treating and disposing of chemicals in waste.

Program Requirements

There were no specific mandatory guidelines developed by the EPA for the 33/50 Program. The companies could set their own reduction commitments that were truly voluntary and not enforceable by law. A company's participation in the program did not preclude it from its responsibilities for complying with all other laws.

The majority of the companies (1,066) set measurable goals or pledges to reduce their releases and transfers of the 17 targeted chemicals, with the companies choosing which of the 17 chemicals they would reduce. Other participants developed goals tied to changes in their production levels, chose alternative baseline years (when the 1988 baseline year was not a representative year for their facility), or set a reduction target for all their TRI reporting without specifying goals for the 33/50 chemicals.

The EPA did not monitor the companies to ensure that their reported figures were accurate. The simple letters sent to the EPA stating their reduction pledges or plans were required to be signed by the chief financial officers (CFOs) of the companies. It was in the interest of the CFOs to reduce costs for their company; therefore, they were interested in the amount of waste reduced or chemicals released from a cost-savings opportunity perspective.

Role of Stakeholders

The EPA, major corporations, industry trade associations, and environmental groups all played a part in developing the 33/50 Program. The stakeholders, not the sponsoring agency (the EPA), were directly involved in the implementation and program design to meet self-determined reduction targets for one or more of the targeted chemicals. The 33/50 Program was free from government regulations, paperwork, penalties, punishments, and lawsuits. Companies participating in the program voluntarily developed their own goals and plans and provided their reduction commitments to the EPA in a simple letter.

Financial Impacts

The EPA funded the administrative costs of the 33/50 Program through Pollution Prevention Incentives for States (PPIS) grants (now referred to as the P2 Grants Program) to the states. The 33/50 Program administrative and technical assistance activities were added to the list of other state P2 programs already funded by the grants.

The voluntary approach of the 33/50 Program avoided the costly process of legislation and the substantial costs of monitoring and enforcement by the regulatory agency. It was also cost-effective for those making pollution reductions because it allowed firms to make the most cost-effective emission reductions.

In addition to costs savings, program participation included possible public recognition by the EPA and special awards for innovators and firms with outstanding P2 achievements. The incentives for a company to participate were the benefits a company derived from a clean environmental record and were one way for a company to indicate that it was environmentally conscientious.

Effectiveness of Program

The 33/50 Program achieved its goals in 1994, a year ahead of schedule. The program proved to be a successful way for the EPA to partner with industries to effectively reduce the releases and transfers of toxic chemicals through voluntary regulation

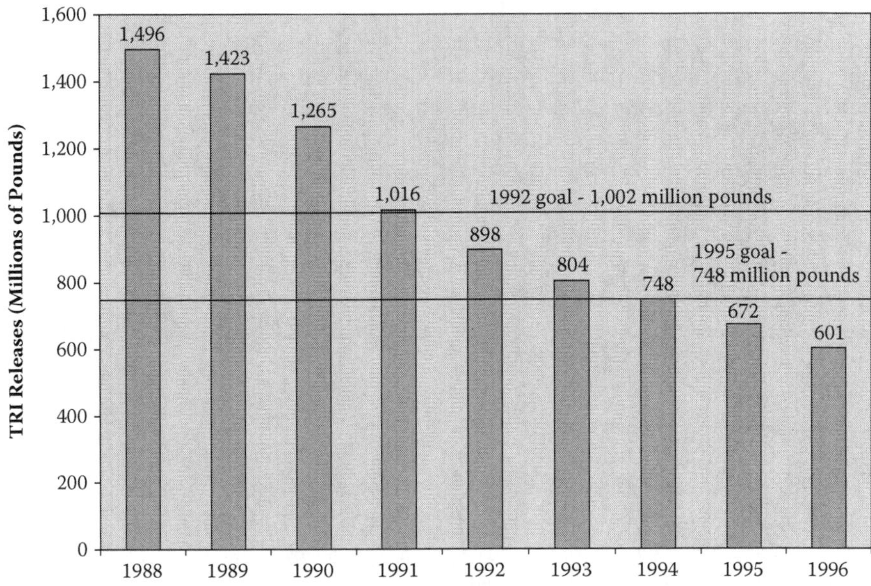

FIGURE 11.1 Releases and transfers of 33/50 Program chemicals (1988–1996). (From U.S. Environmental Protection Agency, *33/50 Program: The Final Record.* Office of Pollution Prevention and Toxics. Washington, DC: EPA-745-R-99-004, 1999.)

instead of using their usual command-and-control approach. In addition, by focusing on the top 600 companies with the largest releases and transfers, the EPA was able to get 64 percent of these companies to participate in the program, compared to less than 14 percent of the smaller companies (EPA, 1999).

The program measured its progress in reducing releases and transfers to treatment, publicly owned treatment works (POTWs), and disposal based on the original TRI reporting categories. To control for changes in the TRI chemical list over time, year-to-year comparisons were based on a consistent list of chemicals reported in all years 1988 to 1996. Figure 11.1 shows the average percentage decrease in releases and transfers between 1988 and 1996.

The 17 targeted chemicals in the program outpaced all other TRI chemicals for reductions in releases and transfers (Figure 11.2 and Table 11.2). Between 1988 and 1995, release of the 33/50 chemicals was reduced by 60 percent, and non-33/50 chemicals were reduced by 36 percent. Reductions of targeted chemicals continued at a higher rate than other TRI chemicals in the year after the 33/50 Program ended.

During the implementation of the 33/50 Program, an international agreement, the Montreal Protocol, was also in effect to phase out ozone-depleting chemicals. The chemicals in the 33/50 Program that would have also been affected by the protocol were carbon tetrachloride and 1,1,1-trichloroethane. Although these chemicals reflected the largest reduction, facilities also reduced releases and transfers of the other 33/50 chemicals by 50 percent from 1988 to 1995.

Some environmentalists criticized the program as not reducing pollution at the source but instead at the point of departure from a facility, arguing that any chemical

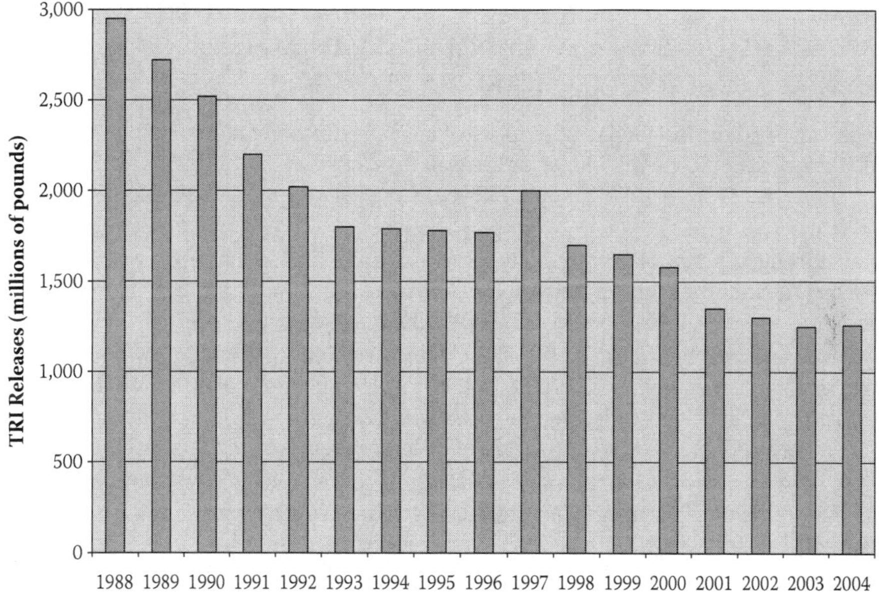

FIGURE 11.2 Total releases of TRI core chemicals in the United States.

TABLE 11.2
Releases and Transfers of 33/50 Program Chemicals versus Other TRI Chemicals, 1988–1996 (millions of pounds)

Year	All TRI Chemicals (Excluding Additions or Deletions)	TRI Chemicals Less 33/50 Chemicals	33/50 Chemicals Only
1988	4,020	2,524	1,496
1990	3,428	2,163	1,265
1995	2,289	1,617	672
1996	2,217	1,616	601

Source: U.S. Environmental Protection Agency. *33/50 Program: The Final Record.* Office of Pollution Prevention and Toxics. Washington, DC: EPA-745-R-99-004, 1999.

Note: Does not include delisted chemicals or chemicals added in 1990, 1991, 1994, and 1995 or transfers to recycling or energy recovery.

is a potential hazard to workers. They also worried that self-reporting by the companies would not provide accurate figures on chemical releases since companies would have an interest in making their numbers look good (U.S. Food and Drug Administration, Center for Devices and Radiological Health 2000).

Researchers from the University of Pittsburgh found that the EPA voluntary 33/50 industrial toxics reduction program, which used the TRI to measure progress, only reduced toxics emissions in politically active communities (Gamper-Rabindran 2006).

WASHINGTON DEPARTMENT OF ECOLOGY PERSISTENT, BIOACCUMULATIVE TOXINS PROGRAM AND OTHER TARGETED (MERCURY) CHEMICALS PROGRAMS

The state of Washington has a number of other toxic chemical reduction programs that target specific chemicals, as detailed in this section.

In August 1998, the Washington Department of Ecology (DOE) released a dioxin source assessment and announced plans to develop a long-term strategy to reduce and eliminate persistent, bioaccumulative and toxic (PBT) substances. Established by the Washington DOE in 2000, the department's PBT program was implemented to phase out all sources of persistent toxic chemicals through specific state actions, focusing on one PBT at a time. PBT chemicals build up through the food chain and accumulate in human and animal tissues. Specifically, in December 2000, the Washington State DOE developed the *Proposed Strategy to Continually Reduce Persistent, Bioaccumulative Toxins (PBTs) in Washington State* (Washington State DOE 2000b). This document outlined strategies and actions to phase out PBTs.

The primary goals established by the department's PBT program to reduce the use and prevent the pollution of PBTs are as follows:

- Development of alternative materials used in products and industrial or manufacturing processes
- Incremental and continuous improvements in pollution reduction through better science and processes, technical assistance, and use of regulatory tools
- Additional guidance and direction to agencies by Washington legislature

Under guidance provided by the state legislature in 2002, the Departments of Ecology and Health targeted mercury as the first priority pollutant in the PBT initiative in the state. Mercury was made the highest priority PBT chemical in Washington due to its widespread use and ease of release into the environment. The Washington State *Mercury Chemical Action Plan* (Peele 2003) outlined strategies to reduce the use and pollution of mercury. This plan was followed by the 2003 Mercury Education Reduction Act (MERA; Chapter 70.95M RCW).

Chapter 173-333 WAC, Persistent Bioaccumulative Toxins Regulations, was finalized in February 2006. The draft multiyear *PBT Chemical Action Plan Schedule* was published in September 2006 and received public comment until December 20, 2006. This plan proposes to develop chemical action plans for lead, polycyclic aromatic hydrocarbons (PAHs), and perfluorooctane sulfonate (PFOS), in that order, from November 2006 to March 2010.

The Washington DOE also produced a plan to examine and reduce the use of polybrominated diphenyl ether (PBDE) flame retardants. A final PBDE action plan was published in January 2006.

Lead and cadmium have been designated metals of concern under WAC 173-333-315. The "metals-of-concern" category was established as an interim category pending completion of the inorganic metals assessment framework process of the EPA. The Washington DOE can prepare chemical action plans for one or more metals of concern.

The most mature of the targeted toxic chemical reduction programs in Washington State is the mercury reduction program. Therefore, the remainder of this section focuses on further analysis of the mercury program.

Basis for the Mercury Reduction Program

It is estimated that approximately 5,000 lb of mercury were released in Washington State in 2001. By 2006, this number had been reduced to approximately 2,300 lb of mercury released. The goal of the mercury program is to virtually eliminate anthropogenic releases by 2015.

Mercury emission levels are estimated by

- Conducting ongoing fish tissue sampling across the state
- Comparing historic and current soil and sediment monitoring
- Conducting studies of mercury levels in landfill gas to determine disposal of mercury-containing wastes
- Maintaining a list of products containing mercury and tracking their disposal and recycling rates

Most mercury releases were found to come from small or medium businesses, schools, and households. The majority of these releases could be prevented through purchasing alternative products that contain no mercury and improving waste separation and disposal methods in dentistry, households, medical facilities, and other areas where mercury is used.

The *Mercury Chemical Action Plan* (Peele 2003) lists and describes the major sources of mercury in the state of Washington. These include mercury release from fossil fuel combustion; mining and manufacturing; use of products containing mercury; and disposal of products containing mercury at end of life. MERA requires labeling of fluorescent lamps, bans the use of elemental mercury in schools, and prohibits sales of mercury. The state developed Memorandums of Understanding (MOUs) between DOE and industry for dental practices, hospitals, and automotive recyclers. They also focused attention on disposal of auto switches, thermostats, and fluorescent lamps. Mercury emission standards were developed for coal-fired power plants. The list of products containing mercury was not sufficiently comprehensive by 2001; therefore, additional products are tracked as information is available. Disposal of utility switches and relays, for example, was not counted in 2001, yet these items were recently estimated to constitute releases of as much as 800 lb of mercury per year.

Program Requirements

Washington state requires the following by law:

- Mercury-containing lamps must be appropriately labeled.
- Washington State government must purchase only low-mercury or mercury-free products.

- Sale of mercury-containing novelties, thermometers, thermostats, manometers, and automobiles with mercury switches is prohibited.
- Schools cannot purchase, use, or possess elemental mercury.

Washington state policy requires that:

- Communities maintain voluntary collection and recycling programs for mercury-containing products, elemental mercury, and mercury waste.
- Collected mercury must go to retort facilities.
- Funding must be provided to local governments for mercury lamp recycling. Thermostat collection bins are funded by either the state or the county. King County provides a rebate for recycled thermostats.

The Washington DOE has three separate MOUs with industry groups. These MOUs target groups with an established mercury waste stream and define best management practices (BMPs) to reduce or eliminate mercury releases, as detailed next.

- The Washington DOE had an MOU with the Washington State Dental Association (WSDA) outlining BMPs for the disposal of mercury and mercury-containing products from dental practices. WSDA is an association with a voluntary membership, with participation 55 percent of Washington dentists. The MOU expired on December 2005, and it is now required through water quality and hazardous waste regulations that dentists install amalgam separators or document that they are not exceeding regulatory discharge standards. The Washington DOE has an MOU with the Washington State Hospital Association (WSHA). WSHA is an association with a voluntary membership of hospitals in Washington and represents over 100 medical facilities in the state. This MOU focuses on mercury and other relevant PBTs in hospitals. The goal of this MOU is to reduce mercury waste generated from hospitals statewide. It also specifies that the DOE and WSHA will cosponsor training and seminars on the topic.
- The Washington DOE has an MOU with the Automotive Recyclers of Washington (AROW) and the End-of-Life Vehicle Solutions Corporation (ELVS), which represents auto manufacturers. AROW is a trade association of licensed auto recyclers in Washington State, and the ELVS is a 501(c)(4) nonprofit social welfare corporation created by the automotive industry to promote the environmental efforts of the industry. The purpose of this MOU is to establish a voluntary motor vehicle mercury switch removal program for end-of-life vehicles.
- Education and outreach are a large part of this program. The Washington State Department of Health provides these services as mandated by the 2003 legislature.

Washington State also regulates mercury releases from wastewater treatment, chloralkali plants, hazardous waste, and municipal solid waste incinerators.

Role of Stakeholders

The Washington State Departments of Health and Ecology worked together with an external mercury advisory committee to develop the Washington State *Mercury Chemical Action Plan* (Peele 2003). The external committee consisted of statewide agriculture, business, environmental, local government, and public health representatives. There has also been a push toward mercury reduction at the local level, with many counties having mercury reduction programs. The *Mercury Chemical Action Plan* lists the identified most common sources of mercury and evaluates potential methods of contamination from each of these sources. After the plan was published, the Washington State legislature passed MERA (Chapter 70.95M RCW) in 2003.

Financial Impacts

Any fiscal impact on the Department of Health or Ecology that results from the implementation of MERA is paid for out of funds that are appropriated by the legislature from the state toxics control account for the implementation of the PBT chemical strategy of the department.

If there is a violation of MERA, it is punishable by a civil penalty not to exceed $1,000 for each violation in the case of a first violation. Repeat violators are liable for a civil penalty of up to $5,000 for each repeat violation. Penalties collected must be deposited in the state toxics control account created in RCW 70.105D.070.

Effectiveness of Program

MERA has caused measurable improvements in the amount of mercury used and released since 2001, as noted in Table 11.3. The goal of the program is to eliminate mercury use by 2015.

TABLE 11.3
Improvements in the Amount of Mercury Used and Released Since 2001 under MERA

Banned under MERA	Goals	Results to 2005
Bulked mercury (schools)	100% (2,600 lb)	90% (2,300 lb)
Thermometers	100% (300 lb)	99% (288 lb)
Thermostats	100% (400 lb)	10% (40 lb)
Manometers	100% (300 lb)	99% (295 lb)
Novelties	No data	No data
Other items		
Button cell batteries	90% (88 lb)	10% (8 lb)
Fluorescent lamps	90% (500 lb)	26% (130 lb)
Amalgam	90% (400 lb)	90% (360 lb)
Switches and relays	90% (800 lb)	10% (80 lb)
Auto switches	90% (4,000 lb)	10% (55 lb)

Source: Washington State Department of Ecology. 2006. Mercury Strategies and Outcome Measures. http://www.ecy.wa.gov/mercury/mercury_outcomes.html. Accessed 11/06.2006.

REGISTRATION, EVALUATION AUTHORIZATION, AND RESTRICTION OF CHEMICALS PROGRAM OF THE EUROPEAN UNION

Chemicals pose a quandary for the European Union (EU). Production of chemicals is the third largest industry in Europe, employing 1.7 million people directly. There are over 100,000 different chemicals used in the European market, and a number of these chemicals have been linked to certain health conditions, such as the connection of asbestos to lung cancer and benzene to leukemia. Of the 100,000 chemicals used in the market, only a small portion has adequate information on carcinogenicity or other toxicity. In addition, new substances are introduced to the market annually. As its name implies, the EU program for Registration, Evaluation, Authorization, and Restriction of Chemicals (REACH) has four basic components for dealing with toxic chemicals.

The first part of the program, registration, is described in Chapter 2 of this book as it focuses on composition reporting. The remaining three parts of the program (evaluation, authorization, and restriction) are based on focusing on impact chemicals; therefore, these are described further in this chapter.

Basis for Program

REACH is based on a number of council directives: 67/548/EEC of June 27, 1967, is related to regulating classification, packaging, and labeling of dangerous substances; 76/769/EEC of July 27, 1976, is associated with restricting marketing and use of dangerous substances; 1999/45/EC of May 31, 1999, is related to the classification, packaging, and labeling of dangerous preparations; and 793/93 of March 23, 1993, is related to the evaluation and control of the risks of existing substances. These directives identified problems of disparities in the laws of individual countries and the need to do more to protect health and the environment from chemicals.

Program Requirements

Evaluation

The evaluation component of the program will require manufacturers or importers of chemicals to supply information on the toxicity of the chemicals. The goal is to establish a single source for toxicity data on existing and new chemicals. The information can be supplied by acceptable testing in countries outside the EU or by evaluating toxicity of the chemical functional groups or toxicity of analogous compounds. The program is also driven by a desire to limit new animal testing. The goal of the program is to expand the knowledge of chemical toxicity and to place the burden of testing and providing that information on the chemical manufacturer or users of a chemical.

The EU wishes to maximize the use of nonanimal testing methods for determining toxicity of new and existing chemicals, including the use of quantitative structure-activity (or structure-property) relationships (QSARs) to determine toxicity based on knowledge of toxicity of similar chemicals or functional groups.

Authorization

Substances with certain hazardous properties that give rise to very high concern will have to be given use-specific permission before they can be employed in particular uses. Evidence demonstrating that the specific use only presents a negligible risk or, in other cases, that the use is acceptable will be considered before granting an authorization. This evidence includes taking into account socioeconomic benefits; lack of "safer" chemicals for the same task; and measures minimizing the exposure of consumers, workers, the general public, and the environment.

An authorization would be granted if the risk to human health or the environment from the use of a substance arising from the intrinsic properties specified in Annex XIV (list of substances subjected to the authorization procedure in the REACH program) is adequately controlled; it is shown that socioeconomic benefits outweigh the risk to human health or the environment; and if there are no suitable alternative substances or technologies.

One goal of authorization is to ensure the good functioning of the internal market while ensuring that the risks from substances of very high concern are properly controlled. Another goal is to ensure that these substances are eventually replaced by suitable alternative substances or technologies if these are economically and technically viable.

The EU recognizes that the authorization program will require considerable resources for industries to comply with the program and for the regulatory community to review applications. It is expected that about 5 or 10 chemicals will be selected each year for the authorization process. The system is also organized such that one application should be prepared by interested parties for each chemical and use.

The following substances may be included:

- Category 1 or 2 carcinogens (Tables 11.4 and 11.5)
- Category 1 or 2 mutagens (Table 11.6)
- Category 1 or 2 substances that are toxic for reproduction (Tables 11.7 and 11.8)
- Substances that are PBTs

The following information is included in an authorization for a substance:

1. The persons to whom the authorization is granted
2. The identity of the substance
3. The use for which the authorization is granted
4. Any conditions under which the authorization is granted
5. The time-limited review period
6. Any monitoring arrangement

Notwithstanding any conditions of an authorization, the holder should ensure that the exposure is reduced to as low a level as is technically and practically possible.

TABLE 11.4
Initial List of Category 1 Carcinogens

Chromium (VI) trioxide	Chloroethylene
Zinc chromates, including zinc potassium chromate	Bis(chloromethyl) ether
Nickel monoxide	Chloromethyl methyl ether
Nickel dioxide	2-Naphthylamine; beta-naphthylamine
Dinickel trioxide	Benzidine; 4,4′-diaminobiphenyl
Nickel sulfide	Biphenyl-4,4′-ylenediamine
Nickel subsulfide	Salts of benzidine
Diarsenic trioxide	Salts of 2-naphthylamine
Arsenic trioxide	Biphenyl-4-ylamine
Arsenic pentoxide	Xenylamine
Arsenic acid and its salts	4-Aminobiphenyl
Lead hydrogen arsenate	Salts of biphenyl-4-ylamine
Butane (containing ≥ 0.1% butadiene)	Salts of xenylamine
Isobutane (containing ≥ 0.1% butadiene)	Salts of 4-aminobiphenyl
1,3-Butadiene; buta-1,3-diene	Coal tar
Benzene	Erionite
Vinyl chloride	Asbestos
Chlorodimethyl ether	

Source: Commission of the European Communities, *White Paper, Strategy for a Future Chemicals Policy*, Brussels, BE. 2001.

Authorizations may be reviewed at any time if the circumstances of the original authorization have changed so they affect the risk to human health, the environment, or the socioeconomic impact or if new information on possible substitutes becomes available.

Applications for authorization may be submitted by the manufacturer, importer, or downstream user of the substance. Applications may be submitted by one or several persons and could include one or several substances for one or more uses. Applications can be filled out for the applicant's own use or for uses for which the applicant intends to place the substance on the market.

An application for authorization shall include the following information:

1. The identity of the substance
2. The name and contact details of the person or persons making the application
3. A request for authorization, specifying for which use the authorization is sought and covering the use of the substance in preparation or the incorporation of the substance in articles, if this is relevant
4. A chemical safety report covering the risks to human health or the environment from the use of the substance arising from the intrinsic properties
5. An analysis of the alternatives considering their risks and the technical and economic feasibility of substitution.

TABLE 11.5
Initial List of Category 2 Carcinogens

Beryllium	Hexachlorobenzene
Beryllium compounds with the exception of aluminum beryllium silicates	1,4-Dichlorobut-2-ene
Beryllium oxide	2,3-Dibromopropan-1-ol
Sulfallate (ISO*)	2,3-Dibromo-1-propanol
2-Chlorallyl diethyldithiocarbamate	Ethylene oxide
Dimethylcarbamoyl chloride	Oxirane
Diazomethane	1-Chloro-2,3-epoxypropane
Hydrazine	Epichlorhydrin
N,N-Dimethylhydrazine	Propylene oxide
1,2-Dimethylhydrazine	1,2-Epoxypropane
Salts of hydrazine	Methyloxirane
Hydrazobenzene	2,2′-Bioxirane
1,2-Diphenylhydrazine	1,2:3,4-diepoxybutane
Hydrazine bis(3-carboxy-4-hydroxybenzensulfonate)	2,3-Epoxypropan-1-ol
Hexamethylphosphoric triamide	Glycidol
Hexamethylphosphoramide	Phenyl glycidyl ether
Dimethyl sulfate	2,3-Epoxypropyl phenyl ether
Diethyl sulfate	1,2-Epoxy-3-phenoxypropane
1,3-Propanesultone	Styrene oxide
Dimethylsulfamoylchloride	(Epoxyethyl)benzene
Potassium dichromate	Phenyloxirane
Ammonium dichromate	Furan
Sodium dichromate	R-2,3-Epoxy-1-propanol
Sodium dichromate, dihydrate	(R)-1-Chloro-2,3-epoxypropane
Chromyl dichloride	4-Amino-3-fluorophenol
Chromic oxychloride	5-Allyl-1,3-benzodioxole
Potassium chromate	Safrole
Calcium chromate	3-Propanolide
Strontium chromate	1,3-Propiolactone
Chromium III chromate	Urethane(INN)
Chromic chromate	Ethyl carbamate
Chromium (VI) compounds, with the exception of barium chromate	Methyl acrylamidomethoxyacetate (containing ≥ 0.1% acrylamide)
Sodium chromate	Methyl acrylamidoglycolate (containing ≥ 0.1% acrylamide)
Cobalt dichloride	Acrylonitrile
Cobalt sulfate	2-Nitropropane
Potassium bromate	2,4-Dinitrotoluene [1]
Cadmium oxide	Dinitrotoluene [2]
Cadmium fluoride	Dinitrotoluene, technical grade
Cadmium chloride	5-Nitroacenaphthene
Cadmium sulfate	2-Nitronaphthalene

continued

TABLE 11.5 (continued)
Initial List of Category 2 Carcinogens

Benzo[a]pyrene	4-Nitrobiphenyl
Benzo[d,e,f]chrysene	Nitrofen (ISO*)
Benzo[a]anthracene	2,4-Dichlorophenyl-4-nitrophenyl ether
Benzo[b]fluoranthene	2-Nitroanisole
Benzo[e]acephenanthrylene	2,6-Dinitrotoluene
Benzo[j]fluoranthene	2,3-Dinitrotoluene
Benzo[k]fluoranthene	3,4-Dinitrotoluene
Dibenz[a,h]anthracene	3,5-Dinitrotoluene
Chrysene	Hydrazine-tri-nitromethane
Benzo[e]pyrene	2,5-Dinitrotoluene
1,2-Dibromoethane	Azobenzene
Ethylene dibromide	Methyl-ONN-azoxymethyl acetate
1,2-Dichloroethane	Methyl azoxy methyl acetate
Ethylene dichloride	Disodium {5-[(4-((2,6-hydroxy-3-((2-hydroxy-5-sulfophenyl) azo)phenyl)azo) (1,1-biphenyl)-4-yl)azo]salicylato(4-)} cuprate(2-)
1,2-Dibromo-3-chloropropane	CI direct brown 95
Bromoethylene	4-o-Tolylazo-o-toluidine; 4-amino-2,3-dimethylazobenzene
Trichloroethylene	o-Aminoazotoluene
Trichloroethene	4-Aminoazobenzene
α-Chlorotoluene	Benzidine-based azo dyes
Benzyl chloride	4,4′-Diarylazobiphenyl dyes
α,α,α-Trichlorotoluene	Disodium 4-amino 3-[[4-[(2,4- diaminophenyl)azo] [1,1-biphenyl]-4-yl]azo]-5-hydroxy-6-(phenylazo) naphtalene-2,7-disulfonate
Benzotrichloride	Tetrasodium 3,3′-[[1,1-biphenyl]-4,4-dylbis(azo)] bis[5-amino-4-hydroxynaphthalene-2,7-disulfonate]
1,3-Dichloro-2-propanol	Disodium 3,3-[[1,1-bifenyl]-4,4dylbis(azo)] bis[4-aminonaphthalene-1-sulfonate)

Source: Commission of the European Communities, *White Paper, Strategy for a Future Chemicals Policy,* Brussels, BE. 2001.

* International Organization for Standards.

In addition, the application may also include

1. A socioeconomic analysis
2. A substitution plan, if appropriate, including research and development and a timetable for proposed actions by the applicant
3. A justification for not considering risks to human health and the environment arising from either
   a. Emissions of a substance from an installation for which a permit was granted or
   b. Discharges of a substance from a point source

TABLE 11.6
Initial List of Category 2 Mutagens (No Category 1 Mutagens Were Initially Listed)

Hexamethylphosphoric triamide	Benzo[d,e,f]chrysene
Hexamethylphosphoramide	1,2-Dibromo-3-chloropropane
Diethyl sulfate	Ethylene oxide
Potassium dichromate	Oxirane
Ammonium dichromate	Propylene oxide
Sodium dichromate	1,2-Epoxypropane
Sodium dichromate, dihydrate	Methyloxirane
Chromyl dichloride	2,2′-Bioxirane
Chromic oxychloride	1,2:3,4-Diepoxybutane
Potassium chromate	Methyl acrylamidomethoxyacetate (containing ≥ 0.1% acrylamide)
Sodium chromate	Methyl acrylamidoglycolate (containing ≥ 0,1% acrylamide)
Cadmium fluoride	Ethyleneimine
Cadmium chloride	Aziridine
Butane (containing ≥ 0.1% butadiene)	1,3,5,-Tris(oxiranylmethyl)-1,3,5-triazine-2,4,6(1H,3H,5H)-trione
Isobutane (containing ≥ 0.1% butadiene)	Triglycidylisocyanurate (TGIC)
1,3-Butadiene buta-1,3-diene	Acrylamide
Benzo[a]pyrene	1,3,5-Tris-[(2S and 2R)-2,3-epoxypropyl]-1,3,5-triazine-2,4,6-(1H,3H,5H)-trione

Source: Commission of the European Communities, *White Paper, Strategy for a Future Chemicals Policy,* Brussels, BE. 2001.

TABLE 11.7
Initial List of Chemicals Toxic to Reproduction, Category 1

Carbon monoxide	Trilead bis(orthophosphate)
Lead hexafluorosilicate	Lead acetate
Lead alkyls	Lead(II) methanesulfonate
Lead azide	C.I. pigment yellow 34
Lead chromate	C.I. pigment red 104
Lead di(acetate)	

Source: Commission of the European Communities, *White Paper, Strategy for a Future Chemicals Policy,* Brussels, BE. 2001.

TABLE 11.8
Initial List of Chemicals Toxic to Reproduction, Category 2

6-(2-Chloroethyl)-6(2-methoxyethoxy)-2,5,7,10-tetraoxa-6-silaundecane; etacelasil
Flusilazole (ISO*); bis(4-fluorophenyl)-(methyl)-(1H-1,2,4-triazol-1-ylmethyl)-silane
A mixture of 4-[[bis-(4-fluorophenyl)-methylsilyl] methyl]-4H-1,2,4-triazole;
1-[[bis-(4-Fluorophenyl)methylsilyl]methyl]-1H-1,2,4-triazole
Nickel tetracarbonyl
Cadmium fluoride
Cadmium chloride
Benzo[a]pyrene
Benzo[d,e,f]chrysene
2-Methoxyethanol
Ethylene glycol monomethyl ether
Methylglycol
2-Ethoxyethanol
Ethylene glycol monoethyl ether
Ethylglycol
2,3-Epoxypropan-1-ol
Glycidol
2-Methoxypropanol
Bis(2-methoxyethyl) ether
R-2,3-epoxy-1-propanol
4,4′-Isobutylethylidenediphenol
2,2-Bis(4′-hydroxyphenyl)-4-methylpentane
2-Methoxyethyl acetate
Ethylene glycol monomethyl ether acetate
Methylglycol acetate
2-Ethoxyethyl acetate
Ethylene glycol monoethyl ether acetate
Ethylglycol acetate
2-Ethylhexyl 3,5-bis(1,1-dimethylethyl)-4-hydroxyphenyl methyl thio acetate
Bis(2-methoxyethyl) phthalate
2-Methoxypropyl acetate
Fluazifop-butyl (ISO*); butyl (RS)-2-[4-(5-trifluoromethyl-2- pyridyloxy)phenoxy]propionate
Vinclozolin (ISO*);
N-3,5-dichlorophenyl-5-methyl-5-vinyl-1,3-oxazolidine-2,4-dione
Methoxyacetic acid
Bis(2-ethylhexyl) phthalate
Di-(2-ethylhexyl) phthalate
DEHP
Dibutyl phthalate
(+/–) Tetrahydrofurfuryl (R)-2-[4-(6-chloroquinoxalin-2-yloxy)phenyloxy]propionate
Binapacryl (ISO*);
2-secbutyl-4,6-dinitrophenyl-3-methylcrotonate
Dinoseb; 6-sec-butyl-2,4-dinitrophenol
Dinoterb; 2-tert-butyl-4,6-dinitrophenol
Salts and esters of dinoterb
Nitrofen (ISO*); 2,4 dichlorophenyl 4-nitrophenyl ether
Methyl-ONN-azoxymethyl acetate
Methyl azoxy methyl acetate
Tridemorph (ISO*);
2,6-dimethyl-4-tridecylmorpholine
Ethylene thiourea
Imidazolidine-2-thione
2-Imidazoline-2-thiol
Cycloheximide
Flumioxazin (ISO*); N-(7-fluoro-3,4-dihydro-3-oxo-4-prop-2-ynyl-2H-1,4-benzoxazin-6-yl) cyclohex-1-ene-1,2-dicarboxamide
(2RS,3RS)-3-(2-Chlorophenyl)-2-(4-fluorophenyl)-[(1H-1,2,4-triazol-1-yl)-ethyl]oxirane
N,N-Dimethylformamide
Dimethyl formamide
N, N-Dimethylacetamide
Formamide
N-Methylacetamide
N-Methylformamide

Source: Commission of the European Communities, *White Paper, Strategy for a Future Chemicals Policy,* Brussels, BE. 2001.

* International Organization for Standards.

PBT Substances

A substance that fulfills all three of the criteria of the following sections is a PBT substance.

Persistent chemicals are those that have

1. A half-life in marine water longer than 60 days,
2. A half-life in fresh- or estuarine water longer than 40 days,
3. A half-life in marine sediment longer than 180 days,
4. A half-life in fresh- or estuarine water sediment longer than 120 days, or
5. A half-life in soil longer than 120 days

Bioaccumulative chemicals have a bioconcentration factor (BCF) higher than 2,000 in aquatic species. Data from freshwater as well as marine water species can be used.

Toxic chemicals are those that have

1. The long-term no-observed effect concentration (NOEC) for marine or freshwater organisms is less than 0.01 mg/L,
2. The substance is classified as carcinogenic (Category 1 or 2), mutagenic (Category 1 or 2), or toxic for reproduction (Category 1, 2, or 3), or
3. There is other evidence of chronic toxicity

Very Persistent, Bioaccumulative, and Toxic Substances

The very persistent, bioaccumulative, and toxic (VPBT) chemicals are those that have

1. A half-life in marine, fresh-, or estuarine water longer than 60 days or
2. A half-life in marine, fresh-, or estuarine water sediment longer than 180 days or in soil longer than 180 days

Bioaccumulative chemicals have a BCF higher than 5,000 in aquatic species.

Restriction

The REACH program has proposed to ban certain uses of chemicals. Some of these chemicals and uses are shown in Table 11.9.

TABLE 11.9
Examples of Banned Uses of Chemicals in Proposed REA CH Legislation

Designation of the Substance, of the Groups of Substances or of the Preparation	Conditions of Restriction
1. Polychlorinated terphenyls (PCTs) — Preparations, including waste oils, with a PCT content higher than 0,005 % by weight.	1. Shall not be used. However, the following use of equipment, installations and fluids which were in service on 30 June 1986 shall continue to be permitted until they are disposed of or reach the end of their service life: (a) closed-system electrical equipment transformers, resistors and inductors; (b) large condensers (≥ 1 kg total weight); (c) small condensers; (d) heat-transmitting fluids in closed-circuit heattransfer installations; (e) hydraulic fluids for underground mining equipment.
2. Chloro-1-ethylene (monomer vinyl chloride)	Shall not be used as aerosol propellant for any use.
4. Tris (2,3 dibromopropyl) phosphate	Shall not be used in textile articles, such as garments, undergarments and linen, intended to come into contact with the skin.
5. Benzene	1. Not permitted in toys or parts of toys as placed on the market where the concentration of benzene in the free state is in excess of 5 mg/kg of the weight of the toy or part of toy. 2. Shall not be used in concentrations equal to, or greater than, 0,1 % by mass in substances or preparations placed on the market. 3. However, paragraph 2 shall not apply to: (a) motor fuels which are covered by Directive 98/70/EC; (b) substances and preparations for use in industrial processes not allowing for the emission of benzene in quantities in excess of those laid down in existing legislation; (c) waste covered by Council Directive 91/689/EEC of 12 December 1991 on hazardous waste (1) and Directive 2006/12/EC.

Source: Regulation (EC) No 1907/2006 Of The European Parliament And Of The Council of 18 December 2006 concerning the Registration, Evaluation, Authorisation and Restriction of Chemicals (REACH), Annex VII, P 167. Council of the European Union, Brussels, BE.

BIBLIOGRAPHY

Arora, S., and T. N. Cason, 1996. Why Do Firms Volunteer to Exceed Environmental Regulations? Understanding Participation in EPA's 33/50 Program. Madison, WI: University of Wisconsin Press, 72(4): 413–432.

Commission of the European Communities. 2001. *White Paper, Strategy for a Future Chemicals Policy.* Brussels, BE.

Commission of the European Communities. 2003. *Commission Staff Working Paper: Regulation of the European Parliament and of the Council Concerning REACH—Extended Impact Assessment.* Brussels, BE.

Council of the European Union. 2006a. *EU Common Ground on REACH.* Brussels, BE.

Council of the European Union. 2006b. *Questions and Answers on REACH.* Brussels, BE.

Council of the European Union. 2006c. *The REACH Proposal Process Description.* Brussels, BE.

Fan, P. L., H. Batchu, H. Chou, W. Gasparac, J. Sandrik, and D. M. Meyer. 2002. Laboratory Evaluation of Amalgam Separators. *JADA,* 133: 577–84.

Gallagher, M. J. 2000. *Proposed Strategy to Continually Reduce Persistent, Bioaccumulative Toxins (PBTs) in Washington State.* Washington State Department of Ecology, with contributions from the Ecology PBT Technical Committee. Olympia, WA: Publication No. 00-03-054.

Gamper-Rabindran, S. 2006. Did EPA's Voluntary Industrial Toxics Program Reduce Emissions? A GIS Analysis of Distributional Impacts and By-Media Analysis of Substitution. *Journal of Environmental Economics and Management,* 52(1): 391–410.

Mercury Education Reduction Act (MERA). Washington State Legislature Revised Code of Washington 2003 c 260 § 5. Olympia, WA: Washington State Department of Ecology.

Multiyear PBT Chemical Action Plan Schedule. Washington Department of Ecology Publication No. 07-07-016, March 2007, Olympia, WA. http://www.ecy.wa.gov/biblio/ 0607025.html.

Peele, C. 2003. *Washington State Mercury Chemical Action Plan.* Department of Ecology Publication No. 03-03-001. Olympia, WA: Department of Health Publication No. 333-051.

Peeler, M. 2006. Apresentação ao Companhia de Tecnologia de Saneamento Ambiental Redução do Mercúrio. Power Point presentation for the Department of Ecology Meeting, Lacy, WA.

U.S. Environmental Protection Agency. 1999. *33/50 Program: The Final Record.* Office of Pollution Prevention and Toxics. Washington, DC: EPA-745-R-99-004.

U.S. Food and Drug Administration, Center for Devices and Radiological Health. 2000. Outside Leveraging: Creating Relationships through Partnerships with Stakeholders—Mini Case Study—EPA's 33/50 Program: From Confrontation to Collaboration. http://www.fda.gov/cdrh/leveraging/2.html (accessed January 20, 2006).

U.S. Geological Survey. 2007. Mineral commodity summaries 2007: U.S. Geological Survey, Washington, DC: United States Government Printing Office.

Washington State Dental Association and the Washington Department of Ecology. 2004. Dental Amalgam Waste Management. http://www.ecy.wa.gov/mercury/mercury_dental_bmps.html (accessed January 20, 2006).

Washington State Department of Ecology. 2000a. *Chapter 173-307 WAC, Pollution Prevention Plans.* Olympia, WA.

Washington State Department of Ecology. 2000b. *Proposed Strategy to Continually Reduce Persistent, Bioaccumulative Toxins (PBTs) in Washington State.* Olympia, WA: Publication Number 00-03-054.

Washington State Department of Ecology. 2006a. *Chapter 173-333 WAC, Persistent Bioaccumulative Toxins Regulation.* Olympia, WA.

Washington State Department of Ecology. 2006b. *Pollution Prevention Plan: User's Guide.* Olympia, WA: Publication Number 02-04-023.

Washington State Department of Ecology. 2006c. *Reducing Toxics in Washington: Progress Report for 2001 through 2003.* Olympia, WA: Publication Number 06-04-009.

Washington State Department of Ecology. 2006d. Washington State Polybrominated Diphenyl Ether (PBDE) Chemical Action Plan: Final Plan. http://www.ecy.wa.gov/biblio/0507048.html (accessed January 20, 2010).

Washington State Department of Ecology. 2006a. Don't Mess With Mercury. http://www.ecy.wa.gov/programs/eap/pbt/hgreductionstrategy.html (accessed November 6, 2006).

Washington State Department of Ecology. 2006b. Hazardous Waste and Toxics Reduction—Mercury Reduction. http://www.ecy.wa.gov/mercury/index.html (accessed November 6, 2006).

Washington State Department of Ecology. 2006c. Mercury Memorandums of Understanding with Industry. http://www.ecy.wa.gov/mercury/mou.html (accessed November 3, 2006).

Washington State Department of Ecology. 2006d. Mercury Strategies and Outcome Measures. http://www.ecy.wa.gov/mercury/mercury_outcomes.html (accessed November 6, 2006).

Washington State Dioxin Source Assessment. July 1998. Publication number 98-320. Olympia, WA: Washington State Department of Ecology.

Washington State Legislature. 2010. Revised Code of Washington (RCW), Chapter 70.95M RCW, Mercury. http://apps.leg.wa.gov/RCW/default.aspx?cite=70.95M (accessed November 6, 2006).

Washington State Legislature. 2006. *WAC Chapter 173-333 Persistent Bioaccumulative Toxins.* Olympia, WA.

12 Use versus Release Reporting

INTRODUCTION

Although the Toxics Release Inventory (TRI) program has been successful in initially reducing the use of toxic chemicals, to continue to build on this success, releases of toxic chemicals contained in the products sold to customers versus just releases to environmental media, actual use of toxic chemicals by manufacturers versus just releases, and worker exposure now need to be addressed. It is important to note that it is expensive to track releases by each source in the factory. As part of the TRI process, the company must determine use of the chemical to determine if the releases of that chemical need to be quantified and reported; however, the data determined on use itself are not reported. This chapter presents a comparison between use and release reporting and programs aimed at reducing use. If we limit use, we will reduce worker exposure to toxic chemicals and releases, both directly to the air, water, and land and indirectly through products sold to customers.

COMPARISON OF USE AND RELEASES

Based on the toxicity units (TUs), in doses per capita, developed in Chapter 10 of this book, chromium, cobalt, lead, and mercury were determined as among the chemicals with the highest TUs. In Chapter 10, TUs were calculated by multiplying the effective toxicity factors (ETFs) in doses per capita-pound by the 2007 release amounts reported in TRI (pounds) to determine the toxicity threshold of a chemical if the 2007 releases of a chemical were distributed evenly, and equal exposure occurred, over the population of the United States. This section presents a comparison between use and releases of some of these chemicals with higher TUs to demonstrate the magnitude of difference between the level of use and releases of chemicals in the United States. This further demonstrates the importance of reporting and subsequently targeting reduction of the use of relatively higher toxicity chemicals versus just release reporting and targeting reduction of releases.

According to the U.S. Geographical Survey Mineral Commodity Summaries, 2007 apparent consumption of chromium in the United States was 629,000 metric tons as compared to the U.S. TRI 2007 release data of 27,398 metric tons (5,338 metric tons of chromium and 22,060 metric tons of chromium compounds). Since chromium is not a material that is consumed or destroyed in processing, release reporting only accounts for less than 5 percent of the use. The rest is either released by users who are not required to report under TRI or is a component of products. Similarly, the 2007 apparent consumption of cobalt in the United States was 9,600 metric tons, as compared to the U.S. TRI 2007 release data of 3,162 metric tons (397 metric tons of cobalt and 2,765 metric tons of cobalt compounds). The 2007 estimated consumption of lead in the United States was 1,540,000 metric tons, as compared to the U.S. TRI 2007 release data of 224,925 metric tons (9,498 metric tons of lead and 215,427 metric tons of lead compounds).

Excluding the quantity of mercury that has accumulated as a result of historical use (reservoir), best estimates from a 2002 study indicated that the total use of mercury in raw materials in the United States was approximately up to 2,143 tons per year, or 28 percent greater than the approximately up to 1,665 tons per year of mercury waste released (air, water, solid) in the United States across various sectors (U.S. Environmental Protection Agency [USEPA] 2002).

MASSACHUSETTS TOXICS USE REDUCTION ACT

The Massachusetts Toxics Use Reduction Act (TURA; 1989) requires that large-quantity users of toxic materials submit an annual Toxics Use Report to the Massachusetts Department of Environmental Protection (MADEP) and participate in a planning process designed to reduce their wastes and use of toxic chemicals.

The MADEP Bureau of Waste Prevention is responsible for implementing TURA through review of toxics use reports, enforcement actions, managing toxics use data, and evaluating the overall progress of the state.

The Administrative Council on TUR consists of seven members representing state agencies and is responsible for environmental protection, public health, occupational safety, and economic development.

Reporting Requirements

Regulated facilities in Massachusetts are required to file a toxics use report if they meet all three of the following criteria:

- The facility employs at least 10 full-time employees.
- The facility conducted business practices activities described by certain SIC (Standard Industrial Classification) codes.
- The facility is a large-quantity toxics user; it processed or used a TUR-regulated chemical in excess of reporting thresholds. A facility is considered a large-quantity toxics user if
  - 25,000 pounds of toxic substances were manufactured or processed during the reporting year or
  - 10,000 pounds of a toxic substance were used during the reporting year or
- 100 lbs, 10 lbs, or 0.1 lbs of persistent, bioaccumulative, and toxic (PBT) chemicals (depending on the specific PBT chemical) were manufactured, processed, or used during the reporting year.

Some examples of where manufacturers installed upgrades to their manufacturing processes to achieve a reduction in their use of toxic chemicals under the

Massachusetts TURA program are presented in Table 12.1. The table also shows cost reductions that have been realized by companies participating in the program.

MAINE TOXICS AND HAZARDOUS WASTE REDUCTION PROGRAM

The state of Maine Department of Environmental Protection (DEP) encourages an integrated approach to toxics use reduction, toxics release reduction, and hazardous waste reduction based on pollution prevention (P2) management strategies as outlined in the state of Maine Toxics Use and Hazardous Waste Reduction Law, or Toxics Law (State of Maine 1989).

- **Toxic Use Reduction:** The state promotes reducing the use of toxic substances through changes in production or other processes or operations, in products, or in raw materials that reduce, avoid, or eliminate the use or production of toxic substances without creating substantial new or increased risk to public health.
- **Toxic Release Reduction:** The state encourages reducing the release of toxics during manufacturing and other processes.
- **Hazardous Waste Reduction:** The state promotes reducing the generation of hazardous waste through use and release reduction techniques employed by the facility.

Reporting Requirements

Facilities in Maine must meet one or more of the following three reporting categories before they are required to report and submit a P2 plan:

- **Hazardous Waste:** A facility reports shipping 2,640 lb of hazardous waste in a calendar year.
- **Toxic Use:** A facility reports their use of extremely hazardous substances to Maine's Emergency Management Agency (EMA).
- **Toxic Release:** a facility reports release of toxic chemicals to the TRI database of the USEPA.

Facilities that ship between 661 and 2,639 lb of hazardous waste are only required to pay a $100 fee but not report or develop a P2 plan.

NEW JERSEY POLLUTION PREVENTION PROGRAM

The New Jersey Department of Environmental Protection Office of Pollution Prevention and Right-to-Know administers the P2 program in New Jersey; this includes P2 planning focused on toxics use reduction. The preparation of P2 plans by industries is one of the primary tools used by New Jersey for toxic use and waste reductions.

Under this program, Part I of the P2 plans can be broken down into six categories, one of which includes process-level inventory data or "the use of each hazardous

TABLE 12.1
Massachusetts Toxic Use Reduction Program Case Studies

Company Name	**Coyne Textile Services**
Company description	Industrial laundry service
Project description	Discontinued acceptance of saturated laundry that was dripping, installed extractor to remove liquids from laundry prior to washing
Capital/installation cost	$60,000
Annual cost savings	Over $25,000
Payback period	Less than 3 yrs
Annual reduction in chemical use	3,500 lb sulfuric acid; 1,200 lb ferric chloride; 15,000 lb potassium hydroxide
Reduction in air emissions	Reduced solvent emissions
Additional benefits	Reduced water use by 2 million gallons per year, increased productivity
Company Name	**Cranston Print Works**
Company description	Prints on cotton and blended fabrics
Project description	Installed liquid carbon dioxide system for wastewater neutralization to reduce the use of sulfuric acid
Capital/installation cost	$115,000
Annual cost savings	$80,000
Payback period	1.5 yrs
Annual reduction in chemical use	2.66 million pounds of sulfuric acid
Additional benefits	Reduced risk of employee exposure, reduced training requirements, improved operational control
Company Name	**Poly-Plating**
Company description	Production of nickel-plated parts
Project description	Installed filtration, recycling, and concentration system for waste reclamation
Capital/installation cost	$225,000
Annual cost savings	$107,000
Payback period	25 mos
Annual reduction in waste	Reduced disposal costs by 91%
Annual reduction in chemical use	Reduced acid purchased by 96%
Additional benefits	Reduced water use and sewage fees by 98%; new company formed to manufacture and sell waste purification equipment
Company Name	**V. H. Blackinton**
Company description	Manufactures metallic uniform insignia (badges)
Project description	Eliminated the use of freon from drying system, replaced brazing furnaces that used ammonia with ones that used hydrogen and nitrogen, and various modifications to plating and wastewater treatment systems
Payback period	3 yrs

TABLE 12.1 (continued)
Massachusetts Toxic Use Reduction Program Case Studies

Company Name	V. H. Blackinton
Annual reduction in chemical use	5,000 lb sulfuric acid; 9,000 lb sodium hydroxide; 1,500 lb sodium hypochlorite; 1,900 lb cyanide; 30,000 lb trichloroethane; 20,000 lb ammonia
Additional benefits	Improved employee safety, reduced water use, reduced reporting burden

Source: Toxics Use Reduction Institute, *TURI Overview, Working to Make Massachusetts Safer for Everyone*. Lowell: University of Massachusetts Lowell, 2003.

substance, the generation of nonproduct output, the amount recycled, and the amount released for each process" (New Jersey Technical Assistance Program 1999).

Part II of the P2 program focuses on toxic substances produced or manufactured in quantities greater than 10,000 lb. The P2 planning is targeted at 90 percent use or nonproduct output or releases and must target PBTs due to their low thresholds.

BIBLIOGRAPHY

Geiser, K., and M. Rossi. 1995. *Toxics Chemical Management in Massachusetts: The Second Report on Further Chemical Restriction Policies*. Lowell: University of Massachusetts Lowell.

The General Laws of Massachusetts. 2007. Chapter 21I. Massachusetts Toxics Use Reduction Act. http://www.mass.gov/legis/laws/mgl/21i-1.htm (accessed July 31, 2006).

Massachusetts Department of Environmental Protection, Bureau of Waste Prevention. 2005. *2005 Toxics Use Reporting Instructions*.

New Jersey Department of Environmental Protection, Office of Pollution Prevention. 2002. *Industrial Pollution Prevention Planning: Meeting Requirements under the New Jersey Pollution Prevention Act.* Trenton, NJ.

New Jersey Technical Assistance Program. 1999. The New Jersey Pollution Prevention Act. http://ycees.njit.edu/njtap/njppa.htm (accessed January 21, 2010).

State of Maine. 1989. *Title 38, Chapter 26, Toxics Use and Hazardous Waste Reduction Act.* Augusta, ME: Maine State Legislature, Office of the Revisor of Statutes.

Toxics Use Reduction Institute. 2003. *TURI Overview, Working to Make Massachusetts Safer for Everyone*. Lowell: University of Massachusetts Lowell.

U.S. Environmental Protection Agency. 2002. *Use and Release of Mercury in the United States*. Washington, DC: EPA/600/R-02/104.

U.S. Geographical Survey. 2009. Commodity Statistics and Information. http://minerals.usgs.gov/minerals/pubs/commodity/ (accessed January 21, 2010).

13 Pollution Prevention Planning

INTRODUCTION

One method of reducing use or releases is through requirements that companies perform pollution prevention (P2) plans and set goals for reducing use or releases. Although the details included in a P2 plan vary from state to state, most plans contain the following basic information:

1. **Baseline inventory:** either an inventory of chemical use or releases or an assessment of the basic processes that use toxic chemicals
2. **Reduction opportunity assessment:** identification of various opportunities to reduce the use or release of toxic chemicals and a cost-benefit analysis
3. **Goal setting:** commitment to reduce toxic use or release; can be in the form of a percentage reduction of toxic chemical use/release or specification of individual projects that will be implemented
4. **Progress reports:** documenting what measures have been implemented to meet the goals in the plan

The following is a review of how several state programs require or encourage facilities to prepare and implement a P2 plan.

MASSACHUSETTS TOXICS USE REDUCTION ACT

The Massachusetts Toxics Use Reduction Act (TURA) (1989) requires that facilities that use large quantities of toxic chemicals prepare a toxic use reduction plan designed to reduce their wastes and use of toxic chemicals.

A toxics use reduction (TUR) plan must evaluate the technical and economic impacts of toxic use reduction opportunities and identify which measures, if any, the facility will implement. Essential elements of a TUR plan include a corporate TUR policy statement, an assessment of which chemicals are used, how much is used and what quantities are generated as waste, a list of available TUR options and evaluations of those that appear to be technically and economically feasible, and a description and implementation schedule for the options that will be pursued (Massachusetts Department of Environmental Protection [MassDEP] 2009a).

Although the facility is not obligated to submit the entire plan to the MassDEP, it is required to be approved by a state-certified TUR planner and a summary of the plan must be submitted to MassDEP. The summary is also made available by MassDEP for public review.

According to MassDEP, many companies, with the support of senior management, establish planning teams and involve the workforce in various steps: analyzing production, tracking the use of materials, auditing health and environmental regulations, and identifying available TUR options. A company that has completed an initial TUR plan and two plan updates may also develop a resource conservation plan addressing water, energy, or materials use (allowed every other planning cycle) or implement an environmental management system (EMS) that addresses toxics in lieu of a TUR plan.

Massachusetts TUR reporting supplements the federal Form R reporting. Facilities submitting a state toxics use report must first obtain federal Toxics Release Inventory (TRI) reporting forms and instructions to complete the state toxics use report.

Facility-level materials accounting data and production unit information are reported annually, with the state protecting confidential business information by combining conventional trade secret protections with the use of data collection and reporting strategies. This approach is designed to allow tracking without the need for public release of sensitive business data.

In the 15-yr period from 1990 to 2005, industries that were subject to TURA reporting in both years reduced toxic chemical use by 40 percent, toxic by-products by 71 percent, toxics shipped in product by 41 percent, on-site releases of toxics to the environment by 91 percent, and transfers of toxics off site for further waste management by 60 percent (MassDEP 2009b).

MAINE TOXICS AND HAZARDOUS WASTE REDUCTION PROGRAM

The state of Maine Department of Environmental Protection (DEP) encourages an integrated approach to TUR, toxics release reduction, and hazardous waste reduction based on P2 management strategies as outlined in the State of Maine *Toxics Use and Hazardous Waste Reduction Law* (Toxics Law) (State of Maine 1989).

- **Toxic Use Reduction:** The state promotes reducing the use of toxic substances through changes in production or other processes or operations and in products or in raw materials that reduce, avoid, or eliminate the use or production of toxic substances without creating substantial new or increased risk to public health.
- **Toxic Release Reduction:** The state encourages reducing the release of toxics during manufacturing and other processes.
- **Hazardous Waste Reduction:** The state promotes reducing the generation of hazardous waste through use and release reduction techniques employed by the facility.

P2 planning in Maine is driven by and is encouraged to be in line with the reduction goals of the facilities. P2 planning development by a facility must also include the employees, management, and technical staff of the facility who will carry out the recommendations of the plan.

The Toxics Law requires regulated facilities to (State of Maine DEP 2006)

- Develop a P2 plan
- Solicit employee input during P2 plan development
- Set specified reduction goals
- Report biannually to the Maine DEP Toxics Program and local municipal officials on the progress made toward meeting reduction goals
- Pay an annual fee to the Maine DEP Toxics Program

WASHINGTON STATE DEPARTMENT OF ECOLOGY POLLUTION PREVENTION PROGRAM

The Washington State P2 program applies to all hazardous substance users and to hazardous waste generators who generate more than 2,640 lb of hazardous waste per year except for those facilities that are primarily treatment, storage, and disposal facilities or recycling facilities. Applicable facilities must prepare a P2 plan, which is then updated at least every 5 yrs. If a facility has an EMS, it can be used instead of submitting P2 planning forms.

Washington P2 plans must consider opportunities based on the following priorities: hazardous substance use reduction and hazardous waste reduction, recycling, and treatment. The plans are required to contain four parts:

- **Part 1:** This section includes a statement expressing the commitment of the facility to the plan and its goals and includes the scope and objectives of the plan. It describes the type of facility and products made or services provided at the facility, including the current production or activity level. It provides the total pounds of extremely hazardous waste and total pounds of dangerous waste reported on Form 4, *Dangerous Waste Annual Report*, for the last calendar year, and if applicable, the total pounds of toxic releases reported on Form R under the Superfund Amendments and Reauthorization Act (SARA) Title III, Section 313, for the same time period. It also includes a description of current reduction, recycling, and treatment activities and compares these with documentation of any reduction efforts that took place before the first plan was submitted.
- **Part 2:** This section includes an identification of hazardous substances used and hazardous wastes generated by the facility; a description of the facility processes; an identification of reduction, recycling, and treatment opportunities; an evaluation of those opportunities; a selection of proposed options; a policy to prevent shifting of risks; performance goals; and an implementation schedule.
- **Part 3:** This section provides a financial description of the plan, which identifies costs and benefits realized from implementing selected opportunities to the extent reasonably possible. Part 3 also includes a description of accounting systems that will be used to identify hazardous substance

use and hazardous waste management costs. Liability, compliance, and oversight costs must be components of these accounting systems.
- **Part 4:** Part four of the plan includes a description of personnel training and employee involvement programs. Each facility required to write a plan is encouraged to advise its employees of the planning process and solicit comments or suggestions from its employees on hazardous substance use and waste reduction opportunities.

Progress reports must be submitted to the department annually. The purpose of the progress report is to provide information on quantities of hazardous waste and hazardous substances or products containing hazardous substances in the prior 12-mo period. These progress reports consist of

- Progress toward any numeric performance goals and a description of reduction, recycling, and treatment opportunities that were implemented
- A description of the process or processes on which each opportunity had an impact
- A description of the quantities, by weight, of hazardous substances or products containing hazardous substances reduced and hazardous waste reduced by each option
- Discussion of any changes in measurement or estimation techniques
- Problems encountered in the implementation process

Hazardous waste generators must develop their plans to fit their individual situations. Once a draft of the plan is complete, the department may review a plan, executive summary, or an annual progress report to determine whether the document is adequate. If a hazardous substance user or hazardous waste generator fails to complete an adequate plan, executive summary, or annual progress report, the department notifies the user or generator of the inadequacy, identifying specific deficiencies. The generator then has at least 90 days to complete a modified plan, executive summary, or annual progress report addressing the specified deficiencies. If this modified document is still considered inadequate, the department will assign penalty fees.

Different facilities have varying success with P2, depending on their processes. Some facilities have been able to bring themselves below the planning thresholds based on changes implemented through the P2 program. Other facilities with older or more specialized processes have not found cost-effective methods or materials to modify their processes.

An analysis performed by the Washington Department of Ecology indicated that between 1990 and 2003, the goal of an overall decrease of 50 percent in the amount of recurrent hazardous waste generated in the state of Washington was met (Figure 13.1). Per capita waste generation has also decreased from 52 lb/yr in 1990 to 17 lb/yr in 2003. These decreases were due to a variety of reasons. Some companies and industries have closed due to economic factors. This inherently decreases hazardous waste generation. Most industries, however, continued to experience economic growth while decreasing their hazardous waste output.

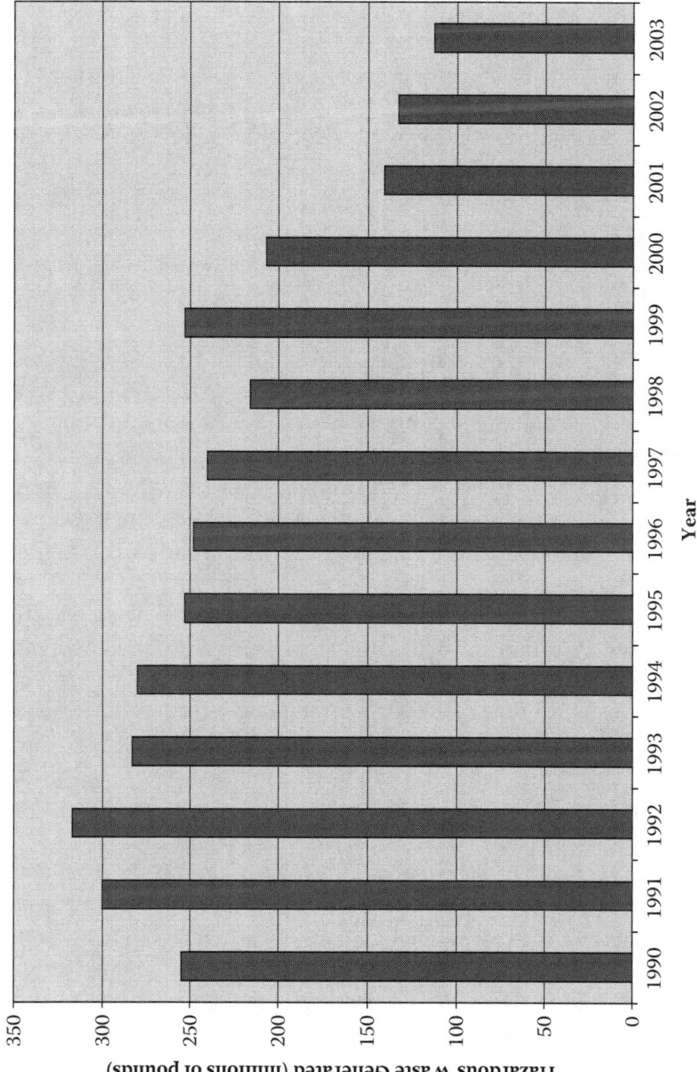

FIGURE 13.1 State of Washington hazardous waste generation, 1990 through 2003.

NEW JERSEY POLLUTION PREVENTION PROGRAM

The New Jersey DEP Office of Pollution Prevention and Right-to-Know administers the P2 program in New Jersey, which includes P2 planning focused on TUR. The preparation of P2 plans by industries is one of the primary tools used by New Jersey for toxic use and waste reductions. The P2 planning is targeted at 90 percent use or nonproduct output (NPO) or releases and must target PBT (persistent, bioaccumulative, and toxic) substances due to their low thresholds. P2 planning does not include waste minimization. Waste minimization is encouraged but is not counted as part of the overall reduction for a facility.

Although the P2 planning is a mandatory component in the NJ P2 program for regulated facilities, the plans prepared by the facilities are voluntarily implemented. Between 500 and 600 facilities are currently subjected to the P2 planning requirements. The plans are approved by an administrative checklist to encourage self-direction. The plans are not reviewed for technical content by the engineers at the New Jersey DEP, but the engineers do review the plans to make sure that the material accounting calculations make sense.

The New Jersey P2 program sets goals of "significant reduction" in the use of toxic (TRI-listed) chemicals and a statewide 50 percent reduction in toxic waste generation. Materials accounting is used in the New Jersey program as a regulatory tool to measure use of toxic substances and to facilitate preparation of mandated P2 plans. Materials accounting data provide a complete view of hazardous substances as they flow through communities and manufacturing operations of the facilities. The P2 planning program is structured so that the technical assistance component is separate from the data collection and enforcement components.

P2 planning in New Jersey is pursuant to the federal TRI reporting requirements and the New Jersey Pollution Prevention Act (P2 Act). Worker and Community Right-to-Know (WCRTK) Act of New Jersey established a regulation process for collecting data on the flow of toxic materials through individual facilities. Under the WCRTK Act, information on releases to the environment, transfers to off-site waste management facilities, NPO (i.e., total production-related waste), and throughput (i.e., use) of over 600 toxic chemicals is reported annually.

The regulations established by the P2 Act were adopted by the state in 1993 and are readopted every 5 yrs. The basis behind the legislation was the findings by the legislators that thousands of tons of hazardous substances are discharged into the environment of the state each year, and the regulatory approaches at the time were fragmented and allowed pollution to be shifted from one environmental medium to another. The legislators also declared that the inherent limitations of the traditional system of P2 should be addressed by a new emphasis on P2 that included the following:

- Reduction in the use of hazardous substances in industrial and manufacturing processes
- Rigorous accounting of the use, generation, and multimedia environmental releases of hazardous substances to identify points at which pollution can be prevented

- Achievement of P2 by a more efficient and rational use of hazardous substances through the use of less-hazardous substances, substitution of substances, or processes less prone to produce pollution
- Implementation of a soundly planned P2 program without adversely affecting the economic health of the state and those employed by the industries that use and discharge hazardous substances

Facilities that use, process, or manufacture chemicals listed in the TRI in quantities greater than 10,000 lb, have 10 or more full-time employees, or submit at least one federal toxic release inventory Form R or Form A to the Environmental Protection Agency (EPA) are required to implement P2 planning and reporting at their facility. Facilities subject to P2 planning are required to develop P2 plans every 5 yrs; these remain at the facility. These plans are required to be certified that they are true, accurate, and complete.

P2 planning documents include a P2 Plan with two parts, a P2 Plan Summary and a P2 Plan Progress Report.

The P2 Plan Part 1 includes an inventory of hazardous chemical use and NPO, including the following:

- The name of a top-level company official responsible for the P2 plan and identification of at least one nonmanager employee representative. The top-level company official must certify that the plan is true, accurate, and complete.
- Identification of the facility-wide chemicals manufactured, stored, or used at the facility and how the substances flow through the facility. Specifically, the plan must identify hazardous substance inputs: chemicals that are stored on the first day of the year, brought into the facility, manufactured, or recycled on site and reused. The plan must also identify hazardous substance outputs: chemicals stored on the last day of the year, consumed, shipped off site as/or in product or coproduct, generated as NPO.
- Identification of each process that involves hazardous substances and a "unit of product" for each process. A *unit of product* is an amount of useful product produced. This unit of product allows the facility to report NPO per unit of useful production. This method of reporting gives the facility a method of determining how successful its P2 measures have been that is independent of fluctuations in production levels.
- An inventory of the use of each hazardous substance, the generation of NPO, the amount recycled, and the amount released for each process.
- An inventory of the wastes generated at each process and how they are handled.
- A financial analysis to evaluate costs for storage or handling; monitoring, tracking, and reporting; treatment; transportation and disposal; hazardous waste manifesting; permit fees; liability; raw materials; and safety and compliance.
- An evaluation of the facility-level reductions and increases (use, NPO, and releases) of TRI substances in comparison to the facility P2 goals.
- An evaluation of process-level reductions and increases (use, NPO, and releases) of TRI substances in comparison to the process-level P2 goals.

P2 Plan Part 2 is an analysis of the P2 options and contains the following:

- Identification of P2 options that target processes and sources that contain TRI substances
- A financial and technical feasibility analysis for implementing the P2 options
- Five-year reduction goals for NPO and use of TRI substances

A P2 Plan Summary is required to be revised and submitted electronically to the New Jersey DEP every 5 yrs and includes a summary of the P2 plan elements, such as the reduction goals set by the facility.

A P2 Plan Progress Report is required to be submitted electronically to the New Jersey DEP annually. The report includes the actual P2 reductions or increases and is integrated with the Release and Pollution Prevention Report (RPPR), which is an annual report submitted to the New Jersey DEP that contains facility-level materials accounting numbers and throughput of TRI substances.

Some other aspects of the program to note are as follows:

- A facility is not penalized if it tries a failed TUR process.
- Releases are available for public review, but use or storage of toxic substances is not.
- A reading room was created by the New Jersey DEP for the public to review facility information.

The data collected by the New Jersey DEP Office of Pollution Prevention and Right-to-Know (P2RTK) is used to create trend analysis, determine enforcement targets, and help set priorities. The trend analysis helps the New Jersey DEP P2RTK to set meaningful metrics through the evaluation of multimedia releases, use, NPO, production, and shipped in/as product. The trend analysis also helps determine the top 10 lists of substances such as carcinogens, PBTs, extraordinary hazardous substances, and facilities with the largest increases and decreases of use, NPO, and releases.

A trend analysis of materials accounting data from 1994 through 2001 was completed in New Jersey in 2004 (New Jersey DEP 2004). The overview findings of the analysis are as follows:

- There was a decrease statewide in NPO and releases of hazardous substances.
- Hazardous substance use increased due to higher statewide industrial production; however, the amount of releases when adjusted for industrial production decreased. In other words, facilities were using hazardous substances more efficiently.
- Hazardous substance use increases as reported in New Jersey were primarily the result of hazardous substances contained in products shipped as or in a product.
- Statewide trends were driven by changes at a few large facilities.
- Although the trend showed an overall increased efficiency of hazardous substance use and a reduction in the release of hazardous substances, certain industries and chemicals did not follow this trend.

Enforcement targeting was determined through a relative risk analysis of the data and by evaluating permit exceedances. This helps to prioritize P2 inspections. The data also help the agency with priority setting through health tracking of cancer, child blood lead levels, and birth defects; multimedia release reports; and the discovery of hot spots and chemicals of concern.

On the other hand, the requirements for the New Jersey P2 plans were laid out in the P2 Act legislation and resulted in some outcomes that did not necessarily lead to P2. For example, some companies produced products that contained nontoxic forms of chemicals listed on the TRI lists, such as nickel, chromium, or cobalt in their alloy forms. Targeting is an option for facilities to streamline their planning requirements. A facility could choose to target the metal component; however, it is possible that other more toxic substances also contributed to 90 percent of use, NPO, or releases and would also be targeted. Targeting does not prevent a facility from looking at all substances.

In addition, in 1996 the New Jersey DEP commissioned a report by Hampshire Research Associates (HRA), *Evaluation of the Effectiveness of Pollution Prevention Planning in New Jersey*. HRA and the New Jersey DEP visited 115 facilities to evaluate P2 planning, including planning costs and cost savings. The report showed the following findings:

- Of the 28 facilities able to calculate their annual cost savings, the annual average cost savings was approximately $254,000, for a total of $7.1 million. Most facilities calculated only the direct savings in raw material costs for their reduction projects.
- Eight of the facilities calculated the costs/amounts of both capital investment and annual cost savings, yielding a total (for all eight facilities) capital investment of $7.2 million and a total annual return of $1.7 million, for an average payback period of 4.2 yrs.
- Potential capital investment to complete all of the projects in the plans of the facilities ranged from zero to $19 million for the 48 facilities visited by HRA.
- Of 115 facilities that could place a dollar figure on the cost of their planning, 49 spent an average of $35,000 each.
- Of 115 facilities that could quantify the amount of time necessary to prepare their plans, 95 spent, on average, 6.5 wks of one person's time to develop the plan. Taking an average salary of $50/h yields an average planning cost of $13,000 per facility.
- Overall, the average cost of planning was less than the average annual savings per facility found through planning.

Lessons Learned by New Jersey Companies

Howmet Castings, a division of Alcoa Aluminum Company, located in Dover, New Jersey, produces a specialty alloy and manufactures turbine blades of various sizes for jet engines and power plant gas turbines. The alloy facility produces a variety of superalloys of chromium, nickel, and cobalt to withstand the high temperatures of gas turbine engines. Alloy scrap is recycled on site.

The processing facility uses a lost wax process to cast a wide variety of complex turbine blades and other critical parts. Using best management practices, the processing facility tracks wasted parts and minimizes waste as much as possible, with the wax and sludge generated during processing sent to a recycling facility in Pennsylvania.

Management of Howmet discussed the New Jersey DEP P2 program with us and how the P2 Program could be improved.

The advantages are as follows:

- Preparing a P2 plan helps to detect toxic use reduction opportunities.
- Hazardous waste that is recycled is credited back to the facility when hazardous waste disposal is reported on Form R.

The disadvantages are as follows:

- Because the New Jersey P2 Program is specifically outlined in the New Jersey law, it is hard to modify and is therefore not flexible for all industries.
- The preparation of a P2 plan requires many hours of labor to complete, and there are also efforts to fill out duplicate mass balance reports for New Jersey in addition to multiple Form Rs that contain similar information under the TRI program.
- The New Jersey P2 program recommends that facilities target for reduction the TRI chemicals that constitute 90 percent of use; however, since chromium, cobalt, and nickel are in the alloy that makes up the product of the company, the plan would target these rather than materials used in production that are much more of a toxic product and should be targeted for reduction. The superalloy is inert and not an environmental or health threat. Lower-volume materials like etching solutions should be targeted.
- Therefore, the law has the unintended consequence of targeting the product of the facility not the toxic chemicals used in its production.

Air Products Polymers, L.P., of Dayton, New Jersey, produces raw materials, primarily vinyl acetate-ethylene emulsion polymers for high-quality binders used for packaging, wood bonding/furniture, pressure-sensitive adhesives, wall and ceiling coverings, flooring, consumer glues, and film laminates.

P2 is part of the corporate culture of Air Products and is implemented voluntarily. The company makes an effort to reduce the use of toxic materials at the source by substituting alternative materials to produce products. Air Products also has a hierarchy for waste minimization, with land disposal at the bottom.

Management of Air Products discussed the New Jersey DEP P2 program with us and how the P2 program could be improved.

- The managers at Air Products suggested that the New Jersey P2 program should be more compliance advantageous to industries by providing incentives for those firms that comply with P2 regulations.
- The managers at Air Products noted that larger companies have P2 programs as part of their corporate culture. Smaller-to-midsize facilities

normally need more assistance developing P2 programs. It was noted that if the larger companies do not help the smaller-to-midsize firms with P2, the New Jersey DEP may create more stringent regulations for the entire industry. Air Products has provided assistance to smaller companies in an effort to decrease the amount of regulations required for the chemical industry.

- Managers at Air Products suggested that reporting be for 2 yrs rather than yearly. It was also felt that the TRI program by itself was the most effective at promoting toxics reductions due to the shame factor.
- Those at Air Products also suggested that less emphasis be placed on material accounting as it can require a lot of resources that could be used to fund actual projects.
- Those at Air Products advised against creating regulations that are too prescriptive for manufacturing facilities. Regulations that are too prescriptive can stifle the manufacturing and economic growth of a facility.

CALIFORNIA: HAZARDOUS WASTE SOURCE REDUCTION AND MANAGEMENT REVIEW ACT OF 1989 (SENATE BILL 14)

In California, Senate Bill (SB) 14 requires that generators that routinely generate more than 12,000 kg of hazardous waste or 12 kg of extremely hazardous waste in a reporting year comply with SB 14. These regulated industries are required to evaluate source reduction opportunities and report on accomplishments every 4 yrs (California EPA 2006).

The regulated facilities are required to stay in compliance with the legislation by preparing a Source Reduction Evaluation Review and Plan, a Hazardous Waste Management Performance Report, and a Summary Progress Report. TUR is not a separate program and is integrated within the P2 planning in California.

A P2 Plan and Summary are required to be prepared every 4 yrs in addition to Hazardous Waste Management Performance Reports. The generator is required to certify that the plan is being implemented unless the selected measures for source reduction are not technically feasible or economically practicable or if attempts to implement the measure result in an adverse impact to hazardous waste reduction goals, product quality, or human health or the environment (State of California 2005).

TEXAS

The Waste Reduction Policy Act of 1991 was adopted by the Texas legislature to prevent pollution in Texas. The Texas Commission on Environmental Quality (TCEQ 2003) adopted the corresponding rule under 30 TAC 335 Subchapter Q. This act requires that large- and small-quantity generators of hazardous waste and TRI Form R reporters

1. Prepare a 5-yr P2 plan
2. Submit an executive summary of the P2 plan
3. Report annually on their activities to prevent pollution

The P2 plan must contain

- A list of all hazardous wastes and TRI chemicals
- The activities that generate the waste or TRI chemicals
- An explanation of P2 projects
- An implementation schedule
- The measurable P2 goals
- An employee awareness program
- A P2 plan executive summary

The TCEQ offers technical assistance with Waste Reduction Policy Act (WRPA) applicability, planning, and reporting requirements, including a planning tool that provides step-by-step assistance in developing and printing a complete P2 plan. The tool includes the option to select from a list of P2 practices that TCEQ has seen work at other facilities (TCEQ 2009).

MINNESOTA

The Minnesota Toxic Pollution Prevention Act (TPPA) requires that most companies subject to TRI Form R reporting must write plans for reducing the emissions of Form R chemicals. P2 plans are also required of companies participating in the Minnesota Pollution Control Agency (MPCA) Environmental Auditing (Green Star Award) Program (Minnesota Technical Assistance Program 2008).

The P2 plan must contain the following:

- A policy statement expressing management support for eliminating or reducing the generation or release of toxic chemicals (pollutants) at the facility
- A description of the current processes generating or releasing toxic chemicals specifying the types, sources, and quantities of toxic chemicals being generated or released by the facility
- A description of the current and past practices used to eliminate or reduce the generation or release of toxic pollutants at the facility and an evaluation of the effectiveness of these practices
- An assessment of the technically and economically feasible options available to eliminate or reduce the generation or release of toxic chemicals at the facility, including options such as changing the raw materials, operating techniques, equipment, and technology; personnel training; and other practices used at the facility
- Objectives and a schedule for achieving those objectives

The TPPA requires companies to express objectives in numeric terms if technically and economically feasible. Otherwise, nonnumeric objectives can be stated; however, they must include a clearly stated list of actions designed to lead to establishing numeric objectives as soon as they become feasible. Facility P2 plans must at a minimum contain objectives for each chemical for which a facility submits a TRI Form R report. The plan must also include

- The rationale for each objective established for the facility.
- A list of options that were considered but were not technically or economically feasible.
- A certification, signed and dated by the facility manager and an officer of the company, attesting to the accuracy of the information in the plan. P2 plans must be updated biennially.

Facilities are also required to submit an annual progress report, which must include

- A summary of each objective established in the plan, including the base year for any objective stated in numeric terms and the schedule for meeting each objective
- A summary of progress made during the past year, if any, toward meeting each objective established in the plan, including the quantity of each toxic pollutant eliminated or reduced
- A statement of the methods through which elimination or reduction has been achieved
- If necessary, an explanation of the reasons objectives were not achieved during the previous year, including identification of any technological, economic, or other impediments the facility faced in its efforts to achieve its objectives
- A certification signed and dated by the facility manager and an officer of the company under penalty of Section 609.63 attesting that a plan meeting the requirements of Section 115D.07 has been prepared and also attesting to the accuracy of the information in the progress report (Minnesota Office of the Revisor of Statutes 2009)

ARIZONA DEPARTMENT OF ENVIRONMENTAL QUALITY

The P2 Planning Program was created by the Arizona legislature in 1991. The program requires all industrial facilities within a certain threshold of hazardous waste generation and toxic substance use to perform a P2 analysis and to file an annual P2 plan. The P2 plan becomes a stand-alone management tool that

- Documents P2 assessments that are performed
- Records the toxic substances used and emissions and wastes generated by facility operations
- Outlines specific P2 opportunities and performance goals within a suggested implementation schedule
- Creates a database for planning and tracking progress

Arizona makes a number of resource tools available to regulated facilities, including technical assistance, partnerships, outreach, a data collection program, and innovation. Although the program had mandatory requirements for facilities, the state of Arizona promoted and enforced the program through deficiency letters that did not impose financial penalties. Through outreach programs, a collaborative effort between the state and regulated facilities was established through the Arizona

Partnership for Pollution Prevention. This program helped facilities gain more information and experience with P2 programs.

According to P2 plans submitted, 911 million pounds of pollution were prevented by over 200 companies that had submitted plans to the state through 2002, including particulates and fugitive emissions, generation of hazardous wastes, and wastewater, in addition to the reduction in the demand for new water (USEPA 2006).

The Arizona Department of Environmental Quality (ADEQ) offers the additional incentive of a 50 percent reduction in hazardous waste generation fees when a company has an approved P2 plan in place. From 1999 through 2001, this translated into an average annual savings to filers of more than $260,000.

According to the ADEQ, the following elements are considered important by the most successful P2 programs:

- Facilities have a clear understanding of their P2 direction:
    - Have a definition of P2
    - Have either a facility or corporate P2 policy
- Facilities have a method for identifying and documenting wastes and emissions.
- Facilities have P2 goals:
    - There are facility or corporate P2 goals
    - Using input solicited from employees and other sources, facility environmental leaders provide input into the corporate and facility goal-setting processes.
    - Corporate P2 directives influence the program.
- Facilities use a champion or facilitator or focal point person to lead the program:
    - Management supports P2 and commits the necessary resources to support P2 activities.
    - P2 is integrated into business planning.
- Environmental considerations are part of the business planning process:
    - Facility P2 goals are part of the business planning process.
    - P2 is used, whenever possible, in anticipation of future compliance requirements.
- Priorities are assigned to waste streams.
- Cross-functional teams are used.
- Sustainable P2 programs are cost effective.
- P2 projects need to meet a rate of return on investment:
    - Facilities use financial and nonfinancial criteria to evaluate projects.
    - Facilities implement some P2 projects that are not cost effective.
- P2 progress is tracked and communicated:
    - Facilities have the ability to measure progress.
    - Facilities periodically publish results against goals.
    - Results are communicated to key people.
- Facilities use quality tools in their P2 program (i.e., team-based quality culture, ISO 9000/14000, use of Pareto principles, total quality management, etc.).

- There is responsibility and accountability for P2 results:
    - Many facilities tie waste and emissions accountability to the generating operation.
- Facility P2 teams know their plant culture and pattern the program to that culture.
- Recognition sustains employees motivation:
    - Immediate recognition of early accomplishments helps establish the P2 program.
    - Facility- or corporate-level recognition programs help sustain employee motivation.
- Company resources support facility P2 programs:
    - Facilities have access to corporate resources for program implementation.
    - Facilities use external resources to aid their P2 program (i.e., corporate engineering, marketing, research, laboratories, outside suppliers).
- Effective communication increases P2 awareness:
    - Have communication process within the facility.
    - Have communication process with the community.
- P2 is integrated into premanufacturing decisions:
    - P2 begins at research, development, and design phases of the product or process life.
    - Facilities work with equipment and raw material suppliers and customers to help identify P2 opportunities for products and processes.
- Facilities use new technology to achieve significant improvement (ADEQ 2006).

VOLUNTARY TECHNICAL ASSISTANCE PROGRAMS

A number of states, including New Hampshire, Rhode Island, Delaware, Iowa, and Colorado, have voluntary P2 programs that provide free, confidential P2 technical assistance without the risk of regulatory enforcement. These programs often include site visits, technical research, training seminars, grants, clearinghouse programs, pilot-scale testing of P2 equipment, laboratory analyses, and economic analyses.

BIBLIOGRAPHY

Arizona Department of Environmental Quality. 2006. Pollution Prevention: About the Program. http://www.azdeq.gov/environ/waste/p2/about.html (accessed December 16, 2009).

California Environmental Protection Agency, Department of Toxic Substances Control, Office of Pollution Prevention and Technology Development. 2006. *Guidance Manual for Complying with the Hazardous Waste Reduction and Management Review Act of 1989.*

California Environmental Protection Agency, Office of Environmental Health Hazard Assessment. 2003. *Proposition 65 in Plain Language.* Sacramento, CA.

Colorado Department of Public Health and Environment. 2009. Pollution Prevention Program. http://www.cdphe.state.co.us/op/p2_program/index.html (accessed December 16, 2009).

Geiser, K., and M. Rossi. 1995. *Toxics Chemical Management in Massachusetts: The Second Report on Further Chemical Restriction Policies.* Lowell: University of Massachusetts Lowell.

The General Laws of Massachusetts. 2006. Chapter 21I. Massachusetts Toxics Use Reduction Act. http://www.mass.gov/legis/laws/mgl/21i-1.htm (accessed July 31, 2006).

Hampshire Research Associates (HRA). Evaluation of the Effectiveness of Pollution Prevention Planning in New Jersey May 1996. A Report prepared under New Jersey Contract Number: P39724. Trenton, NJ: New Jersey Department of Environmental Protection.

Hazardous Waste Source Reduction and Management Review Act of 1989 (Senate Bill 14), Sacramento, CA: California State Legislature.

Iowa Department of Natural Resources. 2009. Pollution Prevention for Business. http://www.iowadnr.gov/waste/p2/index.html (accessed December 16, 2009).

Massachusetts Department of Environmental Protection. 2009a. Toxics and Hazards, About Toxics Use Reduction Planning. http://www.mass.gov/dep/toxics/tura/plan2.htm (accessed December 14, 2009).

Massachusetts Department of Environmental Protection. 2009b. Toxics and Hazards, Toxics Use Reduction Act Overview. http://www.mass.gov/dep/toxics/tura/turaover.htm (accessed December 14, 2009).

Massachusetts Department of Environmental Protection, Bureau of Waste Prevention. 2005. *2005 Toxics Use Reporting Instructions.* Boston, MA.

Minnesota Office of the Revisor of Statutes. 2009. *Minnesota Statutes 115D.08 Progress Reports.* St. Paul, MN.

Minnesota Technical Assistance Program. 2008. Pollution Prevention Planning. http://mntap.umn.edu/prevention/index.htm (accessed December 15, 2009).

New Hampshire Department of Environmental Services. 2009. Pollution Prevention Program. http://des.nh.gov/organization/commissioner/p2au/pps/ppp/ (accessed December 16, 2009).

New Jersey Department of Environmental Protection. 2004. *Industrial Pollution Prevention in New Jersey: A Trends Analysis of Materials Accounting Data from 1994 to 2001 and an Annual Report for 2001.* Trenton, NJ.

New Jersey Department of Environmental Protection, Office of Pollution Prevention. 2002. *Industrial Pollution Prevention Planning: Meeting Requirements under the New Jersey Pollution Prevention Act.* Trenton, NJ.

New Jersey State Assembly. 1991. *Pollution Prevention Act.* Assembly Appropriations Committee Statement, Senate, No. 2220-L. 1991, c. 235. Trenton, NJ.

State of California. 2005. *Chapter 31, Waste Minimization, Article 1, Hazardous Waste Source Reduction and Management Review, §67100.4.(c) Plan and Summary.* Sacramento, CA: Code of Regulations, Office of Administrative Law.

State of Maine. 1989. *Title 38, Chapter 26, Toxics Use and Hazardous Waste Reduction Act.* Augusta, ME: Maine State Legislature, Office of the Revisor of Statutes.

State of Maine Department of Environmental Protection. 2006. *Toxics and Hazardous Waste Reduction Pollution Prevention (P2) Planning Guidebook.* Augusta, ME.

State of Rhode Island. 2009. Pollution Prevention Program. http://www.dem.ri.gov/programs/benviron/assist/pollut.htm (accessed December 16, 2009).

Superfund Amendments and Reauthorization Act (SARA) Title III, Section 313, US Congress, 1986, Washington, DC.

The Texas Commission on Environmental Quality. 2003 (January 30). Texas Administrative Code Title 30, Part 1, Chapter 335, Subchapter Q, Pollution Prevention: Source Reduction and Waste Minimization, Office of the Secretary of State, Austin, TX.

Texas Commission on Environmental Quality. 2009. About the Waste Reduction Policy Act. http://www.tceq.state.tx.us/assistance/P2Recycle/wrpa/wrpa.html (accessed December 15, 2009).

Toxics Use Reduction Institute. 2003. *TURI Overview, Working to Make Massachusetts Safer for Everyone.* Lowell: University of Massachusetts Lowell.

U.S. Environmental Protection Agency. 2006. Arizona Department of Environmental Quality. http://www.epa.gov/opptintr/p2home/pubs/casestudies/azmandatoryp2.htm (accessed December 16, 2009).

Washington State Department of Ecology. 2000. Pollution Prevention Planning. http://www.ecy.wa.gov/programs/hwtr/p2/p3.html (accessed October 31, 2006).

Washington State Department of Ecology. 2001. EPA Pollution Prevention Guide. http://www.ecy.wa.gov/programs/hwtr/P2/p3additional.html (accessed November 6, 2006).

Washington State Department of Ecology. 2006a. Pollution Prevention in Washington State. http://www.ecy.wa.gov/programs/hwtr/p2/whatisp2.html (accessed October 31, 2006).

Washington State Department of Ecology. 2006b. *Reducing Toxics in Washington, Progress Report for 2001 through 2003. Hazardous Waste and Toxics Reduction Program.* Publication Number 06-04-009. Olympia, WA.

Washington State Department of Ecology. 2010. Form 4 Dangerous Waste Annual Report, accessible through TurboWaste.Net https://fortress.wa.gov/ecy/turbowaste/Login/Splash.aspx. Olympia, WA (accessed May 23, 2010).

Washington State Hazardous Waste and Toxics Reduction Program. 2005. *Ecology Information Document: Cost Analysis for Pollution Prevention*. Publication Number 95-400.

Washington State Legislature. 2000. Chapter 173-307 WAC, Pollution Prevention Plans. http://apps.leg.wa.gov/WAC/default.aspx?cite=173–307 (accessed October 31, 2006).

14 Technical Assistance

INTRODUCTION

By providing technical assistance to companies to implement cost-effective technologies for reducing or replacing toxic chemicals, toxic chemical use reduction can be accomplished faster and more efficiently than if individual companies were required to do it on their own. This is especially true for small businesses. Technical assistance consists of providing direct assistance to companies and providing a clearinghouse for industry sectorwide toxic chemical use reduction methods. This chapter provides examples of the technical assistance program component of various existing toxic chemical use reduction programs.

MASSACHUSETTS TOXICS USE REDUCTION ACT

The Massachusetts Toxics Use Reduction Act (TURA) requires that large-quantity users of toxic materials participate in a planning process designed to reduce their wastes and use of toxic chemicals (General Laws of Massachusetts 2007). Technical assistance is provided by two programs.

The Office of Technical Assistance (OTA) is a nonregulatory agency with the Executive Office of Environmental Affairs that provides free, confidential, on-site technical and compliance consultation to manufacturers, businesses, institutions, and other toxic chemical users in Massachusetts to help reduce or eliminate their use of toxic chemicals or the generation of hazardous by-products. It also develops toxics use reduction (TUR) technologies and sponsors workshops and conferences focusing on TUR for specific industries.

The Massachusetts Toxics Use Reduction Institute (TURI), through the University of Massachusetts Lowell campus, is a multidisciplinary research, education, and policy center established by the Massachusetts TURA and is considered to be the education and research component of the TURA program. The institute sponsors and conducts research, organizes education and training programs, and provides technical support to governments to promote reduction in the use of toxic chemicals or the generation of toxic chemical by-products in industry and commerce (TURI 2003).

MAINE TOXICS AND HAZARDOUS WASTE REDUCTION PROGRAM

The Office of Innovation and Assistance (OIA) offers free compliance, technical, and pollution prevention (P2) assistance through three main programs: P2 Program, Small Business Technical Assistance Program, and the Toxics and Hazardous Waste Reduction Program. The OIA provides on-site technical P2 and small business assistance, manages innovative P2 programs, conducts literature searches and financial analysis of P2 projects, and assists in identifying cleaner technologies.

TECHNICAL RESOURCES FOR ENGINEERING EFFICIENCY PROGRAM

The Technical Resources for Engineering Efficiency (TREE) program began in 1998, when it was recognized that a system was needed to work with companies in a proactive manner to keep them from falling out of environmental compliance and to assist in reducing resource use. It is a voluntary program offered by the Washington Department of Ecology and was designed to look at the overall operations of a facility and identify specific areas that can be altered to improve both environmental performance and industrial efficiency. This involves conducting site visits, interviews, and research and writing reports. Once potential improvements are identified, it is up to the company to implement the changes. The suggested changes are targeted at inefficiencies and waste minimization at the facility. Examples of suggested changes include reusing water, chemicals, or other materials; chemical substitutions; combining multiple processes into one process; changing the order of processes in an operation; improving housekeeping and maintenance; working with outside organizations; and implementing other systematic and organizational changes.

Basis for Program

The TREE program is a voluntary process designed to benefit individual companies. Assistance is targeted to industrial facilities that do not have their own engineering staff available. TREE has worked with facilities consisting of 4 to 500 employees. If a company feels that its organization could profit from an analysis, it is their responsibility to contact the TREE team. Once a company contacts the Washington Department of Ecology, they meet for an initial evaluation. During this meeting, they discuss site specifics and the expectations of both the company and the TREE team. Based on this meeting and any other information provided by the company, the team decides whether their resources match the needs of the company and if they will be able to increase the efficiency of operations at the company. The company also decides if it is interested in pursuing the evaluation or project. Some companies that might wish to take part in this program will not be able to do so. The needs of the company have to match the resources of the team for an analysis to be effective.

Program Requirements

Once a company is selected to work with the program, it is assigned a team of three to five TREE members. The composition of the team varies depending on the specific issues at the company.

The TREE team has personnel with expertise in chemical engineering, water and wastewater treatment, solid waste minimization, water resources, air emission treatment, and spill prevention. The company that is undergoing an assessment provides man-hours and any necessary material analyses for the assessment along with any costs associated with the eventual improvements. The Washington Department of Ecology provides the man-hours of its staff. This assessment is not an environmental audit. If the TREE team identifies an environmental violation while on site, that information cannot be used in enforcement. TREE will inform the company of a violation and suggest ways to solve the issue. The TREE team will generally spend 300–350 hours working on each project. This includes time in the facility, generally 3 to 7 days; travel; research; and report writing. The assessment consists of both on-site and off-site analyses. The team inventories what the company is doing and collects any necessary information on the specific process units contained in the facility. It looks at chemicals that are the most toxic and the most used. Some of the industries are common; some are unique. The team also identifies possible changes through interviews. One of the most valuable sources of information is interviews with facility staff. Workers at the facility are intricately familiar with its processes. This includes not only management personnel.

At the end of the analysis, the team compiles a report for the facility. It is the decision of the facility whether it will implement the changes. Not all changes are implemented. This can be due to organizational or management changes within the company, budget constraints of the company, or other decision-making factors within the company.

After the analysis is complete, an estimate is developed to compare the one-time investment costs to the annual saving incurred by implementing all the changes. Although the specific numbers vary, the initial investments are generally recouped within a year. The one-time investment cost does not include man-hours spent on the analysis. The company specifies the return on investment that it would prefer, and TREE provides only those suggestions that will meet that payback period. The payback period that the companies specify is typically 18 to 24 months.

At 1 yr and 2 yrs after the analysis, follow-up interviews are conducted; qualitative information is provided. No follow-up quantitative analysis is performed.

Role of Stakeholders

The program is an effective networking tool between the Washington Department of Ecology and individual industries. It affects both groups in a positive way by building relationships and increasing environmental awareness. It also positively affects the population of the state of Washington by reducing the pollution in their environment.

Financial Impacts

The TREE program does not have its own source of funding. Participants from the Washington Department of Ecology come from various departments within the agency. Each department essentially "donates" the labor and expenses, which include local travel and some minimal equipment, from its existing budgets. The cost to participating industries is limited to the labor hours required to interact with TREE staff. The facility is not required to implement the recommendations from the program. However, the recommendations typically involve a capital investment that provides a payback within a few years. For example, the program has been in existence since 1998, and as of 2007, the program had identified potential savings representing over $1 million per year for Washington businesses.

During the analysis phase, the team evaluates methods to combine processes and reduce the use of toxic and prolific chemicals. To receive buy-in from the company, the team has to demonstrate that the changes are not only environmentally beneficial but also economically beneficial. The estimated savings of past analyses have shown up to a sevenfold return on investment within the first year. Because it is up to the companies to implement their own changes, the suggestions have to be realistic in scope.

Effectiveness of Program

The environmental and occupational health impacts based on historical recommendations have stemmed from the reduction in the use of water and the generation of hazardous waste.

The TREE team aims for a 10–15 percent design level in its evaluations. Because this is a free program, the program does not have the resources to analyze all the systems and potential systems in a facility along with all the potential side effects of any change. If the company feels that it needs this service, it can hire an outside engineering firm or vendor.

The use of an outside organization leads to a fresh look for the facilities in the program. In addition, because the program is voluntary, the assessors offer an unbiased viewpoint to the process. For each company, the analysis of the needs that is performed before the project begins focuses this program on those companies that will benefit the most. However, as the program develops, it will be helpful to expand the expertise and the marketing of the program. Smaller businesses, which are the ones that are most likely to need a program like this, are the ones that are the least likely to hear about it.

The TREE program initially consisted of two part-time staff members. Today, there is the equivalent of one to one-and-a-half full-time employees dedicated to this program. It is also expanding its scope. The Washington Department of Ecology added an air quality expert to its staff; this person will help the program expand to help companies with air quality issues. The program also feels that it would benefit from an energy expert. The fact that the program continues to grow is a testament to its success.

LEAN AND THE ENVIRONMENT PROGRAM (WASHINGTON STATE)

Lean manufacturing is the reduction of waste and the increase of efficiency in a production operation. These changes are based on economic benefits to a company and minimize any activity that does not directly contribute to the product or service that is manufactured or provided. That covers more than just materials; it could be down- or wait time, rent and utilities for unnecessary facility space, or dealing with defects. Once a production operation is streamlined, environmental benefits occur indirectly. Due to the fact that lean manufacturing does not have a specific environmental focus, it can miss opportunities for environmental source reduction.

In 2006, the U.S. Environmental Protection Agency (EPA) funded a Lean and the Environment pilot project to develop a nonregulatory approach for managing P2 and waste management at manufacturing companies to improve and sustain their competitiveness and profitability and reduce the need for regulatory compliance. The Washington Department of Ecology worked with the EPA and Washington Manufacturing Services (WMS) to provide lean environmental assistance and training to facilities in Washington State. WMS, the Washington State center for the federal Manufacturing Extension Partnership program under the National Institute of Standards and Technology (NIST MEP) is a not-for-profit organization with the goal of increasing the competitiveness of Washington manufacturers.

This pilot program has three participating companies. These companies manufacture cabinets, fiberglass tubs, and paint. This program was designed to take the principals of lean manufacturing and apply them to these companies in a way that also improves environmental management.

The goals of the pilot program are to

- Develop a collaborative partnership between the Washington State Department of Ecology and WMS
- Integrate environmental tools with lean practices
- Identify and reduce material and resource wastes and risks at three pilot companies

Basis for Program

Companies were chosen for the program based on several criteria. They had to agree to the publicity associated with the pilot program. The names of the companies and an analysis of the process at each facility will be included in both Washington Department of Ecology and WMS documentation. The companies will also be expected to testify to the state legislature regarding the effectiveness of the program. Each company chosen had to have processes and products in their operating systems that would benefit significantly from both lean and environmental improvements. If the company could only make changes in one of these areas, then it would not be considered for the program. Size and industrial classification were not directly considered but instead were factored into the potential changes. The Washington Department of Ecology and WMS contacted likely candidates within Washington and encouraged them to apply, then chose the top three applicants based on the criteria listed.

Program Requirements

Lean manufacturing is based on the idea of *kaizen*, a Japanese word adopted into English that refers to the continuous, incremental improvement of production activities. Kaizen is broken down into three phases:

Phase 1: Planning and preparation
Phase 2: Implementation
Phase 3: Follow-up

Value stream mapping (VSM) is used during the implementation phase of the kaizen process. VSM is a lean process-mapping method for understanding the sequence of activities and information flows used to produce a product or deliver a service. It involves site visits to the facility and interviews of facility personnel. Lean practitioners use VSM to

- Identify major sources of non-value-added time in a value stream
- Envision a less-wasteful future state
- Develop an implementation plan for future lean activities

Lean and the Environment work at the cabinet manufacturing facility included a week-long VSM workshop and three week-long kaizen events. Operations in the facility were broken down into two separate processes, with a team assigned to each process. All events were conducted between May and August 2006. During VSM and kaizen events, WMS facilitated the events and conducted overall project management for the site. Ecology staff was on site for every day of the lean events, and a minimum of one ecology person participated on each team at all times. The team leaders and the rest of the team participants were facility staff.

The VSM workshop at the cabinet manufacturing facility was conducted over 5 days and included four components: (1) training, (2) analysis and mapping of the "current state" of the finishing department and the production line, (3) development of "future state" value stream maps and associated implementation plans, and (4) report-out presentations from both teams and a debrief meeting. The workshop was extended by 2 days (VSM workshops are generally 3 days) to support additional process mapping and analysis of environmental wastes and costs. About 30 facility staff attended the training. Systems at the facility were broken into two specific processes, with a team for each. Eight or nine facility staff participated in each of the two teams during the rest of the week.

The kaizen events occurred over three separate weeks, with several weeks in between each kaizen. The first kaizen began with a day of training on lean methods useful for kaizen events. From that point, teams brainstormed, planned, and prioritized on actionable items for each week and accomplished the changes using a set of lean and environmental tools. At the end of the project, the cabinet manufacturing facility had implemented many of the identified changes and planned to implement more in the future.

It is assumed that the other two facilities will have similar timelines and participation requirements.

Role of Stakeholders

This program is a collaborative effort. It is a voluntary program that exists because of the potential benefit that it will provide to the state and to private facilities. It has developed through a partnership among the EPA, the Washington Department of the Environment, and WMS. There is a steering group consisting of Ecology and WMS staff, and an advisory/support group consisting of EPA staff. Each project has a project manager and a project supervisor.

The role of WMS is to provide training and assist with assessment and improvement projects. The role of the Department of Ecology is to identify environmental wastes and risks and P2 opportunities and provide regulatory assistance. These two groups, along with the EPA, provide some level of funding. The role of facility staff is to attend and participate in all analysis and implement the suggestions during and after the kaizen events.

Financial Impacts

This program is partially funded through an EPA demonstration grant. This provides funds to each of the 10 EPA regions for regional EPA offices and state environmental organizations to develop new, groundbreaking environmental projects.

For work at the cabinet manufacturing company, the Department of Ecology and the EPA spent \$58,000 on consulting time, personnel training, and overhead. Of this money, \$21,700 came from the Department of Ecology, \$31,600 came from the Pollution Prevention Grant, and \$4,700 came from EPA Innovations Office funding. The Pollution Prevention Grant is comprised of 50 percent EPA funding and 50 percent Department of Ecology funding. The cabinet manufacturing company spent a total of \$258,000. This was broken down into \$42,000 spent on labor for the lean events and \$6,000 spent on WMS services, with the rest spent on capital investments.

Because of the streamlining process that is inherent to lean manufacturing, operating costs at a facility will decrease once the program is in place. This could be due to fewer staff, a shorter processing time, less undesired product, or many other reasons. It also results in a reduced waste stream, which lowers processing, regulatory, and disposal costs. As of September 2006, the cabinet manufacturing facility was expected to save \$1,090,947 annually in cost, time, material, and environmental expenditures from actions implemented during the pilot project (May–August 2006) and was expected to save an additional \$465,618 annually from actions pending implementation.

Effectiveness of Program

The Washington State Department of Ecology Toxics Reduction Advisory Committee published a findings and recommendations document in December 2008. Between 1992 and 2006, the P2 program in Washington has reduced hazardous waste and hazardous substance use by 200 million pounds, saving over \$400 million, 11 times more than the total revenue generated from the P2 planning fee during the same time period. For every dollar businesses invested in P2, most observed a \$6 return through cost savings and efficiencies.

However, there are still some limitations to this program. It is a workload-intensive program that relies on an ongoing partnership. Additional funding sources may be required for the program to continue in the event that EPA funding is no longer available. It is dependent on facilities that have a potential for both lean and environmental improvement.

Because lean manufacturing does not specify that it will focus on environmental improvements, it sometimes does not take advantage of these opportunities. If environmental regulatory requirements are not considered in the kaizen process, the changes may not satisfy these requirements. New environmental risks or hazardous waste streams may be generated, and P2 opportunities may be lost. This program ensures that these opportunities are utilized.

Company Examples

Some examples of where manufacturers installed upgrades to their manufacturing processes to achieve a reduction in their use of toxic chemicals under the Lean and the Environment program are presented in Table 14.1. The table also shows cost reductions that have been realized by companies participating in the program.

NEW JERSEY TECHNICAL ASSISTANCE PROGRAM

The New Jersey Technical Assistance Program (NJTAP) was founded in 1990 and is currently implemented through the New Jersey Program for Manufacturing Excellence (NJME), a component of the Center for Advanced Energy Systems (CAES) at Rutgers University. NJME provides free technical assistance services to industries participating in the New Jersey P2 program. Services provided include manufacturing plant assessments that identify energy efficiency, P2, and waste minimization opportunities as well as workshops or seminars on various P2 topics, such as energy efficiency, and training on how to complete the mandated New Jersey P2 plans that are required to be submitted to the New Jersey Department of Environmental Regulation Office of Pollution Prevention (NJ DEP OPP).

The NJME conducts on-site assessments for approximately 20 facilities each year. The assessments are performed by two experienced engineers and up to three well-trained undergraduate or graduate students. During the on-site assessments, opportunities to save energy, minimize waste, prevent pollution, and improve productivity are identified.

Basis for Program

The establishment of the NJTAP was based on prescriptive legislation, the New Jersey Pollution Prevention Act, which was passed by New Jersey lawmakers in 1991. When New Jersey lawmakers created the New Jersey P2 Act, they purposely incorporated a technical assistance component into the act. This technical assistance program was not to be implemented by the NJ DEP OPP but by a nonregulatory institution that would be able to provide innovative and creative P2 solutions to industries seeking assistance with toxics reductions and P2 at their facilities. The lawmakers also specified the amount of funding, $200,000, that would be provided annually for the technical assistance program. Under the NJ P2 Act, the program was originally carried out by the Hazardous Substance Management Research Center at the New Jersey Institute of Technology. Since 2002, the program has being executed out of the NJME at Rutgers University.

Program Requirements

Although the NJME tries to focus its efforts on small-to-medium facilities, all manufacturing facilities in New Jersey are eligible, with priority given to TRI

TABLE 14.1
Lean and the Environment Program

Company Name	**Canyon Creek Cabinet Company**
Company description	Wood cabinet manufacturer
Project description	Through lean Kaizen events, employees identified production efficiency measures and improved and upgraded staining operations to reduce toxic chemical use.
Project costs to company	$258,000
Financial incentive provided	$58,000 in agency consulting costs
Annual cost savings	$1.2 million
Reduction in hazardous substance use	68,720 lb
Reduction in hazardous waste generation	84,400 lb
Reduction in air emissions	55,130 lb of solvent
Additional benefits	Increased productivity, increased quality, reduced regulatory burden, reduced worker exposure, improved ergonomics, increased employee morale, reduced floor space needed for production, reduced solid (nonhazardous waste) production
Company Name	**Columbia Paint and Coatings**
Company description	Paint and coating manufacturer
Project description	Through lean Kaizen events, employees identified production efficiency measures.
Project costs to company	$17,100
Financial incentive provided	$54,000 in agency consulting costs
Annual cost savings	$138,600
Reduction in hazardous substance use	15,000 lb paint/yr
Reduction in hazardous waste generation	2,820 lb/yr
Additional benefits	Increased productivity, increased quality, improved employee safety, improved staff morale, reduced wastewater generation

Source: Washington Lean and Environment Project Final Report, Publication Number 07-04-033, Revised August 2008, Olympia, WA: Washington State Department of Ecology.

facilities. When a facility contacts NJME for help with their P2 strategies, it is first asked to fill out a short questionnaire to help NJME prepare for the on-site assessment and to provide its specific concern to NJME regarding the facility. Facilities are also requested to provide recent energy and utility bills so that possible suggestions for energy conservation and waste reductions can also be provided with the assessment.

The on-site plant assessment is usually completed in a single day and involves the following activities:

- NJME meets with the plant personnel.
- A tour of the facility is conducted.

- A brainstorming session and discussions with plant personnel are conducted to identify savings opportunities.
- Engineering measurement and observations are noted.
- A short summary meeting at the end of the day is completed to make sure NJME has all the information needed to prepare the assessment report.

The assessment report is provided to the facility within 60 days of the on-site assessment and contains an analysis of utility bills, engineering calculations, associated dollar savings, implementation costs, and vendor information. NJME then follows up with the facility within 6 to 12 mo to check on the status of the implementation of their recommendations. It is not required that a facility carry out the recommendations provided by NJME.

Role of Stakeholders

Because the establishment of the NJTAP was primarily a directive in the P2 Act, the stakeholders, the industries and the NJ DEP, had little input into who would implement the NJTAP. Instead, the legislators decided that a nonregulatory institution would implement the program. Both New Jersey Institute of Technology (NJIT) and NJME have directly assisted the New Jersey industries with their P2 issues with no involvement from the NJ DEP. The role of NJME is to assist the industries and provide recommendations for P2 solutions. The advantage of using Rutgers University to implement the NJTAP is that it is not a regulatory institution; therefore, it is not required to enforce laws and regulations. In contrast, New Jersey law requires NJ DEP employees to report permit violations observed during facility site visits.

The use of non-NJ DEP personnel to conduct on-site assessments of facilities creates a more comfortable situation for the industries seeking P2 assistance. Industries are more inclined to seek help knowing that they will not be fined or cited for permit violations noted during an on-site assessment of their facility.

Financial Impacts

The NJ DEP OPP receives a budget, known as the Pollution Prevention Fund, from the state, as specified by the NJ P2 Act. The amount of monies funded to the NJTAP through the Pollution Prevention Fund was also specified in the NJ P2 Act. The NJTAP receives $200,000 annually from the NJ DEP OPP. Ten percent of the funding goes to the university and the department, and the remainder is used for the salaries, students, tuition, supplies, and services. The funding is received by the grants accounting office of the university, which sets up the account for NJME. NJME typical staff includes the following:

- A full-time center director and program manager (approximately 40 h/wk each)
- Two full-time project engineers, who split their time working on the NJME (approximately 10–20 h/wk each) along with other programs at CAES

- Three graduate students (approximately 10 h/wk each)
- Four undergraduate students (approximately 10 h/wk each)

It is estimated that 20 assessments per year are able to be performed with the funding received from NJ DEP OPP, at a cost of $10,000 per assessment.

An on-site assessment conducted by NJME usually yields energy and waste minimization recommendations with a payback period of less than 2 yrs. Also, because the services are nonregulatory, confidential, and cost free, and there is no obligation to act on any of the recommendations, companies are more likely to use the NJME as a technical resource for P2 and energy cost-savings solutions.

Effectiveness of Program

Since the inception of NJTAP in 1991, over 450 site visits to manufacturing facilities have been completed by both NJIT and NJME. The ability for the NJTAP to offer a facility a variety of services is a major advantage of the program and causes it to be highly effective. For example, a representative from Howmet Castings visited the NJME at Rutgers University for a class on filling out the New Jersey P2 Plan electronic application. When there, he heard the presentation on how Rutgers had assisted another company in finding a method for regenerating ferric chloride etching solutions. He invited Rutgers to evaluate the etching line at Howmet Castings, take samples, and come up with an analysis of the feasibility of regenerating the ferric chloride etching solution of the facility. Rutgers is currently performing this study, and he is impressed by it. If the technology proposed by Rutgers works and is cost effective, he hopes to reduce the waste and materials purchases of the facility. Hence, the Rutgers program has been helpful.

Students who work on the P2 projects for NJME are also able to gain real-world experience from the usually theoretical curriculum. They are able to experience a wide range of industries, which have often offered the students jobs on graduation. Those in the NJME program have published papers; student design projects have been requested by and funded by clients; and graduate students have been able to prepare a thesis through this program.

CALIFORNIA TECHNICAL ASSISTANCE PROGRAM

California's Office of Pollution Prevention and Technology and Development (OPPTD) implements outreach and education programs in the form of training sessions and technical forums that target hazardous waste generators, consultants, government employees, and the general public. The training sessions are presented in cooperation with industry associations, public interest groups, academic institutions, and other agencies of the state, federal, and local governments. The OPPTD also produces publications and videos, which are updated as technology improves and new strategies are developed (California EPA 2002).

BIBLIOGRAPHY

California Environmental Protection Agency, Department of Toxic Substances Control, Office of Pollution Prevention and Technology Development. 2002. *Guidance Manual for Complying with the Hazardous Waste Reduction and Management Review Act of 1989*. Sacramento, CA.

California Environmental Protection Agency, Office of Environmental Health Hazard Assessment. 2003. *Proposition 65 in Plain Language*. Sacramento, CA.

Center for Advanced Energy Systems. 2010. About NJME, Programs and Services. http://www.njme.rutgers.edu (accessed January 21, 2009).

Coleman, L. 2006. TREE Project: In-depth Pollution Prevention. PowerPoint presentation for the Brazilian Technology Centre for Environment Conservation. September 11, 2006.

Geiser, K., and M. Rossi. 1995. *Toxics Chemical Management in Massachusetts: The Second Report on Further Chemical Restriction Policies*. Lowell: University of Massachusetts Lowell.

General Laws of Massachusetts. (enacted 1989, amended 2006). Chapter 21I, Massachusetts Toxics Use Reduction Act. http://www.mass.gov/legis/laws/mgl/21i-1.htm (accessed May 23, 2009).

Massachusetts Department of Environmental Protection, Bureau of Waste Prevention. 2005. *2005 Toxics Use Reporting Instructions*. Boston, MA.

New Jersey State Assembly. 1991. *Pollution Prevention Act*. Assembly Appropriations Committee Statement, Senate, No. 2220-L. 1991, c. 235.

New Jersey Technical Assistance Program for Industrial Pollution Prevention. 1999. About P2. http://www.cees.njit.edu/njtap/aboutp2.htm (accessed January 21, 2009).

State of California. 2005. *Chapter 31, Waste Minimization, Article 1, Hazardous Waste Source Reduction and Management Review, §67100.4.(c) Plan and Summary*. Sacramento, CA.

State of Maine. 1989. *Title 38, Chapter 26, Toxics Use and Hazardous Waste Reduction Act*. Augusta, ME: Maine State Legislature, Office of the Revisor of Statutes.

State of Maine Department of Environmental Protection. 2006. *Toxics and Hazardous Waste Reduction Pollution Prevention (P2) Planning Guidebook*. Augusta, ME.

Toxics Reduction Advisory Committee Findings and Recommendations, December 2008 Publication no. 08-04-029. Olympia, WA: Washington State Department of Ecology.

Toxics Use Reduction Institute. 2003. *TURI Overview: Working to Make Massachusetts Safer for Everyone*. Lowell: University of Massachusetts Lowell.

U.S. Environmental Protection Agency. 2009. Lean Manufacturing and the Environment. http://www.epa.gov/lean/index.htm (accessed May 23, 2009).

Vicklund, J. H. 2006. Washington Manufacturing Services. Lean Environment Project. PowerPoint presentation for the Northwest Environmental Summit.

Washington Lean and Environment Project Final Report, Publication Number 07-04-033, Revised August 2008. Olympia, WA: Washington State Department of Ecology.

Washington Manufacturing Service. 2006. Solutions for Manufacturing Excellence. http://www.wamfg.org/default.htm (accessed November 2, 2009).

Washington State Department of Ecology. 2004. *Hazardous Waste Shoptalk—A Publication for Hazardous Waste Generators*. Vol. 14, No. 3., Publication No. 04-04-003. Olympia, WA.

Washington State Department of Ecology. 2006. Design for Lean and the Earth, a Great Value! http://www.ecy.wa.gov/programs/hwtr/shoptalkonline/2006_spring_summer/free_lean_mfg.html (accessed November 2, 2009). Olympia, WA.

Washington State Department of Ecology, Technical Resources for Engineering Efficiency (TREE) Team. 2002. *Waste Reduction Assessment for Encompass Materials Group, Ltd.* Olympia, WA.

Washington State Department of Ecology, Technical Resources for Engineering Efficiency (TREE) Team. 2004. *Focus on TREE*. Publication No. 00-04-021. Olympia, WA.

Washington State Department of Ecology, Technical Resources for Engineering Efficiency (TREE) Team. 2006. TREE Team. http://www.ecy.wa.gov/programs/hwtr/TREE/index.html (accessed November 7, 2009).

Washington State Department of Ecology and Washington Manufacturing Services. 2006. *Lean and Environment Case Study: Canyon Creek Cabinet Company.* Pollution Prevention Resources Center and Ross and Associates Environmental Consulting, Ltd. Publication Number 06-04-024. Olympia, WA.

Washington State Department of Ecology and Washington Manufacturing Services. Washington's Lean and Environment Project. Power Point Presentation. September 11, 2006.

15 Market-Based Approaches to Environmental Protection

INTRODUCTION

Initially, environmental protection programs were based on a "command-and-control" strategy. An analysis was first performed to determine how much pollution the environment could absorb without causing excessive harm. This level of pollutants was then allocated to dischargers through a permit process. Discharges up to the permitted level were accepted, whereas discharges exceeding the permitted level were permit violations subject to an administrative or criminal penalty. This approach requires considerable resources, first to determine what the acceptable level of pollution is for each local area and second for allocation by a bureaucracy of the acceptable level of pollution equitably among the various dischargers. It also takes a considerable effort to police each of these dischargers.

In addition, this approach may not lead to the most efficient attainment of the goal as it assumes that each discharger can achieve its permit limit with an efficient utilization of its resources. More important, once the individual permit limit is achieved, there is no incentive to achieve a lower discharge. There is no penalty, but likewise there is no reward for lowering the discharge below the permitted level. Perversely, there may not be an additional penalty for massive failure to achieve a limit as compared to barely missing a permit limit.

If any discharge of a pollutant is considered unacceptable, regulations have been enacted that banned the use of a given chemical. This approach was taken with DDT (dichlorodiphenyltrichloroethane), which was found to bioaccumulate and threaten reproduction of raptors, as publicized in Rachel Carson's 1962 book *Silent Spring*, and the banning of ozone-depleting chemicals such as chlorofluorocarbons.

In recent years, alternative market-based approaches for achieving environmental protection have evolved. This chapter reviews these approaches and how they might apply to a program aimed at reducing or eliminating the use of toxic chemicals.

COMMAND WITHOUT CONTROL

In command-and-control environmental protection, the overall limit on pollution is determined and allocated to each discharger through individual permits. Some dischargers may require little effort or money to achieve this limit, whereas others may require significant effort and money. Another approach is to set limits on the overall pollutant load to the environment but allow individuals to collectively remain below this overall established limit, that is, command without control or a market-based approach.

This is the approach used in setting a limit across a factory rather than through limits on individual discharge points within the factory. The overall effect of acceptable overall emissions is achieved but leaves it to the individual to find the most efficient way of achieving the limit, for example, by installing a scrubber on a major source of pollutants rather than finding and eliminating a number of much smaller fugitive emissions sources.

According to an innovative Progressive Policy Institute (PPI) policy report, it is important to adopt standards and regulations that are stringent in their goals but flexible in their means of achieving those goals (Swift 2000). Laws and policies that command but not control (i.e., "second-generation" policy) encourage the redesign of industrial processes to produce less waste instead of relying on costly equipment to clean pollutants at the end of the production process. According to the PPI, the following should be noted when establishing pollution prevention policies:

1. Established policy should continuously stimulate and reward technological innovation. Laws and regulations should be written such that they prevent pollution and at the same time drive innovation, instead of the command-and-control approach in which specific technologies to control pollution (i.e., at the end of a discharge pipe or smokestack) are mandated, often by a costly permitting process. The traditional command-and-control approaches also require reductions in rates of end-of-pipe single pollutants, as opposed to continuous innovation toward an overall cleaner process resulting in comprehensive pollutant discharge reductions. Second-generation approaches would take the following into account:
    - Implement standards that specify a set of desired environmental outcomes rather than end-of-pipe rate reductions in single pollutants
    - Eliminate outdated mandates in federal and state laws requiring specific technologies
    - Favor upstream pollution prevention (vs. downstream)
    - Establish "cap-and-trade" programs
    - Reform hazardous waste laws to promote reuse and recycling

    By offering more choices in how standards are achieved, it has been shown at both the federal and state levels that strict environmental standards can be maintained and even exceeded over time. For example, the Clean Air Act sulfur dioxide (SO_2) emissions trading program achieves major reductions using a cap on total emissions. The program also eliminates individual permit review of technology.
2. By creating the opportunity for technology innovation, and thereby avoiding the permitting process and associated regulator lack of familiarity with the new technology and time delays, along with reducing the number of enforcing jurisdictions, the chance for developing and marketing environmental technologies increases. The traditional laws and regulations that

have led to the traditional scenarios, such as technology restriction, costly and timely permitting, and numerous enforcing jurisdictions, have resulted in far less funding for environmental technology than for telecommunications, health, and general industrial sectors. In addition, a declining trend of funding for environmental technology has been seen within just the past 10 to 20 years.

As noted in the PPI report (Swift 2000), the following case study presents the benefits of applying second-generation policy to mercury reduction: More than half of the mercury releases to the environment are from the intentional and typically nonessential use of mercury in processes and products (e.g., in older chloralkali plants [160 tons], wiring [57 tons], dentistry [40 tons], lamps [29 tons], and measurement instruments [24 tons]). Substitutes are available for most products, and recycling programs that do exist only capture a small percentage of the mercury that is used. Current regulations to control mercury pollution derived from these uses focus on air emissions from waste incinerators, which is expensive and fails to address major releases through product breakage, leakage, and disposal. There have been some focused efforts to reduce mercury at its source (e.g., elimination of mercury in paints and most batteries and through some industry-driven volunteer programs), however, not on a comprehensive scale due to the control-oriented Clean Air Act regulations for air toxics. By focusing more on source reductions of mercury by all intentional users instead of focusing on emissions reductions as is done today, the environmental effectiveness would be more permanent, resulting in 100 percent elimination of mercury waste (vs. none) and would be significantly more cost effective as the regulation of waste incinerators imposes costs of $500 to $3,000 per pound of mercury reduced.

CAP AND TRADE

A wider application of the cap-and-trade approach is the United States Environmental Protection Agency (USEPA) acid rain program, which allocates emissions of oxides of nitrogen and sulfur that are combined with water to produce nitric and sulfuric acid across the dischargers in the United States. An overall "cap," or total emission level that all companies can collectively discharge, is set, and the cap is reduced over time through phases in the program. Subsequently, discharge limits are allocated to individuals across their operation as a whole instead of for individual pieces of equipment within the operation. The program lets companies decide how to achieve their allocation. For example, a company can convert some power plants to burning low-sulfur coal or can install scrubbers on one plant that achieves much lower levels of emissions than required and use that reduction as a credit against another plant with higher emissions. Finally, companies have been able to achieve lower overall emissions and sell excess emission credits to other firms. Buyers of the emission credits have found that the cost of buying credits is less expensive than installing and operating new pollution control equipment. This type of program harnesses the power of the marketplace. For industry, the program provides increased flexibility and a financial incentive to reduce air pollution beyond what laws and traditional command-and-control rules require. For the public, the program translates into cleaner air, efficiently achieved, because the public ultimately pays for pollution control through electric rates.

There are limits to cap and trade. First, a cap-and-trade approach is most effective for pollutants that are dispersed and have widespread rather than local impacts. Acid rain is a national or regional problem caused by widespread dispersion of acid-forming gases by a relatively small number of coal-fired electricity-generating stations and metal refineries. Trade across a region or even across the country can result in controlling local impacts. This is in contrast to toxic wastewater discharge, which has significant local impacts that are not taken care of by buying a credit from a distant company.

The success of the cap-and-trade approach to acid rain has led to proposals to use cap-and-trade approaches for combating greenhouse gases. As such, it would appear to be an ideal candidate for this approach since it is a problem with international impact, rather than local to any emission source. A complication with a traditional cap-and-trade program is that each emission source must be equitably allocated a specific cap, which takes considerable regulatory staff time for something with even relatively few chemicals and sources such as acid rain. While power plants and factories are significant individual dischargers of greenhouse gases, virtually everyone is a source of these emissions through our individual use of fuels for transportation and for heating individual homes.

A cap-and-trade system for toxic chemical use would be even more difficult as caps would be required for each toxic chemical and would need to be allocated to the diverse group that uses those chemicals. Finally, trade markets would need to be established for each of those chemicals.

POLLUTION TAXES OR FEES

Another market-based approach to preventing pollution is to charge taxes or fees based on the environmental impact of the discharge. As an example, for greenhouse gas, this would amount to a "carbon tax," or fee charged per pound of carbon in fuel that, when burned, would emit a given amount of greenhouse gas. This approach is simpler to administer than a cap-and-trade program because, for example, the sources of fuel would be regulated rather than all of the individual users of the fuel. In doing so, this would cover all users of the fuel (as the cost of the tax to the fuel suppliers would be reflected in the cost of the fuel to end users), including individuals, and not just the major emitters. Similarly, in the case of a toxic chemical use reduction program, the total impact cost of chemical use would be borne by the individuals enjoying the benefits of using the chemicals rather than borne by the public through medical or other toxic impacts.

There are several advantages of pollution taxes or fees to the enforcing agency over a direct control or command-and-control approach (Humboldt State University 1997). Money is generated for the program by pollution taxes or fees generate money for the program as opposed to money being spent on enforcement. Because enforcement under this market-based approach is simply based on the payment of taxes or fees, it is relatively straightforward, more immediate, and less costly compared to enforcing rules and regulations under the command-and-control approach.

In addition, program efficiency in both time and cost is gained, as opposed to a command-and-control approach, because firms have flexibility in deciding how to implement emission reductions. Specifically, those who are able to reduce pollution the cheapest will likely do so first and without specific and timely enforcement program monitoring and prosecution.

From the industry perspective, firms subject to pollution taxes or fees have an incentive to develop cleaner technologies, thus reducing their "cost of pollution" over time. Firms that pay the tax or fee and pass their costs on to customers may ultimately be confronted with a falling demand for their products provided there are cheaper, less-polluting, substitutes available. This in turn will provide firms with an incentive to alter their products and production processes to meet the altering demand (Groosman 1999).

From the industry perspective, firms mostly view market-based approaches, such as pollution taxes, to impose much greater costs on them than command-and-control policies (i.e., standards and voluntary agreements). In addition, although this holds true for any type of nationally implemented environmental regulation, from the industry perspective, implementing an environmental tax may damage the competitive position of domestic industries in comparison with international competitors. This is viewed as a temporary comparative disadvantage, however, as the imposed tax may form an incentive for companies to improve, for example, quality and the use of new technologies. In addition, a country that introduces national policies to protect the environment will be more competitive when international environmental regulation is implemented (Groosman 1999). The economic impact on the affected companies or individuals is reduced if the fee program is revenue neutral, with the fees collected returned to the users in rebates to reward reductions in emissions or grants or loans to be used to invest in equipment designed to reduce the pollution. One proposed use of a carbon tax would be a rebate to individual families, not to buy the fuel but to be invested in more fuel-efficient vehicles.

New Jersey enacted a fee system with money collected going to a Pollution Prevention Fund and not to the general treasury. Funding for pollution prevention is dedicated for that function and cannot be used for other purposes.

BIBLIOGRAPHY

Carson, R. 1962. *Silent Spring*. Boston: Houghton Mifflin.

Groosman, B. 1999. *2500 Pollution Tax*. Ghent, Belgium: Center for Environmental Economics and Management Faculty of Economics and Applied Economics, University of Ghent.

Humboldt State University. 1997. Environmental Economics: Pollution. http://sorrel.humboldt.edu/~economic/econ104/pollute/ (accessed January 21, 2010).

Swift, B. 2000. How Environmental Laws Can Discourage Pollution Prevention. http://www.ppionline.org/ppi_ci.cfm?knlgAreaID=116&subsecID=150&contentID=1159 (accessed January 21, 2010).

16 A Program to Reduce Toxic Chemical Use

INTRODUCTION

In this chapter, we propose a program for reducing the use of toxic chemicals. This program is an initial proposal that builds on the existing Toxics Release Inventory (TRI) program, and utilizes lessons learned from this program; similar programs overseas; and other U.S. state-based toxic chemical use reduction programs aimed at reducing the impacts of toxic chemicals.

The impetus and intention behind proposing the program in this book is to spur thinking among other professionals, especially federal-level policy makers regarding the next level of toxic chemical use reduction, with the ultimate goal of instituting policy updates that build on and continue the success of the TRI and other associated programs. Regardless of how successful past programs have been, to continue progress in any field, especially those concerning public health, it is the responsibility of professionals like us to reflect on what has worked and what needs to be changed based on the world as it has evolved. Our proposal is certainly not set in stone; in fact, we welcome healthy discussion and suggestions on our specific proposed program elements and what modifications, if any, would maximize the effectiveness of the next-level U.S.-based toxic chemical use reduction program. As this is an initial proposal to generate a new way of thinking and move toward certain toxic chemical use reduction policy adjustments, we also recognize that if select or even all program elements proposed in this chapter are deemed appropriate, additional refinement of details will be required as part of actual program implementation to convert these ideas on paper into actual policy. Some thoughts regarding the additional conceptual-level steps that could be taken toward this objective to further refine our initial thoughts over time as part of a formal program, as warranted, are also provided as part of this chapter.

Based on an evaluation of the various existing programs that have resulted in reducing the use of and associated exposure to toxic chemicals, we suggest that the following objectives be used to guide the development of the next-level U.S.-based toxic chemical use reduction program:

1. A toxic chemical use reduction program should target those chemicals that have the greatest adverse impact.
2. Users of products containing toxic chemicals should be made aware of, and be made responsible for obtaining the information for, the chemicals being used, the associated concentrations in the products, and the relative risks to

their health and well-being. By *user*, we mean end-use consumers as well as manufacturers or maintenance personnel who directly use the chemicals in the manufacturing process or in other products or maintain them. Currently in the United States, neither labels nor Material Safety Data Sheets (MSDSs) contains the detailed information needed by businesses and the public alike. For example, U.S. MSDSs do not contain information regarding engineered nanoparticles or their structure and potential impacts, as described by other countries and the American Society for Testing and Materials (ASTM).

3. We need a simple yardstick for quantifying the relative toxicity of each chemical that is used for producing or maintaining products.
4. We need separate accounting of the different forms of chemicals with varying toxicity. For instance, hexavalent chromium is a carcinogen, whereas metallic chromium is relatively inert. Fumes or dust of nickel and cobalt are toxic, but these chemicals are not toxic in larger particle sizes. Increasing use of nanoparticles, such as nanoscale titanium dioxide particles in sunscreen, have raised the potential for chemicals with normally low toxicity to have greater toxicity or environmental impact (Environmental Protection Agency [EPA] 2009c).
5. We need a system that accounts for worker exposure to toxic chemicals in manufacturing and quantifies amounts shipped in products in addition to the amounts reported as released to the environment.
6. The total impact cost of chemical use should be borne by the user of the chemicals, rather than by the public through medical or other toxic impacts. Including these costs would result in a free-market incentive to replace toxic chemicals with ones that are more benign but allow the continued use of chemicals that have a high ratio of benefit to total cost.
7. Any money collected from a system of loading the toxicity costs on toxic chemical use should be used to fund efforts to reduce the use of toxic chemicals or to mediate the medical costs imposed on the public by the use of these chemicals.
8. Users of toxic chemicals should be required to evaluate alternatives for reducing or eliminating their use.
9. There is value in public support for research in developing methods for eliminating the use of toxic chemicals and disseminating the technical information to the companies that could implement process and chemical changes.

These objectives were used to develop specific elements that could be components of the next U.S.-based toxic chemical use reduction program:

1. **Target Impact Chemicals:** A program that treats all chemicals as bad and targets none risks accomplishing nothing, whereas targeting the chemicals with the greatest potential toxic impact will make reduction feasible and provide the most benefit for expenditure of resources.
2. **Chemical Composition Reporting:** The concentration and physiochemical characteristics of toxic chemicals should be listed in products used in manufacturing, in maintenance, and by the public. This will facilitate

monitoring of their toxic chemical use by manufacturers and inform the public of the presence, concentration, and characteristics of these chemicals in the products they use.

3. **Chemical Toxicity Rating:** An ongoing effort should be made to determine the toxicity, mobility, persistence, and bioconcentration factors of the toxic chemicals that are used and to refine factors to assign to these chemicals. One effect of assigning toxicity values to chemicals is to have users switch to other chemicals, many of which do not currently have associated toxicity data. In addition, we need to move away from the practice of inventorying under a single category name the different forms of a chemical that have widely varying toxicities.
4. **Chemical Use Reporting:** Companies should monitor and report use or production of toxic chemicals in addition to reporting on releases to the environment. The companies that are currently required to report releases already must measure use to determine if they need to continue measuring and reporting releases. Use reporting is much easier to implement than release reporting and accounts for worker exposure, release in products (to end users), and release to the environment. Chemical use reporting should be expanded beyond the companies presently in the TRI program.
5. **Public Disclosure:** To incentivize toxic chemical use reduction, publish successes (positive public publicity) and chemical use inventory disclosure (via the Internet).
6. **Toxic Chemical Use Fee:** Base a fee on toxic chemical use multiplied by an effective toxicity factor (ETF) to place the toxic impact costs for the use of the chemical at the point where the chemical is used. This will provide an economic incentive for reduction while allowing its use for high-value uses. The fee would be proportionate to the level of toxicity adjusted by factors to account for the mobility, persistence, and tendency to bioconcentrate of the chemical.
7. **Incentives:** Any money collected as toxic chemical use fees would be used to fund research and development of alternative chemicals and processes and provide incentives to chemical users to pay for changing processes. Incentives could be in the form of low-interest loans or grants to companies to change manufacturing processes. Money could also be used to pay for the health impacts caused by the current use of toxic chemicals.
8. **Chemical Use Reduction Planning:** Companies that use toxic chemicals should be required to evaluate process changes or product reformulations that would reduce or eliminate use of toxic chemicals. Planning would consist of inventorying use of chemicals, evaluating alternative processes and materials, funding cost-effective projects, and setting toxic chemical use reduction goals.
9. **Technical Assistance:** It is in the public interest to reduce use of toxic chemicals. By having a central organization to provide information and technical assistance to implement cost-effective technologies for reducing or replacing toxic chemicals, toxic chemical use reduction can be accomplished faster and more efficiently than if individual companies were required to do it on their

own. This is especially true for small businesses. This can be accomplished by providing direct technical assistance and by providing a clearinghouse for industry sectorwide toxic chemical use reduction methods.

Further details of these nine specific proposed program elements are provided in the remainder of this chapter.

TARGET IMPACT CHEMICALS

Based on the analysis of the TRI releases in 2007 (Figure 16.1), any program should start with targeting the toxic chemicals that have the most potential toxic impact. The first 10 chemicals (of the 650 chemicals reported as part of the TRI program) represent over 99.98 percent of the estimated toxicity impact.

CHEMICAL COMPOSITION REPORTING

When a chemical product is supplied to a company user, it is typically supplied with an MSDS. An MSDS provides basic information on the properties and potential hazards of the material, how to use it safely, and what to do if there is an emergency associated with the chemical. An MSDS describes hazards associated with the material by effects such as

- Health hazards (e.g., skin contact with strong acids will cause burns)
- Fire hazards (e.g., propane burns easily and may explode)
- Reactivity hazards (e.g., mixing ammonia and household bleach will result in the release of a harmful gas)

An MSDS does not contain sufficient information on the concentrations or physiochemical characteristics of toxic chemicals contained in the product for an individual to determine the degree of toxicity or the total amount of the chemical used in the product. When MSDSs contain composition information, the information is usually limited to reporting concentrations of 1 percent or higher. More composition information is needed as a starting point in any program designed to reduce or eliminate the use of toxic chemicals.

The TRI program places the burden of inventorying toxic chemicals on the users of those chemicals. The result of the TRI program was to require product suppliers to reveal the concentrations of TRI chemicals in products sold to those companies required to report under the TRI program. This composition reporting was one of the more difficult parts of the TRI program as it was based on each user agreeing to supply the information provided by each of the suppliers of that company. Composition information is generally considered to be proprietary and is usually well protected from competitors. Think of the secret formula for Coca-Cola, for instance.

As a result of the TRI program, the hurdle of providing toxic chemical composition in products has been lowered. Chemical composition information for toxic chemicals needs to be provided to all users and end users of products containing these chemicals.

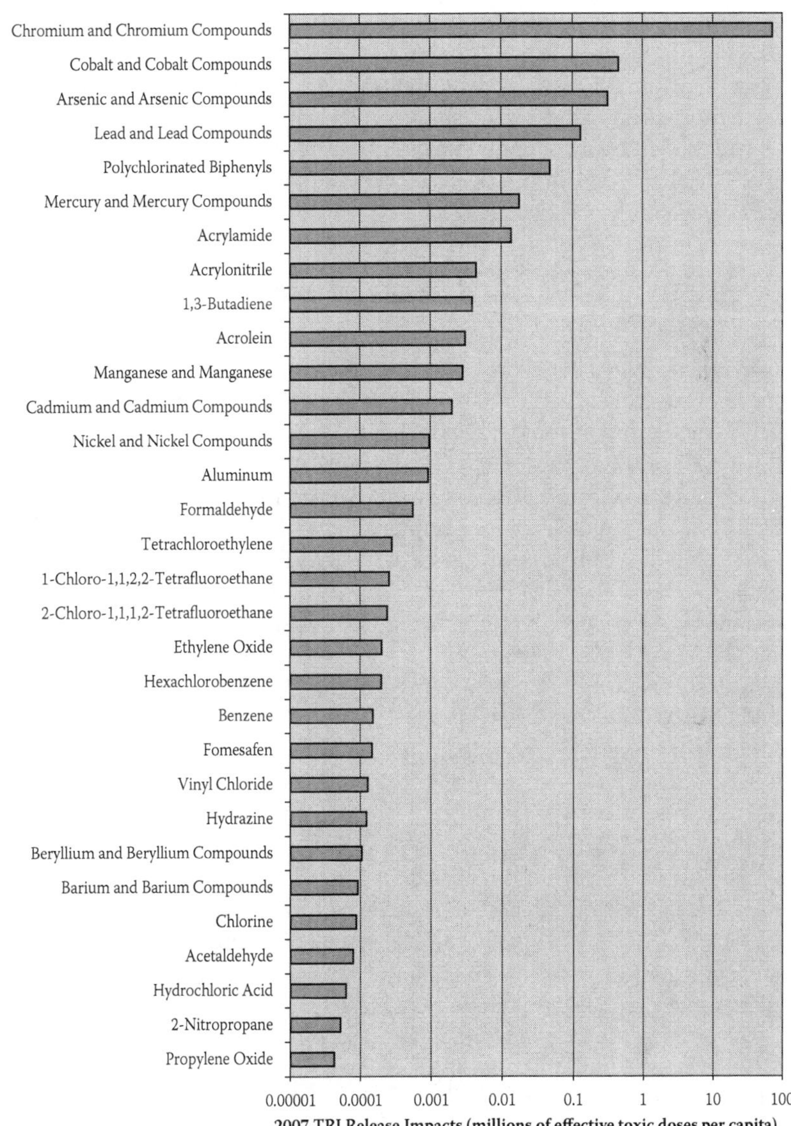

FIGURE 16.1 Analysis of the TRI releases in 2007.

It is not feasible, or even necessary, to report on trace concentrations of toxic chemicals. In other programs (e.g., European Registration, Evaluation, and Authorization of Chemicals [REACH]), a chemical component does not need to be reported if it is present in less than 0.1 percent concentration by weight. It would make sense to establish the threshold concentration for reporting based on relative toxicity factors. For example, for extremely toxic compounds, such as hexavalent chromium and dioxin, a lower concentration reporting limit or threshold would be in order.

Based on the ETFs developed in Chapter 10 of this book, we have proposed concentration reporting thresholds that could be used to report composition relative to ETFs. If an individual is exposed over the course of a year to 1 lb of a product that contains a toxic chemical at a concentration of 1 part per million (ppm), and if the chemical in this product has an ETF of 0.01 (1.00E-02) doses/capita-lb (assuming a total U.S. capita of 306 million people), the total doses that the person would be exposed to would be

$$\frac{1 \text{ lb toxic chemical}}{1{,}000{,}000} \times \frac{0.01 \text{ doses}}{\text{capita-lb}} \times 306{,}000{,}000 \text{ capita} = \frac{3 \text{ total doses}}{\text{yr}}$$

In essence, an annual equal exposure to the toxic chemical at a concentration of 1 ppm in 1 lb of the product could result in three times the acceptable threshold beyond which an adverse effect of some type could occur in every single person in the United States. An ETF resulting in this level of doses per year (on a total capita basis), or this type of potential adverse effect, warrants a lower reporting limit. Therefore, the reporting limit was set at 1 ppm for any toxic chemical with an ETF greater than 0.01 doses/capita-lb and progressively set at higher values for lower ETFs.

Specifically, as shown in Table 16.1, for ETFs higher than 0.01 (1.00E-02) doses/capita-lb, the concentration reporting threshold was set to 1 ppm. For ETFs between 1.00E-03 and 1.00E-02 doses/capita-lb, the concentration reporting threshold was set to 10 ppm, and so on, with the highest concentration reporting threshold, 10,000 ppm (1 percent), set for ETFs less than 1.00E-05 doses/capita-lb ETFs (represent those ETFs that could not be established for a compound due to lack of input data).

CHEMICAL TOXICITY RATING

In this book, we utilized available information on toxicity, mobility, persistence, and bioconcentration for the 650-plus chemicals in the TRI program. There will need to be an ongoing effort to collect and analyze similar data on other chemicals used, particularly new chemicals that are developed to replace toxic chemicals currently in use. For example, when trichloroethylene (TCE) was identified as having toxic issues, facilities that used TCE in vapor degreasers replaced TCE with trichloroethane (TCA). At the time, there was little information on the toxicity of TCA. After a significant conversion had taken place, data were developed that showed that TCA also had toxicity issues, and there was a need to modify processes. Consequently, we will need to have an ongoing program to evaluate chemical toxicity, with changes in the targeted chemicals list based on changes in use as new chemicals are developed.

One option for those chemicals that do not have corresponding toxicity data available is the application of quantitative structure-activity, or structure-property, relationships (QSARs) (Nikolova and Jaworska 2004). QSAR is the process by which a chemical structure is quantitatively correlated with a defined process, such as chemical reactivity or biological activity. More specifically, QSARs represent predictive models, mathematic relationships or quantitative structure-activity relationships, derived from application of statistical tools correlating quantitative desirable

TABLE 16.1
Proposed Chemical Usage Concentration Reporting Thresholds for TRI Chemicals

Chemical	ETF (Doses/ Capita-lb)	Reporting Threshold
(1,1′-Biphenyl)-4,4′-diamine, 3,3′-dimethyl-	5.51E–04	100 ppm
1-(3-Chloroallyl)-3,5,7-triaza-1-azoniaadamantane chloride	—	1%
1,1,1,2-Tetrachloro-2-fluoroethane (Hcfc-121a)	—	1%
1,1,1,2-Tetrachloroethane	7.64E–05	0.10%
1,1,1-Trichloroethane	1.21E–09	1%
1,1,1-Trifluoro-2,2-dichloroethane	—	1%
1,1,2,2-Tetrachloro-1-fluoroethane	—	1%
1,1,2,2-Tetrachloroethane	5.52E–05	0.10%
1,1,2-Trichloroethane	2.26E–05	0.10%
1,1′-Bi(ethylene oxide)	—	1%
1,1-Dichloro-1-fluoroethane	—	1%
1,1-Dichloroethane	7.17E–06	1%
1,1-Dichloroethylene	4.59E–08	1%
1,1-Dimethyl hydrazine	2.51E–03	10 ppm
1,1′-Methylenebis(4-isocyanatobenzene)	1.02E–06	1%
1,2,3-Trichloropropane	8.71E–04	100 ppm
1,2,4-Trichlorobenzene	4.08E–06	1%
1,2,4-Trimethylbenzene	3.49E–07	1%
1,2-Butylene oxide	6.05E–08	1%
1,2-Dibromo-3-chloropropane (Dbcp)	7.58E–03	10 ppm
1,2-Dibromoethane	4.58E–04	100 ppm
1,2-Dichloro-1,1,2-trifluoroethane	—	1%
1,2-Dichloro-1,1-difluoroethane	—	1%
1,2-Dichlorobenzene	1.04E–08	1%
1,2-Dichloroethane	3.80E–05	0.10%
1,2-Dichloroethylene	6.34E–08	1%
1,2-Dichloropropane	1.75E–05	0.10%
1,2-Diphenylhydrazine	6.20E–05	0.10%
1,3-Butadiene	2.12E–03	10 ppm
1,3-Dichloro-1,1,2,2,3-pentafluoropropane	—	1%
1,3-Dichlorobenzene	—	1%
1,3-Dichloropropene (mixed isomers)	3.70E–05	0.10%
1,4-Dichloro-2-butene	4.07E–03	10 ppm
1,4-Dichlorobenzene	1.19E–05	0.10%
1,4-Dioxane	1.79E–05	0.10%
1-Bromo-1-(bromomethyl)-1,3-propanedicarbonitrile	—	1%
1-Chloro-1,1,2,2-tetrafluoroethane	2.92E–02	1 ppm
1-Chloro-1,1-difluoroethane	6.82E–10	1%
2,2′,6,6′-Tetrabromo-4,4′-isopropylidenediphenol	6.32E–08	1%
2,3-Dichloropropene	2.43E–08	1%

continued

TABLE 16.1 (continued)
Proposed Chemical Usage Concentration Reporting Thresholds for TRI Chemicals

Chemical	ETF (Doses/ Capita-lb)	Reporting Threshold
2,4,5-Trichlorophenol	2.97E–08	1%
2,4,6-Trichlorophenol	1.13E–05	0.10%
2,4,6-Trinitrophenol	—	1%
2,4-D	1.85E–09	1%
2,4-D 2-Ethylhexyl ester	9.24E–05	0.10%
2,4-D Butoxyethyl ester	3.77E–05	0.10%
2,4-D Butyl ester	1.16E–08	1%
2,4-D Sodium salt	6.67E–05	0.10%
2,4-Db	1.45E–08	1%
2,4-Diaminoanisole	—	1%
2,4-Diaminotoluene	1.18E–03	10 ppm
2,4-Dichlorophenol	5.77E–08	1%
2,4-Dimethylphenol	5.62E–09	1%
2,4-Dinitrophenol	1.86E–08	1%
2,4-Dinitrotoluene	9.09E–05	0.10%
2,6-Dinitrotoluene	9.59E–09	1%
2,6-Xylidine	1.23E–05	0.10%
2-Acetylaminofluorene	1.14E–04	100 ppm
2-Aminonaphthalene	2.09E–04	100 ppm
2-Chlor-1,3-butadiene	4.68E–07	1%
2-Chloro-1,1,1,2-tetrafluoroethane	4.33E–04	100 ppm
2-Chloro-1,1,1-trifluoroethane	0.00E+00	1%
2-Chloroacetophenone	2.05E–05	0.10%
2-Mercaptobenzothiazole	1.53E–06	1%
2-Methyllactonitrile	1.56E–07	1%
2-Methylpyridine	7.43E–05	0.10%
2-Nitrophenol	7.25E–06	1%
2-Nitropropane	1.79E–03	10 ppm
2-Phenylphenol	1.42E–07	1%
3,3-Dichloro-1,1,1,2,2-pentafluoropropane	—	1%
3,3′-Dichlorobenzidine	1.49E–03	10 ppm
3,3′-Dichlorobenzidine dihydrochloride	4.78E–05	0.10%
3,3′-Dichlorobenzidine sulfate	1.49E–03	10 ppm
3,3′-Dimethoxybenzidine	8.34E–08	1%
3,3′-Dimethoxybenzidine dihydrochloride	8.88E–07	1%
3-Chloro-2-methyl-1-propene	1.83E–05	0.10%
3-Chloropropionitrile	—	1%
3-Iodo-2-propynyl butylcarbamate	1.17E–09	1%
4,4′-Diaminodiphenyl ether	2.56E–07	1%
4,4′-Isopropylidenediphenol	7.27E–10	1%

TABLE 16.1 (continued)
Proposed Chemical Usage Concentration Reporting Thresholds for TRI Chemicals

Chemical	ETF (Doses/ Capita-lb)	Reporting Threshold
4,4′-Methylenebis(2-chloroaniline)	1.92E–04	100 ppm
4,4′-Methylenebis(N,N-dimethyl)benzenamine	1.33E–05	0.10%
4,4′-Methylenedianiline	2.40E–04	100 ppm
4,6-Dinitro-O-cresol	1.62E–07	1%
4-Aminoazobenzene	7.65E–07	1%
4-Aminobiphenyl	2.37E–03	10 ppm
4-Dimethylaminoazobenzene	4.08E–04	100 ppm
4-Nitrophenol	1.25E–07	1%
5-Nitro-O-anisidine	1.43E–05	0.10%
5-Nitro-O-toluidine	1.71E–06	1%
Abamectin	2.75E–06	1%
Acephate	3.62E–06	1%
Acetaldehyde	6.89E–06	1%
Acetamide	3.10E–07	1%
Acetone	1.49E–10	1%
Acetonitrile	3.86E–08	1%
Acetophenone	4.09E–10	1%
Acifluorfen, sodium salt	1.16E–06	1%
Acrolein	1.81E–03	10 ppm
Acrylamide	2.22E–03	10 ppm
Acrylic acid	1.04E–06	1%
Acrylonitrile	6.18E–04	100 ppm
Alachlor	9.87E–06	1%
Aldicarb	1.24E–07	1%
Aldrin	5.96E–03	10 ppm
Allyl alcohol	5.86E–06	1%
Allyl amine	7.48E–08	1%
Allyl chloride	2.29E–05	0.10%
alpha-Naphthylamine	2.09E–04	100 ppm
Aluminum	2.28E–05	0.10%
Aluminum oxide (fibrous forms)	—	1%
Aluminum phosphide	3.72E–05	0.10%
Ametryn	7.54E–09	1%
Amitraz	4.65E–08	1%
Amitrole	2.00E–08	1%
Ammonia	8.87E–10	1%
Ammonium nitrate (solution)	—	1%
Ammonium sulfate (solution)	—	1%
Aniline	2.91E–05	0.10%

continued

TABLE 16.1 (continued)
Proposed Chemical Usage Concentration Reporting Thresholds for TRI Chemicals

Chemical	ETF (Doses/ Capita-lb)	Reporting Threshold
Anthracene	6.55E–11	1%
Antimony and antimony compounds	7.49E–08	1%
Arsenic and arsenic compounds	3.19E–03	10 ppm
Asbestos (friable)	3.80E–10	1%
Atrazine	1.08E–06	1%
Auramine	—	1%
Barium and barium compounds	3.71E–07	1%
Bendiocarb	—	1%
Benfluralin	1.06E–09	1%
Benomyl	2.32E–09	1%
Benzal chloride	1.34E–10	1%
Benzene	1.76E–05	0.10%
Benzidine	3.27E–02	1 ppm
Benzo(Ghi)perylene	2.30E–13	1%
Benzoic trichloride	1.32E–04	100 ppm
Benzoyl chloride	1.13E–10	1%
Benzoyl peroxide	5.70E–12	1%
Benzyl chloride	2.20E–05	0.10%
Beryllium and beryllium compounds	1.18E–04	100 ppm
Bifenthrin	6.13E–08	1%
Biphenyl	5.15E–10	1%
Bis(2-chloro-1-methylethyl) ether	1.55E–05	0.10%
Bis(2-chloroethoxy)methane	1.54E–08	1%
Bis(2-chloroethyl) ether	3.13E–04	100 ppm
Bis(2-ethylhexyl) adipate	1.40E–07	1%
Bis(2-ethylhexyl)phthalate	6.14E–08	1%
Bis(chloromethyl) ether	6.36E–02	1 ppm
Bis(tributyltin) oxide	2.74E–06	1%
Boron trichloride	—	1%
Boron trifluoride	4.40E–06	1%
Bromacil	1.60E–10	1%
Bromine	—	1%
Bromochlorodifluoromethane	8.73E–11	1%
Bromotrifluoromethane	6.04E–11	1%
Bromoxynil	1.67E–09	1%
Bromoxynil octanoate	1.34E–08	1%
Brucine	—	1%
Butyl acrylate	2.06E–10	1%
Butyl benzyl phthalate	2.21E–07	1%
Butyraldehyde	2.00E–11	1%

TABLE 16.1 (continued)
Proposed Chemical Usage Concentration Reporting Thresholds for TRI Chemicals

Chemical	ETF (Doses/ Capita-lb)	Reporting Threshold
C.I. acid red 114	—	1%
C.I. basic green 4	—	1%
C.I. basic red 1	—	1%
C.I. direct blue 218	2.39E–14	1%
C.I. disperse yellow 3	—	1%
C.I. food red 15	—	1%
C.I. solvent orange 7	—	1%
C.I. solvent yellow 14	—	1%
C.I. solvent yellow 3	—	1%
Cadmium and cadmium compounds	5.11E–04	100 ppm
Calcium cyanamide	—	1%
Camphechlor	6.58E–03	10 ppm
Captan	1.15E–08	1%
Carbaryl	9.72E–10	1%
Carbofuran	4.88E–09	1%
Carbon disulfide	8.61E–09	1%
Carbon tetrachloride	1.21E–04	100 ppm
Carbonyl sulfide	7.96E–12	1%
Carboxin	1.54E–10	1%
Catechol	2.46E–12	1%
Chloramben	7.74E–09	1%
Chlordane	1.63E–11	1%
Chlorendic acid	9.56E–12	1%
Chlorimuron ethyl	1.12E–07	1%
Chlorine	1.52E–05	0.10%
Chlorine dioxide	2.12E–05	0.10%
Chloroacetic acid	1.41E–07	1%
Chlorobenzene	4.15E–08	1%
Chlorobenzilate	9.30E–05	0.10%
Chlorodifluoromethane	2.28E–10	1%
Chloroethane	2.26E–10	1%
Chloroform	4.83E–05	0.10%
Chloromethane	3.07E–08	1%
Chloromethyl methyl ether	1.10E–03	10 ppm
Chlorophenols	8.22E–06	1%
Chloropicrin	3.69E–06	1%
Chlorotetrafluoroethane	1.80E–03	10 ppm
Chlorothalonil	1.39E–07	1%
Chlorotrifluoromethane	1.04E–11	1%

continued

TABLE 16.1 (continued)
Proposed Chemical Usage Concentration Reporting Thresholds for TRI Chemicals

Chemical	ETF (Doses/ Capita-lb)	Reporting Threshold
Chlorpyrifos methyl	1.03E–08	1%
Chlorsulfuron	1.54E–08	1%
Chromium, hexavalent	1.17E+00	1 ppm
Cobalt and cobalt compounds (fumes or dust)	6.37E–02	1 ppm
Copper and copper compounds	3.55E–10	1%
Creosotes	3.33E–16	1%
Cresol (mixed isomers)	1.26E–09	1%
Crotonaldehyde	3.14E–04	100 ppm
Cumene	1.50E–09	1%
Cumene hydroperoxide	3.17E–09	1%
Cupferron	—	1%
Cyanazine	3.77E–05	0.10%
Cyanide compounds	1.28E–06	1%
Cycloate	—	1%
Cyclohexane	1.36E–09	1%
Cyclohexanol	3.78E–10	1%
Cyfluthrin	6.63E–05	0.10%
Cyhalothrin	3.87E–07	1%
Dazomet	3.84E–13	1%
Dazomet, sodium salt	3.84E–13	1%
Decabromodiphenyl oxide	7.74E–08	1%
Desmedipham	2.33E–13	1%
Diallate	1.48E–05	0.10%
Diaminotoluene (mixed isomers)	—	1%
Diazinon	1.35E–07	1%
Dibenzofuran	—	1%
Dibromotetrafluoroethane (Halon 2402)	—	1%
Dibutyl phthalate	3.87E–10	1%
Dicamba	3.62E–09	1%
Dichloran	—	1%
Dichlorobenzene (mixed isomers)	1.19E–08	1%
Dichlorobromomethane	8.98E–05	0.10%
Dichlorodifluoromethane	4.30E–08	1%
Dichlorofluoromethane	1.47E–10	1%
Dichloromethane	4.50E–06	1%
Dichlorotetrafluoroethane (Cfc-114)	1.32E–07	1%
Dichlorotrifluoroethane	4.68E–12	1%
Dichlorpentafluoro-propane	9.03E–12	1%
Dichlorvos	4.02E–05	0.10%
Dicofol	—	1%

TABLE 16.1 (continued)
Proposed Chemical Usage Concentration Reporting Thresholds for TRI Chemicals

Chemical	ETF (Doses/ Capita-lb)	Reporting Threshold
Dicyclopentadiene	1.28E–07	1%
Diethanolamine	2.32E–12	1%
Diethyl phthalate	1.45E–10	1%
Diethyl sulfate	4.18E–13	1%
Diflubenzuron	5.81E–09	1%
Diglycidyl resorcinol ether (Dgre)	1.82E–12	1%
Dihydrosafrole	—	1%
Diisocyanates	5.11E–06	1%
Dimethipin	1.73E–09	1%
Dimethoate	2.06E–07	1%
Dimethyl chlorothiophosphate	—	1%
Dimethyl phthalate	1.15E–10	1%
Dimethyl sulfate	—	1%
Dimethylamine	4.28E–07	1%
Dimethylamine dicamba	1.76E–08	1%
Dimethylcarbamoyl chloride	—	1%
Dinitrobutyl phenol	2.13E–08	1%
Dinitrotoluene (mixed isomers)	1.82E–05	0.10%
Dinocap	—	1%
Di-N-propylnitrosamine	1.19E–03	10 ppm
Dioxin and dioxin-like compounds	5.80E–02	1 ppm
Diphenylamine	3.99E–10	1%
Dipotassium endothall	4.50E–10	1%
Dipropyl isocinchomeronate	—	1%
Direct black 38	2.15E–03	10 ppm
Disodium cyanodithioimidocarbonate	—	1%
Dithiobiuret	—	1%
Diuron	7.32E–09	1%
Dodine	2.90E–08	1%
D-trans-Allethrin	—	1%
Epichlorohydrin	2.60E–06	1%
Ethoprop	—	1%
Ethyl acrylate	5.10E–06	1%
Ethyl chloroformate	—	1%
Ethyl dipropylthiocarbamate	5.32E–09	1%
Ethylbenzene	1.01E–06	1%
Ethylene	—	1%
Ethylene glycol	3.74E–09	1%
Ethylene glycol monoethyl ether	7.01E–09	1%

continued

TABLE 16.1 (continued)
Proposed Chemical Usage Concentration Reporting Thresholds for TRI Chemicals

Chemical	ETF (Doses/ Capita-lb)	Reporting Threshold
Ethylene glycol monomethyl ether	1.46E–07	1%
Ethylene oxide	6.45E–04	100 ppm
Ethylene thiourea	4.77E–06	1%
Ethylenebisdithiocarbamic acid, salts, and esters	—	1%
Ethyleneimine	—	1%
Famphur	—	1%
Fenarimol	8.78E–03	10 ppm
Fenbutatin oxide	0.00E+00	1%
Fenoxycarb	7.34E–07	1%
Fenpropathrin	4.65E–09	1%
Fenthion	—	1%
Fenvalerate	—	1%
Fluazifop-butyl	—	1%
Fluometuron	1.65E–09	1%
Fluorine	5.67E–07	1%
Fluoroacetic acid, sodium salt	5.81E–06	1%
Fluorouracil	—	1%
Fluvalinate	4.65E–09	1%
Folpet	1.47E–08	1%
Fomesafen	2.09E–03	10 ppm
Formaldehyde	2.53E–05	0.10%
Formic acid	5.04E–07	1%
Freon 113	9.37E–10	1%
gamma-Lindane	8.75E–04	100 ppm
Glycol ethers	1.79E–10	1%
Heptachlor	1.86E–02	1 ppm
Hexachloro-1,3-butadiene	3.57E–03	10 ppm
Hexachlorobenzene	4.53E–03	10 ppm
Hexachlorocyclopentadiene	3.46E–06	1%
Hexachloroethane	3.47E–05	0.10%
Hexachlorophene (Hcp)	1.06E–03	10 ppm
Hexamethylphosphoramide	—	1%
Hexazinone	4.38E–09	1%
Hydramethylnon	7.00E–12	1%
Hydrazine	7.19E–03	10 ppm
Hydrazine sulfate	7.19E–03	10 ppm
Hydrochloric acid	1.23E–07	1%
Hydrofluoric acid	5.84E–08	1%
Hydrogen cyanide	1.58E–06	1%
Hydroquinone	1.47E–05	0.10%

TABLE 16.1 (continued)
Proposed Chemical Usage Concentration Reporting Thresholds for TRI Chemicals

Chemical	ETF (Doses/ Capita-lb)	Reporting Threshold
Iron pentacarbonyl	—	1%
Isobutyraldehyde	—	1%
Isodrin	—	1%
Isofenphos	—	1%
Isopropyl alcohol	8.77E–11	1%
Isosafrole	—	1%
Lactofen	6.10E–09	1%
Lead and lead compounds	2.60E–04	100 ppm
Linuron	2.91E–08	1%
Lithium carbonate	8.94E–09	1%
Malathion	5.42E–10	1%
Maleic anhydride	1.51E–07	1%
Malononitrile	2.00E–06	1%
Maneb	2.32E–08	1%
Manganese and manganese compounds	1.15E–05	0.10%
M-Cresol	1.97E–09	1%
M-Dinitrobenzene	4.81E–07	1%
Mecoprop	1.12E–07	1%
Mercury and mercury compounds	2.55E–03	10 ppm
Merphos	5.52E–07	1%
Methacrylonitrile	2.14E–06	1%
Metham sodium	—	1%
Methanamine, N-methyl-N-nitroso	3.49E–02	1 ppm
Methanol	2.76E–11	1%
Methoxone	1.72E–07	1%
Methoxychlor	5.67E–08	1%
Methyl acrylate	4.39E–09	1%
Methyl bromide	1.12E–06	1%
Methyl chlorocarbonate	—	1%
Methyl ethyl ketone	3.16E–10	1%
Methyl hydrazine	3.36E–03	10 ppm
Methyl iodide	—	1%
Methyl isobutyl ketone	1.08E–09	1%
Methyl isocyanate	4.23E–05	0.10%
Methyl isothiocyanate	1.16E–09	1%
Methyl methacrylate	9.96E–10	1%
Methyl parathion	3.79E–08	1%
Methyl tert-butyl ether	6.47E–07	1%
Methylene bromide	2.12E–08	1%

continued

TABLE 16.1 (continued)
Proposed Chemical Usage Concentration Reporting Thresholds for TRI Chemicals

Chemical	ETF (Doses/ Capita-lb)	Reporting Threshold
Metribuzin	9.18E–10	1%
Michler's ketone	—	1%
Molinate	5.81E–08	1%
Molybdenum trioxide	—	1%
Monochloropentafluoroethane	—	1%
M-Phenylenediamine	3.64E–08	1%
M-Xylene	3.82E–10	1%
Myclobutanil	1.98E–10	1%
N,N-Dimethylaniline	2.06E–08	1%
N,N-Dimethylformamide	5.65E–08	1%
Nabam	—	1%
Naled	6.18E–10	1%
Naphthalene	6.16E–06	1%
N-Butyl alcohol	6.19E–10	1%
N-Dioctyl phthalate	—	1%
N-Ethyl-N-nitrosourea	2.27E–03	10 ppm
N-Hexane	6.66E–08	1%
Nickel and nickel compounds (fumes or dust)	2.50E–05	0.10%
Nicotine and salts	—	1%
Nitrapyrin	—	1%
Nitrate compounds	1.13E–09	1%
Nitric acid	0.00E+00	1%
Nitrilotriacetic acid	—	1%
Nitrobenzene	7.68E–06	1%
Nitrofen	—	1%
Nitroglycerin	6.71E–07	1%
N-Methyl-2-pyrrolidone	—	1%
N-Methylolacrylamide	3.22E–03	10 ppm
N-Nitrosodiethylamine	6.06E–02	1 ppm
N-Nitrosodi-N-butylamine	7.73E–04	100 ppm
N-Nitrosodiphenylamine	2.21E–06	1%
N-Nitrosomethylvinylamine	—	1%
N-Nitroso-N-methylurea	2.81E–02	1 ppm
N-Nitrosopiperidine	3.04E–03	10 ppm
Norflurazon	2.52E–10	1%
O-Anisidine	6.15E–06	1%
O-Cresol	2.01E–09	1%
O-Dinitrobenzene	1.45E–07	1%
O-Phenylenediamine	3.84E–06	1%
O-Phenylphenate, sodium	—	1%

TABLE 16.1 (continued)
Proposed Chemical Usage Concentration Reporting Thresholds for TRI Chemicals

Chemical	ETF (Doses/ Capita-lb)	Reporting Threshold
Oryzalin	2.32E–09	1%
Osmium oxide Oso4 (T-4)	—	1%
O-Toluidine	2.10E–05	0.10%
O-Toluidine hydrochloride	1.23E–05	0.10%
Oxydemeton methyl	—	1%
Oxydiazon	1.74E–08	1%
Oxyfluorfen	2.41E–08	1%
O-Xylene	5.15E–10	1%
Ozone	—	1%
P-Anisidine	1.43E–05	0.10%
Paraldehyde	—	1%
Paraquat	8.31E–08	1%
Parathion	8.58E–09	1%
P-Chloroaniline	1.03E–05	0.10%
P-Cresidine	—	1%
P-Cresol	1.78E–08	1%
P-Dinitrobenzene	9.31E–08	1%
Pebulate	2.32E–09	1%
Pendimethalin	1.36E–09	1%
Pentachlorobenzene	6.81E–06	1%
Pentachloroethane	6.95E–05	0.10%
Pentachlorophenol	6.15E–06	1%
Peracetic acid	—	1%
Perchloromethyl mercaptan	—	1%
Permethrin	4.27E–09	1%
Phenanthrene	—	1%
Phenol	9.66E–09	1%
Phenothrin	—	1%
Phenytoin	—	1%
Phosgene	9.39E–06	1%
Phosphine	2.25E–05	0.10%
Phosphorus (yellow or white)	9.05E–07	1%
Phospohoric acid	6.14E–08	1%
Phthalic anhydride	4.97E–09	1%
Picloram	3.69E–10	1%
Piperonyl butoxide	—	1%
Pirimiphos methyl	1.16E–08	1%
P-Nitroaniline	3.30E–07	1%
P-Nitrosodiphenylamine	2.16E–06	1%

continued

TABLE 16.1 (continued)
Proposed Chemical Usage Concentration Reporting Thresholds for TRI Chemicals

Chemical	ETF (Doses/ Capita-lb)	Reporting Threshold
Polychlorinated alkanes (C10-C13)	9.18E–12	1%
Polychlorinated biphenyls	2.30E–02	1 ppm
Polycyclic aromatic compounds	9.63E–07	1%
Potassium bromate	—	1%
Potassium dimethyldithiocarbamate	—	1%
Potassium N-methyldithiocarbamate	—	1%
P-Phenylenediamine	4.10E–10	1%
Profenofos	—	1%
Prometryn	1.56E–08	1%
Pronamide	5.43E–10	1%
Propachlor	2.39E–09	1%
Propane sultone	—	1%
Propanil	1.29E–09	1%
Propargite	4.53E–07	1%
Propargyl alcohol	1.58E–07	1%
Propetamphos	—	1%
Propiconazole	1.28E–08	1%
Propionaldehyde	2.08E–07	1%
Propoxur	6.49E–09	1%
Propylene	—	1%
Propylene oxide	1.23E–04	100 ppm
Propyleneimine	—	1%
P-Xylene	4.61E–10	1%
Pyridine	3.68E–07	1%
Quinoline	1.00E–04	100 ppm
Quinone	—	1%
Quintozene	3.40E–05	0.10%
Quizalofop-ethyl	4.65E–10	1%
Resmethrin	3.87E–09	1%
S,S,S-Tributyltrithiophosphate	1.56E–05	0.10%
Saccharin	—	1%
Safrole	8.52E–05	0.10%
sec-Butyl alcohol	1.04E–10	1%
Selenium and selenium compounds	8.10E–07	1%
Sethoxydim	2.21E–05	0.10%
Silver and silver compounds	7.35E–10	1%
Simazine	4.53E–07	1%
Sodium azide	1.78E–08	1%
Sodium dicamba	3.07E–08	1%
Sodium dimethyldithiocarbamate	—	1%

TABLE 16.1 (continued)
Proposed Chemical Usage Concentration Reporting Thresholds for TRI Chemicals

Chemical	ETF (Doses/ Capita-lb)	Reporting Threshold
Sodium hydroxide (solution)	—	1%
Sodium nitrite	—	1%
Strychnine	2.32E–08	1%
Styrene	4.74E–10	1%
Styrene oxide	—	1%
Sulfuric acid	5.42E–12	1%
Sulfuryl fluoride	—	1%
Sulprofos	—	1%
Tebuthiuron	5.51E–10	1%
Temephos	5.81E–09	1%
Terbacil	8.93E–09	1%
Terephthalic acid	1.16E–10	1%
tert-Butyl alcohol	2.46E–10	1%
Tetrachloroethylene	1.24E–04	100 ppm
Tetrachlorvinphos	9.02E–07	1%
Tetracycline hydrochloride	—	1%
Tetramethrin	—	1%
Thallium and thallium compounds	5.83E–07	1%
Thiabendazole	—	1%
Thioacetamide	—	1%
Thiobencarb	4.33E–10	1%
Thiodicarb	3.87E–08	1%
Thiophanate-methyl	1.88E–10	1%
Thiosemicarbazide	—	1%
Thiourea	1.35E–05	0.10%
Thiram	9.21E–10	1%
Thorium dioxide	—	1%
Titanium tetrachloride	2.90E–04	100 ppm
Toluene	9.55E–10	1%
Toluene diisocyanate (mixed isomers)	6.99E–07	1%
Toluene-2,4-diisocyanate	2.00E–05	0.10%
Toluene-2,6-diisocyanate	2.08E–06	1%
trans-1,3-Dichloropropene	2.68E–05	0.10%
trans-1,4-Dichloro-2-Butene	4.29E–03	10 ppm
Triadimefon	3.87E–09	1%
Triallate	1.28E–08	1%
Tribenuron methyl	—	1%
Tribromomethane	1.22E–06	1%
Tributyltin methacrylate	3.87E–07	1%

continued

TABLE 16.1 (continued)
Proposed Chemical Usage Concentration Reporting Thresholds for TRI Chemicals

Chemical	ETF (Doses/ Capita-lb)	Reporting Threshold
Trichlorfon	—	1%
Trichloroacetyl chloride	—	1%
Trichloroethylene	6.38E–06	1%
Trichlorofluoromethane	2.88E–08	1%
Triclopyr triethylammonium salt	1.45E–09	1%
Triethylamine	2.27E–07	1%
Trifluralin	3.02E–06	1%
Triforine	—	1%
Triphenyltin chloride	—	1%
Triphenyltin hydroxide	—	1%
Tris(2,3-dibromopropyl) phosphate	—	1%
Trypan blue	—	1%
Urethane	—	1%
Vanadium and vanadium compounds	6.39E–10	1%
Vinclozolin	4.65E–09	1%
Vinyl acetate	3.48E–09	1%
Vinyl bromide	1.98E–05	0.10%
Vinyl chloride	3.36E–04	100 ppm
Warfarin and salts	8.04E–09	1%
Xylene (mixed isomers)	1.99E–08	1%
Zinc and zinc compounds	1.74E–08	1%
Zineb	2.32E–09	1%

and undesirable biological activity of chemicals with numerical descriptors representative of molecular physiochemical properties or structure. This mathematical expression can then be used to predict the biological response of other chemical structures. QSARs are currently being applied in areas that include toxicity prediction, risk assessments, and regulatory decision making (e.g., used in the EU REACH program) in addition to drug discovery. Obtaining a good-quality QSAR model depends on numerous factors, such as the quality of biological data, the choice of descriptors and statistical methods, and validation of the models. QSAR modeling should ultimately lead to statistically robust models capable of making accurate and reliable predictions of biological activities of new compounds.

In addition, the Office of Pollution Prevention and Toxics of the USEPA released a tool called the analog identification methodology (AIM) in the interest of promoting the use of safer chemical alternatives and promoting the design of safer chemicals (EPA 2009a, 2009b). Specifically, AIM was developed to assist in identifying closely related chemical structures, or analogs, for which experimental toxicity data may be publicly available to help determine the hazards of select input chemicals for which toxicity data are not available.

It is not necessary that we restrict toxicity data to that generated in the United States. Part of the European REACH program is to collect toxicity data from throughout the world and evaluate and determine the most acceptable values for toxicity to use in regulating toxic chemicals. The current U.S. toxicity data cover a range of over 12 orders of magnitude. It is better to have toxicity data that may be off an order of magnitude than to have no values at all and assume that a substance is not toxic.

Another point to consider is that the existing TRI chemicals list combines the different forms of chemicals that have widely varying toxicities. For instance, the category *chromium and chromium compounds* consists of three forms of chromium: hexavalent chromium, trivalent chromium, and metallic or elemental chromium. Hexavalent chromium is a potent carcinogen and is made from trivalent chromium-bearing ores. With very rare exceptions, metallic or trivalent chromium does not convert to hexavalent in the environment. Trivalent chromium is an essential nutrient and has a much lower toxicity. Elemental chromium is nontoxic and is used in stainless steel and highly inert products. As another example, nickel and cobalt fumes and gases are carcinogens, but the metal salts and nonaerosolized metallic cobalt and nickel are much less toxic. It would make sense to provide use reporting, and efforts to reduce use, on hexavalent chromium and gaseous or micronized nickel or cadmium. Each of the other chemicals that are categorized as highly toxic should be evaluated to ensure that the toxic forms, or forms that can be converted to toxic forms, are the forms that are individually reported for use.

Lastly, with time, an additional ecological risk factor could be incorporated into our relative chemical-specific toxicity factors to adjust for impacts on other species or other associated environmental impacts.

CHEMICAL USE REPORTING

We have noted that the existing TRI program only requires reporting on releases of the chemicals to the environment, and that this does not account for the exposure of workers to chemicals or inclusion of the chemicals in products. Companies that report on a chemical under the TRI program are required to calculate use to determine if they need to calculate releases for each of many potential emission points. Use calculation is inherently much easier with suppliers providing composition information on chemicals and purchasing agents typically keeping extensive information on amounts of chemicals purchased. Use is a simple matter of multiplying composition data by the annual use of each product. Release calculations add to the complexity of tracking each chemical to its point of use and calculating release using emission factors for each chemical and use. Reporting on use would require little additional effort and still provide a measure of all releases and worker exposure.

PUBLIC DISCLOSURE

Public reporting of the data from the TRI database has been effective in spurring reduction in the release of the reported chemicals. It notifies neighbors of the releases and therefore puts local pressure to reduce the releases. Public disclosure also highlights which companies are nationally known to release toxic chemicals, resulting in an unfavorable image of the products of these companies. Public disclosure leads to accountability to the public; that is, the public can identify the highest emitters. This provides incentive for companies to reduce toxic chemical use to avoid being among the top emitters. Publishing company successes of reduced releases of toxic chemicals would also generate favorable publicity for the companies.

Similarly, if toxic chemical use or the concentrations of toxic chemicals in products are reported, there will be pressure to reduce. Successful reduction efforts would generate favorable publicity.

TOXIC CHEMICAL USE FEE

There are several advantages for a toxic chemical use reduction program based on toxic chemical use fees. First, the only economic reason for selecting a particular chemical is the purchase price of the chemical. This does not include the downstream cost of the adverse health effects of the chemical use. By paying a fee for use that is based on the effective toxicity of the chemical, the economic analysis is then based on a better measure of the total cost of the chemical, and less costly chemicals will be favored. Moreover, a fee approach will discourage the use of chemicals for lower-value uses or uses for which there is a less-costly (purchase plus toxic chemical use fee) alternative. This also allows continued high-value uses for which there are no safer alternatives while paying for the downstream impacts. The result is a free-market reduction in overall use of a chemical while preserving use for higher-value uses, that is, those uses that are associated with high socioeconomic benefits coupled with a current lack of a safer chemical.

Finally, a toxic chemical use fee would have the effect of reducing the overall exposure of the chemicals used without the requirement to set individual limits. A toxic chemical use fee is easier and more efficient to administer than a command-and-control type of regulatory program. Can you imagine the complexity of a regulatory agency determining what an acceptable level of use would be for each chemical, for each process, for each company in the United States?

As the cost of toxic chemicals will be increased to include the total cost of using the chemicals, there will be an economic incentive to develop lower-toxicity alternatives or processes that do not require the use of higher-toxicity chemicals.

Any toxic chemical use fee collection program should be revenue neutral, with fees collected used for the following:

- Administration of the program
- Research on developing alternative chemicals and processes to reduce the use of highly toxic chemicals
- Grants or low-interest loans for companies to implement process changes needed to reduce or eliminate use of toxic chemicals
- Providing technical assistance to companies in evaluating or implementing projects to reduce use of toxic chemicals
- Funding of medical programs aimed at those adversely impacted by toxic chemical use

Typically, when toxic chemicals need to be replaced, not only does the chemical need to be replaced but also the processes employing the chemicals require upgrading to accommodate the process change. By ensuring that the toxic chemical use fee collection program is revenue neutral by reinvesting even a part of the revenue collected into the more proactive companies, those companies can implement the necessary upgrades in a way that minimizes the impacts of U.S. competitiveness in the world market (i.e., allows for the necessary upgrades with the least impact on additional costs having to be passed on to the end consumer).

Table 16.2 is a list of one possible set of toxic chemical use fees based on setting fees proportional to the ETFs (i.e., higher fee per pound of chemical for those chemicals with relatively higher toxic impacts). We made the following assumptions, solely for the purpose of proposing one way of how a toxic chemical use fee system could be established, but before such a policy is adopted, a more formal and rigorous analysis would be required:

- We assumed that there is 1/10 of 1 percent (0.1 percent) of actual human inhalation or ingestion exposure of a toxic chemical (with the remainder of the chemical released into the environment or remaining in a product and not inhaled or ingested).
- We assumed that 50 percent of the inhalation or ingestion exposure that does occur causes carcinogenic effects (vs. noncarcinogenic effects), with carcinogenic effects posing more potential adverse effects on the quality of human life (therefore, we based the toxic chemical use fee on carcinogenic effects).
- While no cancer is desirable, there is a threshold of cancer risk that is generally acceptable, and according to the USEPA, this is usually from 1 excess cancer case per 100,000 people to 1 excess cancer case per 1 million people over a 70-yr lifetime. For this analysis, we factored in 1 excess cancer risk per million people.
- We assumed that the strict monetary impact of cancer is \$1 million per individual affected.

Based on these assumptions, for 1 lb of a toxic chemical (exposure over a lifetime) that has an ETF of 1 dose/capita-lb (assume a total U.S. capita of 306 million people), the toxic chemical use fee per pound would be

$$\frac{\$1{,}000{,}000}{\text{cancer}} \times 1 \text{ lb chemical} \times \frac{1 \text{ dose}}{\text{capita-lb}} \times 0.1\% \times \frac{1}{2} \times \frac{1 \text{ cancer}}{1{,}000{,}000} \times \frac{306{,}000{,}000 \text{ capita}}{70 \text{ doses/lifetime}}$$

$$= \frac{\$2{,}186}{\text{lb chemical}}$$

Based on this rough analysis, the toxic chemical use fee structure could be based on multiplying the ETF times \$2,000. We also assumed that the maximum fee would be set at \$100/lb. Fees would range from \$0.01 to \$100/lb and would be levied on 153 of the TRI chemicals with ETFs greater than 2.0E-06. ETFs lower than this threshold are relatively innocuous and are not worth the additional effort to collect

TABLE 16.2
Potential Chemical Usage Fees for TRI Chemicals

Chemical	ETF (Doses/ Capita–lb)	Fee ($/lb)
(1,1′-Biphenyl)-4,4′-diamine, 3,3′-dimethyl-	5.51E–04	$1.10
1-(3-Chloroallyl)-3,5,7-triaza-1-azoniaadamantane chloride	—	$0.00
1,1,1,2-Tetrachloro-2-fluoroethane (Hcfc-121a)	—	$0.00
1,1,1,2-Tetrachloroethane	7.64E–05	$0.15
1,1,1-Trichloroethane	1.21E–09	$0.00
1,1,1-Trifluoro-2,2-dichloroethane	—	$0.00
1,1,2,2-Tetrachloro-1-fluoroethane	—	$0.00
1,1,2,2-Tetrachloroethane	5.52E–05	$0.11
1,1,2-Trichloroethane	2.26E–05	$0.05
1,1′-Bi(ethylene oxide)	—	$0.00
1,1-Dichloro-1-fluoroethane	—	$0.00
1,1-Dichloroethane	7.17E–06	$0.01
1,1-Dichloroethylene	4.59E–08	$0.00
1,1-Dimethyl hydrazine	2.51E–03	$5.02
1,1′-Methylenebis(4-isocyanatobenzene)	1.02E–06	$0.00
1,2,3-Trichloropropane	8.71E–04	$1.74
1,2,4-Trichlorobenzene	4.08E–06	$0.01
1,2,4-Trimethylbenzene	3.49E–07	$0.00
1,2-Butylene oxide	6.05E–08	$0.00
1,2-Dibromo-3-chloropropane (Dbcp)	7.58E–03	$15.15
1,2-Dibromoethane	4.58E–04	$0.92
1,2-Dichloro-1,1,2-trifluoroethane	—	$0.00
1,2-Dichloro-1,1-difluoroethane	—	$0.00
1,2-Dichlorobenzene	1.04E–08	$0.00
1,2-Dichloroethane	3.80E–05	$0.08
1,2-Dichloroethylene	6.34E–08	$0.00
1,2-Dichloropropane	1.75E–05	$0.04
1,2-Diphenylhydrazine	6.20E–05	$0.12
1,3-Butadiene	2.12E–03	$4.24
1,3-Dichloro-1,1,2,2,3-pentafluoropropane	—	$0.00
1,3-Dichlorobenzene	—	$0.00
1,3-Dichloropropene (mixed isomers)	3.70E–05	$0.07
1,4-Dichloro-2-butene	4.07E–03	$8.14
1,4-Dichlorobenzene	1.19E–05	$0.02
1,4-Dioxane	1.79E–05	$0.04
1-Bromo-1-(bromomethyl)-1,3-propanedicarbonitrile	—	$0.00
1-Chloro-1,1,2,2-tetrafluoroethane	2.92E–02	$58.45
1-Chloro-1,1-difluoroethane	6.82E–10	$0.00
2,2′,6,6′-Tetrabromo-4,4′-isopropylidenediphenol	6.32E–08	$0.00
2,3-Dichloropropene	2.43E–08	$0.00
2,4,5-Trichlorophenol	2.97E–08	$0.00
2,4,6-Trichlorophenol	1.13E–05	$0.02

TABLE 16.2 (continued)
Potential Chemical Usage Fees for TRI Chemicals

Chemical	ETF (Doses/ Capita–lb)	Fee ($/lb)
2,4,6-Trinitrophenol	—	$0.00
2,4-D	1.85E–09	$0.00
2,4-D 2-Ethylhexyl ester	9.24E–05	$0.18
2,4-D Butoxyethyl ester	3.77E–05	$0.08
2,4-D Butyl ester	1.16E–08	$0.00
2,4-D Sodium salt	6.67E–05	$0.13
2,4-Db	1.45E–08	$0.00
2,4-Diaminoanisole	—	$0.00
2,4-Diaminotoluene	1.18E–03	$2.35
2,4-Dichlorophenol	5.77E–08	$0.00
2,4-Dimethylphenol	5.62E–09	$0.00
2,4-Dinitrophenol	1.86E–08	$0.00
2,4-Dinitrotoluene	9.09E–05	$0.18
2,6-Dinitrotoluene	9.59E–09	$0.00
2,6-Xylidine	1.23E–05	$0.02
2-Acetylaminofluorene	1.14E–04	$0.23
2-Aminonaphthalene	2.09E–04	$0.42
2-Chlor-1,3-butadiene	4.68E–07	$0.00
2-Chloro-1,1,1,2-tetrafluoroethane	4.33E–04	$0.87
2-Chloro-1,1,1-trifluoroethane	0.00E+00	$0.00
2-Chloroacetophenone	2.05E–05	$0.04
2-Mercaptobenzothiazole	1.53E–06	$0.00
2-Methyllactonitrile	1.56E–07	$0.00
2-Methylpyridine	7.43E–05	$0.15
2-Nitrophenol	7.25E–06	$0.01
2-Nitropropane	1.79E–03	$3.57
2-Phenylphenol	1.42E–07	$0.00
3,3-Dichloro-1,1,1,2,2-pentafluoropropane	—	$0.00
3,3′-Dichlorobenzidine	1.49E–03	$2.99
3,3′-Dichlorobenzidine dihydrochloride	4.78E–05	$0.10
3,3′-Dichlorobenzidine sulfate	1.49E–03	$2.99
3,3′-Dimethoxybenzidine	8.34E–08	$0.00
3,3′-Dimethoxybenzidine dihydrochloride	8.88E–07	$0.00
3-Chloro-2-methyl-1-propene	1.83E–05	$0.04
3-Chloropropionitrile	—	$0.00
3-Iodo-2-propynyl butylcarbamate	1.17E–09	$0.00
4,4′-Diaminodiphenyl ether	2.56E–07	$0.00
4,4′-Isopropylidenediphenol	7.27E–10	$0.00
4,4′-Methylenebis(2-chloroaniline)	1.92E–04	$0.38
4,4′-Methylenebis(N,N-dimethyl)benzenamine	1.33E–05	$0.03

continued

TABLE 16.2 (continued)
Potential Chemical Usage Fees for TRI Chemicals

Chemical	ETF (Doses/ Capita–lb)	Fee ($/lb)
4,4′-Methylenedianiline	2.40E–04	$0.48
4,6-Dinitro-O-cresol	1.62E–07	$0.00
4-Aminoazobenzene	7.65E–07	$0.00
4-Aminobiphenyl	2.37E–03	$4.75
4-Dimethylaminoazobenzene	4.08E–04	$0.82
4-Nitrophenol	1.25E–07	$0.00
5-Nitro-O-anisidine	1.43E–05	$0.03
5-Nitro-O-toluidine	1.71E–06	$0.00
Abamectin	2.75E–06	$0.01
Acephate	3.62E–06	$0.01
Acetaldehyde	6.89E–06	$0.01
Acetamide	3.10E–07	$0.00
Acetone	1.49E–10	$0.00
Acetonitrile	3.86E–08	$0.00
Acetophenone	4.09E–10	$0.00
Acifluorfen, sodium salt	1.16E–06	$0.00
Acrolein	1.81E–03	$3.61
Acrylamide	2.22E–03	$4.43
Acrylic acid	1.04E–06	$0.00
Acrylonitrile	6.18E–04	$1.24
Alachlor	9.87E–06	$0.02
Aldicarb	1.24E–07	$0.00
Aldrin	5.96E–03	$11.93
Allyl alcohol	5.86E–06	$0.01
Allyl amine	7.48E–08	$0.00
Allyl chloride	2.29E–05	$0.05
alpha-Naphthylamine	2.09E–04	$0.42
Aluminum	2.28E–05	$0.05
Aluminum oxide (fibrous forms)	—	$0.00
Aluminum phosphide	3.72E–05	$0.07
Ametryn	7.54E–09	$0.00
Amitraz	4.65E–08	$0.00
Amitrole	2.00E–08	$0.00
Ammonia	8.87E–10	$0.00
Ammonium nitrate (solution)	—	$0.00
Ammonium sulfate (solution)	—	$0.00
Aniline	2.91E–05	$0.06
Anthracene	6.55E–11	$0.00
Antimony and antimony compounds	7.49E–08	$0.00
Arsenic and arsenic compounds	3.19E–03	$6.38
Asbestos (friable)	3.80E–10	$0.00
Atrazine	1.08E–06	$0.00

TABLE 16.2 (continued)
Potential Chemical Usage Fees for TRI Chemicals

Chemical	ETF (Doses/ Capita–lb)	Fee ($/lb)
Auramine	—	$0.00
Barium and barium compounds	3.71E–07	$0.00
Bendiocarb	—	$0.00
Benfluralin	1.06E–09	$0.00
Benomyl	2.32E–09	$0.00
Benzal chloride	1.34E–10	$0.00
Benzene	1.76E–05	$0.04
Benzidine	3.27E–02	$65.44
Benzo(Ghi)perylene	2.30E–13	$0.00
Benzoic trichloride	1.32E–04	$0.26
Benzoyl chloride	1.13E–10	$0.00
Benzoyl peroxide	5.70E–12	$0.00
Benzyl chloride	2.20E–05	$0.04
Beryllium and beryllium compounds	1.18E–04	$0.24
Bifenthrin	6.13E–08	$0.00
Biphenyl	5.15E–10	$0.00
Bis(2-chloro-1-methylethyl) ether	1.55E–05	$0.03
Bis(2-chloroethoxy)methane	1.54E–08	$0.00
Bis(2-chloroethyl) ether	3.13E–04	$0.63
Bis(2-ethylhexyl) adipate	1.40E–07	$0.00
Bis(2-ethylhexyl)phthalate	6.14E–08	$0.00
Bis(chloromethyl) ether	6.36E–02	$100.00
Bis(tributyltin) oxide	2.74E–06	$0.01
Boron trichloride	—	$0.00
Boron trifluoride	4.40E–06	$0.01
Bromacil	1.60E–10	$0.00
Bromine	—	$0.00
Bromochlorodifluoromethane	8.73E–11	$0.00
Bromotrifluoromethane	6.04E–11	$0.00
Bromoxynil	1.67E–09	$0.00
Bromoxynil octanoate	1.34E–08	$0.00
Brucine	—	$0.00
Butyl acrylate	2.06E–10	$0.00
Butyl benzyl phthalate	2.21E–07	$0.00
Butyraldehyde	2.00E–11	$0.00
C.I. acid red 114	—	$0.00
C.I. basic green 4	—	$0.00
C.I. basic red 1	—	$0.00
C.I. direct blue 218	2.39E–14	$0.00
C.I. disperse yellow 3	—	$0.00
C.I. food red 15	—	$0.00

continued

TABLE 16.2 (continued)
Potential Chemical Usage Fees for TRI Chemicals

Chemical	ETF (Doses/ Capita–lb)	Fee ($/lb)
C.I. solvent orange 7	—	$0.00
C.I. solvent yellow 14	—	$0.00
C.I. solvent yellow 3	—	$0.00
Cadmium and cadmium compounds	5.11E–04	$1.02
Calcium cyanamide	—	$0.00
Camphechlor	6.58E–03	$13.15
Captan	1.15E–08	$0.00
Carbaryl	9.72E–10	$0.00
Carbofuran	4.88E–09	$0.00
Carbon disulfide	8.61E–09	$0.00
Carbon tetrachloride	1.21E–04	$0.24
Carbonyl sulfide	7.96E–12	$0.00
Carboxin	1.54E–10	$0.00
Catechol	2.46E–12	$0.00
Chloramben	7.74E–09	$0.00
Chlordane	1.63E–11	$0.00
Chlorendic acid	9.56E–12	$0.00
Chlorimuron ethyl	1.12E–07	$0.00
Chlorine	1.52E–05	$0.03
Chlorine dioxide	2.12E–05	$0.04
Chloroacetic acid	1.41E–07	$0.00
Chlorobenzene	4.15E–08	$0.00
Chlorobenzilate	9.30E–05	$0.19
Chlorodifluoromethane	2.28E–10	$0.00
Chloroethane	2.26E–10	$0.00
Chloroform	4.83E–05	$0.10
Chloromethane	3.07E–08	$0.00
Chloromethyl methyl ether	1.10E–03	$2.19
Chlorophenols	8.22E–06	$0.02
Chloropicrin	3.69E–06	$0.01
Chlorotetrafluoroethane	1.80E–03	$3.60
Chlorothalonil	1.39E–07	$0.00
Chlorotrifluoromethane	1.04E–11	$0.00
Chlorpyrifos methyl	1.03E–08	$0.00
Chlorsulfuron	1.54E–08	$0.00
Chromium and chromium compounds	1.17E+00	$100.00
Cobalt and cobalt compounds	6.37E–02	$100.00
Copper and copper compounds	3.55E–10	$0.00
Creosotes	3.33E–16	$0.00
Cresol (mixed isomers)	1.26E–09	$0.00
Crotonaldehyde	3.14E–04	$0.63
Cumene	1.50E–09	$0.00

TABLE 16.2 (continued)
Potential Chemical Usage Fees for TRI Chemicals

Chemical	ETF (Doses/ Capita–lb)	Fee ($/lb)
Cumene hydroperoxide	3.17E–09	$0.00
Cupferron	—	$0.00
Cyanazine	3.77E–05	$0.08
Cyanide compounds	1.28E–06	$0.00
Cycloate	—	$0.00
Cyclohexane	1.36E–09	$0.00
Cyclohexanol	3.78E–10	$0.00
Cyfluthrin	6.63E–05	$0.13
Cyhalothrin	3.87E–07	$0.00
Dazomet	3.84E–13	$0.00
Dazomet, sodium salt	3.84E–13	$0.00
Decabromodiphenyl oxide	7.74E–08	$0.00
Desmedipham	2.33E–13	$0.00
Diallate	1.48E–05	$0.03
Diaminotoluene (mixed isomers)	—	$0.00
Diazinon	1.35E–07	$0.00
Dibenzofuran	—	$0.00
Dibromotetrafluoroethane (Halon 2402)	—	$0.00
Dibutyl phthalate	3.87E–10	$0.00
Dicamba	3.62E–09	$0.00
Dichloran	—	$0.00
Dichlorobenzene (mixed isomers)	1.19E–08	$0.00
Dichlorobromomethane	8.98E–05	$0.18
Dichlorodifluoromethane	4.30E–08	$0.00
Dichlorofluoromethane	1.47E–10	$0.00
Dichloromethane	4.50E–06	$0.01
Dichlorotetrafluoroethane (Cfc-114)	1.32E–07	$0.00
Dichlorotrifluoroethane	4.68E–12	$0.00
Dichlorpentafluoro-propane	9.03E–12	$0.00
Dichlorvos	4.02E–05	$0.08
Dicofol	—	$0.00
Dicyclopentadiene	1.28E–07	$0.00
Diethanolamine	2.32E–12	$0.00
Diethyl phthalate	1.45E–10	$0.00
Diethyl sulfate	4.18E–13	$0.00
Diflubenzuron	5.81E–09	$0.00
Diglycidyl resorcinol ether (Dgre)	1.82E–12	$0.00
Dihydrosafrole	—	$0.00
Diisocyanates	5.11E–06	$0.01
Dimethipin	1.73E–09	$0.00
Dimethoate	2.06E–07	$0.00

continued

TABLE 16.2 (continued)
Potential Chemical Usage Fees for TRI Chemicals

Chemical	ETF (Doses/ Capita–lb)	Fee ($/lb)
Dimethyl chlorothiophosphate	—	$0.00
Dimethyl phthalate	1.15E–10	$0.00
Dimethyl sulfate	—	$0.00
Dimethylamine	4.28E–07	$0.00
Dimethylamine dicamba	1.76E–08	$0.00
Dimethylcarbamoyl chloride	—	$0.00
Dinitrobutyl phenol	2.13E–08	$0.00
Dinitrotoluene (mixed isomers)	1.82E–05	$0.04
Dinocap	—	$0.00
Di-N-propylnitrosamine	1.19E–03	$2.37
Dioxin and dioxin-like compounds	5.80E–02	$100.00
Diphenylamine	3.99E–10	$0.00
Dipotassium endothall	4.50E–10	$0.00
Dipropyl isocinchomeronate	—	$0.00
Direct black 38	2.15E–03	$4.30
Disodium cyanodithioimidocarbonate	—	$0.00
Dithiobiuret	—	$0.00
Diuron	7.32E–09	$0.00
Dodine	2.90E–08	$0.00
D-trans-Allethrin	—	$0.00
Epichlorohydrin	2.60E–06	$0.01
Ethoprop	—	$0.00
Ethyl acrylate	5.10E–06	$0.01
Ethyl chloroformate	—	$0.00
Ethyl dipropylthiocarbamate	5.32E–09	$0.00
Ethylbenzene	1.01E–06	$0.00
Ethylene	—	$0.00
Ethylene glycol	3.74E–09	$0.00
Ethylene glycol monoethyl ether	7.01E–09	$0.00
Ethylene glycol monomethyl ether	1.46E–07	$0.00
Ethylene oxide	6.45E–04	$1.29
Ethylene thiourea	4.77E–06	$0.01
Ethylenebisdithiocarbamic acid, salts, and esters	—	$0.00
Ethyleneimine	—	$0.00
Famphur	—	$0.00
Fenarimol	8.78E–03	$17.56
Fenbutatin oxide	0.00E+00	$0.00
Fenoxycarb	7.34E–07	$0.00
Fenpropathrin	4.65E–09	$0.00
Fenthion	—	$0.00
Fenvalerate	—	$0.00
Fluazifop-butyl	—	$0.00

TABLE 16.2 (continued)
Potential Chemical Usage Fees for TRI Chemicals

Chemical	ETF (Doses/ Capita–lb)	Fee ($/lb)
Fluometuron	1.65E–09	$0.00
Fluorine	5.67E–07	$0.00
Fluoroacetic acid, sodium salt	5.81E–06	$0.01
Fluorouracil	—	$0.00
Fluvalinate	4.65E–09	$0.00
Folpet	1.47E–08	$0.00
Fomesafen	2.09E–03	$4.19
Formaldehyde	2.53E–05	$0.05
Formic acid	5.04E–07	$0.00
Freon 113	9.37E–10	$0.00
gamma-Lindane	8.75E–04	$1.75
Glycol ethers	1.79E–10	$0.00
Heptachlor	1.86E–02	$37.23
Hexachloro-1,3-butadiene	3.57E–03	$7.13
Hexachlorobenzene	4.53E–03	$9.05
Hexachlorocyclopentadiene	3.46E–06	$0.01
Hexachloroethane	3.47E–05	$0.07
Hexachlorophene (Hcp)	1.06E–03	$2.13
Hexamethylphosphoramide	—	$0.00
Hexazinone	4.38E–09	$0.00
Hydramethylnon	7.00E–12	$0.00
Hydrazine	7.19E–03	$14.38
Hydrazine sulfate	7.19E–03	$14.37
Hydrochloric acid	1.23E–07	$0.00
Hydrofluoric acid	5.84E–08	$0.00
Hydrogen cyanide	1.58E–06	$0.00
Hydroquinone	1.47E–05	$0.03
Iron pentacarbonyl	—	$0.00
Isobutyraldehyde	—	$0.00
Isodrin	—	$0.00
Isofenphos	—	$0.00
Isopropyl alcohol	8.77E–11	$0.00
Isosafrole	—	$0.00
Lactofen	6.10E–09	$0.00
Lead and lead compounds	2.60E–04	$0.52
Linuron	2.91E–08	$0.00
Lithium carbonate	8.94E–09	$0.00
Malathion	5.42E–10	$0.00
Maleic anhydride	1.51E–07	$0.00
Malononitrile	2.00E–06	$0.00
Maneb	2.32E–08	$0.00

continued

TABLE 16.2 (continued)
Potential Chemical Usage Fees for TRI Chemicals

Chemical	ETF (Doses/ Capita–lb)	Fee ($/lb)
Manganese and manganese compounds	1.15E–05	$0.02
M-Cresol	1.97E–09	$0.00
M-Dinitrobenzene	4.81E–07	$0.00
Mecoprop	1.12E–07	$0.00
Mercury and mercury compounds	2.55E–03	$5.09
Merphos	5.52E–07	$0.00
Methacrylonitrile	2.14E–06	$0.00
Metham sodium	—	$0.00
Methanamine, N-methyl-N-nitroso	3.49E–02	$69.89
Methanol	2.76E–11	$0.00
Methoxone	1.72E–07	$0.00
Methoxychlor	5.67E–08	$0.00
Methyl acrylate	4.39E–09	$0.00
Methyl bromide	1.12E–06	$0.00
Methyl chlorocarbonate	—	$0.00
Methyl ethyl ketone	3.16E–10	$0.00
Methyl hydrazine	3.36E–03	$6.71
Methyl iodide	—	$0.00
Methyl isobutyl ketone	1.08E–09	$0.00
Methyl isocyanate	4.23E–05	$0.08
Methyl isothiocyanate	1.16E–09	$0.00
Methyl methacrylate	9.96E–10	$0.00
Methyl parathion	3.79E–08	$0.00
Methyl tert-butyl ether	6.47E–07	$0.00
Methylene bromide	2.12E–08	$0.00
Metribuzin	9.18E–10	$0.00
Michler's ketone	—	$0.00
Molinate	5.81E–08	$0.00
Molybdenum trioxide	—	$0.00
Monochloropentafluoroethane	—	$0.00
M-Phenylenediamine	3.64E–08	$0.00
M-Xylene	3.82E–10	$0.00
Myclobutanil	1.98E–10	$0.00
N,N-Dimethylaniline	2.06E–08	$0.00
N,N-Dimethylformamide	5.65E–08	$0.00
Nabam	—	$0.00
Naled	6.18E–10	$0.00
Naphthalene	6.16E–06	$0.01
N-Butyl alcohol	6.19E–10	$0.00
N-Dioctyl phthalate	—	$0.00
N-Ethyl-N-nitrosourea	2.27E–03	$4.54
N-Hexane	6.66E–08	$0.00

TABLE 16.2 (continued)
Potential Chemical Usage Fees for TRI Chemicals

Chemical	ETF (Doses/ Capita–lb)	Fee ($/lb)
Nickel and nickel compounds	2.50E–05	$0.05
Nicotine and salts	—	$0.00
Nitrapyrin	—	$0.00
Nitrate compounds	1.13E–09	$0.00
Nitric acid	0.00E+00	$0.00
Nitrilotriacetic acid	—	$0.00
Nitrobenzene	7.68E–06	$0.02
Nitrofen	—	$0.00
Nitroglycerin	6.71E–07	$0.00
N-Methyl-2-pyrrolidone	—	$0.00
N-Methylolacrylamide	3.22E–03	$6.45
N-Nitrosodiethylamine	6.06E–02	$100.00
N-Nitrosodi-N-butylamine	7.73E–04	$1.55
N-Nitrosodiphenylamine	2.21E–06	$0.00
N-Nitrosomethylvinylamine	—	$0.00
N-Nitroso-N-methylurea	2.81E–02	$56.15
N-Nitrosopiperidine	3.04E–03	$6.08
Norflurazon	2.52E–10	$0.00
O-Anisidine	6.15E–06	$0.01
O-Cresol	2.01E–09	$0.00
O-Dinitrobenzene	1.45E–07	$0.00
O-Phenylenediamine	3.84E–06	$0.01
O-Phenylphenate, sodium	—	$0.00
Oryzalin	2.32E–09	$0.00
Osmium oxide Oso4 (T-4)	—	$0.00
O-Toluidine	2.10E–05	$0.04
O-Toluidine hydrochloride	1.23E–05	$0.02
Oxydemeton methyl	—	$0.00
Oxydiazon	1.74E–08	$0.00
Oxyfluorfen	2.41E–08	$0.00
O-Xylene	5.15E–10	$0.00
Ozone	—	$0.00
P-Anisidine	1.43E–05	$0.03
Paraldehyde	—	$0.00
Paraquat	8.31E–08	$0.00
Parathion	8.58E–09	$0.00
P-Chloroaniline	1.03E–05	$0.02
P-Cresidine	—	$0.00
P-Cresol	1.78E–08	$0.00
P-Dinitrobenzene	9.31E–08	$0.00
Pebulate	2.32E–09	$0.00

continued

TABLE 16.2 (continued)
Potential Chemical Usage Fees for TRI Chemicals

Chemical	ETF (Doses/ Capita–lb)	Fee ($/lb)
Pendimethalin	1.36E–09	$0.00
Pentachlorobenzene	6.81E–06	$0.01
Pentachloroethane	6.95E–05	$0.14
Pentachlorophenol	6.15E–06	$0.01
Peracetic acid	—	$0.00
Perchloromethyl mercaptan	—	$0.00
Permethrin	4.27E–09	$0.00
Phenanthrene	—	$0.00
Phenol	9.66E–09	$0.00
Phenothrin	—	$0.00
Phenytoin	—	$0.00
Phosgene	9.39E–06	$0.02
Phosphine	2.25E–05	$0.05
Phosphorus (yellow or white)	9.05E–07	$0.00
Phospohoric acid	6.14E–08	$0.00
Phthalic anhydride	4.97E–09	$0.00
Picloram	3.69E–10	$0.00
Piperonyl butoxide	—	$0.00
Pirimiphos methyl	1.16E–08	$0.00
P-Nitroaniline	3.30E–07	$0.00
P-Nitrosodiphenylamine	2.16E–06	$0.00
Polychlorinated alkanes (C10-C13)	9.18E–12	$0.00
Polychlorinated biphenyls	2.30E–02	$45.91
Polycyclic aromatic compounds	9.63E–07	$0.00
Potassium bromate	—	$0.00
Potassium dimethyldithiocarbamate	—	$0.00
Potassium N-methyldithiocarbamate	—	$0.00
P-Phenylenediamine	4.10E–10	$0.00
Profenofos	—	$0.00
Prometryn	1.56E–08	$0.00
Pronamide	5.43E–10	$0.00
Propachlor	2.39E–09	$0.00
Propane sultone	—	$0.00
Propanil	1.29E–09	$0.00
Propargite	4.53E–07	$0.00
Propargyl alcohol	1.58E–07	$0.00
Propetamphos	—	$0.00
Propiconazole	1.28E–08	$0.00
Propionaldehyde	2.08E–07	$0.00
Propoxur	6.49E–09	$0.00
Propylene	—	$0.00
Propylene oxide	1.23E–04	$0.25

TABLE 16.2 (continued)
Potential Chemical Usage Fees for TRI Chemicals

Chemical	ETF (Doses/ Capita–lb)	Fee ($/lb)
Propyleneimine	—	$0.00
P-Xylene	4.61E–10	$0.00
Pyridine	3.68E–07	$0.00
Quinoline	1.00E–04	$0.20
Quinone	—	$0.00
Quintozene	3.40E–05	$0.07
Quizalofop-ethyl	4.65E–10	$0.00
Resmethrin	3.87E–09	$0.00
S,S,S-Tributyltrithiophosphate	1.56E–05	$0.03
Saccharin	—	$0.00
Safrole	8.52E–05	$0.17
sec-Butyl alcohol	1.04E–10	$0.00
Selenium and selenium compounds	8.10E–07	$0.00
Sethoxydim	2.21E–05	$0.04
Silver and silver compounds	7.35E–10	$0.00
Simazine	4.53E–07	$0.00
Sodium azide	1.78E–08	$0.00
Sodium dicamba	3.07E–08	$0.00
Sodium dimethyldithiocarbamate	—	$0.00
Sodium hydroxide (solution)	—	$0.00
Sodium nitrite	—	$0.00
Strychnine	2.32E–08	$0.00
Styrene	4.74E–10	$0.00
Styrene oxide	—	$0.00
Sulfuric acid	5.42E–12	$0.00
Sulfuryl fluoride	—	$0.00
Sulprofos	—	$0.00
Tebuthiuron	5.51E–10	$0.00
Temephos	5.81E–09	$0.00
Terbacil	8.93E–09	$0.00
Terephthalic acid	1.16E–10	$0.00
tert-Butyl alcohol	2.46E–10	$0.00
Tetrachloroethylene	1.24E–04	$0.25
Tetrachlorvinphos	9.02E–07	$0.00
Tetracycline hydrochloride	—	$0.00
Tetramethrin	—	$0.00
Thallium and thallium compounds	5.83E–07	$0.00
Thiabendazole	—	$0.00
Thioacetamide	—	$0.00
Thiobencarb	4.33E–10	$0.00
Thiodicarb	3.87E–08	$0.00

continued

TABLE 16.2 (continued)
Potential Chemical Usage Fees for TRI Chemicals

Chemical	ETF (Doses/ Capita–lb)	Fee ($/lb)
Thiophanate-methyl	1.88E–10	$0.00
Thiosemicarbazide	—	$0.00
Thiourea	1.35E–05	$0.03
Thiram	9.21E–10	$0.00
Thorium dioxide	—	$0.00
Titanium tetrachloride	2.90E–04	$0.58
Toluene	9.55E–10	$0.00
Toluene diisocyanate (mixed isomers)	6.99E–07	$0.00
Toluene-2,4-diisocyanate	2.00E–05	$0.04
Toluene-2,6-diisocyanate	2.08E–06	$0.00
Trans-1,3-dichloropropene	2.68E–05	$0.05
Trans-1,4-dichloro-2-butene	4.29E–03	$8.59
Triadimefon	3.87E–09	$0.00
Triallate	1.28E–08	$0.00
Tribenuron methyl	—	$0.00
Tribromomethane	1.22E–06	$0.00
Tributyltin methacrylate	3.87E–07	$0.00
Trichlorfon	—	$0.00
Trichloroacetyl chloride	—	$0.00
Trichloroethylene	6.38E–06	$0.01
Trichlorofluoromethane	2.88E–08	$0.00
Triclopyr triethylammonium salt	1.45E–09	$0.00
Triethylamine	2.27E–07	$0.00
Trifluralin	3.02E–06	$0.01
Triforine	—	$0.00
Triphenyltin chloride	—	$0.00
Triphenyltin hydroxide	—	$0.00
Tris(2,3-dibromopropyl) phosphate	—	$0.00
Trypan blue	—	$0.00
Urethane	—	$0.00
Vanadium and vanadium compounds	6.39E–10	$0.00
Vinclozolin	4.65E–09	$0.00
Vinyl acetate	3.48E–09	$0.00
Vinyl bromide	1.98E–05	$0.04
Vinyl chloride	3.36E–04	$0.67
Warfarin and salts	8.04E–09	$0.00
Xylene (mixed isomers)	1.99E–08	$0.00
Zinc and zinc compounds	1.74E–08	$0.00
Zineb	2.32E–09	$0.00

minimal fees. Toxic chemical use fees would be levied for use of the toxic form of the compound. For chromium, this would be hexavalent chromium. For cobalt, it would be for its use as a fine grain or use that would generate a fume.

INCENTIVES

Changing a process to reduce or eliminate use of a toxic chemical can be expensive. For instance, converting a typical 275,000-ton/yr chloralkali plant, a producer of chlorine and sodium hydroxide, from mercury cells to the newer membrane cells that do not use mercury cost approximately $112 million in 2006 dollars. These conversions are needed but are hard to justify when competing for other projects that could expand the market for a company. By using toxic chemical use fees to provide low-interest loans, the fees can be leveraged, along with the prospects of reducing the fees to provide incentive for companies to invest in changes. Grants associated with fee revenue to cover a portion of conversion costs would also leverage the fees.

Another way to leverage the fees and provide an incentive for change would be to reward companies that demonstrated a reduction in the total effective toxicity associated with their toxic chemical use from year to year. One possible method would be to establish a credit to be earned based on a factor multiplied by the effective toxicity reduction percentage. For example, if the factor was set at 2, then a company that reduced its effective toxicity by 50 percent from the previous year would earn a 100 percent credit multiplied by the fees that would be imposed for the year or, in this example, end up not paying fees for that year. This type of system, which would be reset each year, rewards reductions in toxic chemical use each year based on the ability to reduce the effective toxicity compared to the previous year, encouraging an ongoing reduction in subsequent years.

CHEMICAL USE REDUCTION PLANNING

Many companies are not aware of the opportunities they have to reduce use of toxic chemicals. As noted in this book, in the case of the current TRI program, the established program chemical use thresholds require "behind-the-scene" quantifications of chemical uses to determine if reporting requirements on releases of the chemical to the environment are triggered. In our proposed toxic chemical use reduction program, the same TRI program chemical use thresholds would remain; however, they would be used to determine when chemical use reduction planning requirements are triggered. Requiring that companies using toxic chemicals above use thresholds (the same use thresholds that the current TRI program requires) evaluate alternatives to reduce use and perform a cost-benefit analysis would make the existing costs and benefits of conversion available to managers.

In one case example, one of us was performing a pollution prevention analysis of the missile division of Martin Marietta. The plant was required to clean the missile surface with a virgin cloth and TCE. Barrels of solvent-contaminated cloth were disposed by incineration, costing thousands of dollars per week. When this was pointed out to the plant manager, he found that most of the rags were being produced by staff using the convenient solvent and cloths for operations that did not require TCE, and he directed that these materials only be used for the specified purpose. Just 1 wk later, production of waste rags had dropped by over 80 percent.

Pressure to reduce mercury releases from chloralkali plants has provided an impetus to replace mercury cells with newer mercury-free membrane cells. This has resulted in 110 plants making the switch, eliminating over 50 tons/yr in mercury releases. Much of the cost in the switch has been recovered in improved energy efficiency and increased capacity.

Toxic chemical use reduction planning consists of collecting data on chemicals used and processes in which they are used; evaluating alternative, less-toxic chemicals and processes that could reduce higher-toxicity chemical use; determining technical and economic feasibility of implementing changes; performing a cost-benefit analysis; and funding of cost-effective projects that will reduce use.

As part of planning, a chemical use reduction plan would be developed and include, at a minimum, the following:

- Identification of reduction opportunities
- Identification of cost-effective projects
- Establishment of toxic chemical use reduction goals, including time frames
- Documentation of the progress of meeting the established goals

Under a toxic chemical use reduction program, a full plan would be kept on site at the company headquarters at all times, and a summary of the plan would be submitted to a regulatory agency. The purpose of submitting the plan summary and having the plan available on site is to confirm that a plan is developed and that the process identified in the plan is implemented, not for technical review of the plan.

TECHNICAL ASSISTANCE

One study found that larger firms are generally more successful at pollution prevention efforts because they integrate the pollution prevention processes into other existing management activities (e.g., quality teams). Small and midsize firms generally have fewer internal resources to complete pollution prevention activities and rely more "on external resources for identifying pollution prevention options. They tend to look at published literature, trade associations, vendors and technical assistance programs" (McLees 1995).

A technical assistance program should include three components:

1. Compliance assistance
2. Technology clearinghouse
3. Plant-specific technical assistance

Compliance assistance would focus on helping companies to understand and comply with the requirements of the toxic chemical use reduction program. As part of this assistance, data on toxicity of chemicals would be provided, along with reporting requirements and assistance in the general procedures for performing a pollution prevention opportunity assessment.

A technology clearinghouse would provide information on toxic chemical use reduction methods and best management practices in an industry sector-by-sector basis. These efforts would be based on the most toxic chemicals, similar to the state of Washington mercury effort, which provided best management practice training to hospitals and dentists on methods to eliminate mercury use.

Individual company technical assistance could be accomplished by similar types of programs successfully implemented in New Jersey and Massachusetts, where an institute, funded by fees collected from the regulated community, provides company-specific pollution prevention assessments and analysis of alternatives to reduce toxic chemical use. These have been most useful for small businesses that lack specialized technical staff. By having technical assistance provided by an entity that is not charged with enforcement of environmental regulations, the technical assistance staff can have access to the information needed to perform toxic chemical use reduction analyses without compromising a regulatory enforcement imperative.

BIBLIOGRAPHY

Mahan, S., and J. Savitz. 2007. Cleaning Up: Taking Mercury-Free Chlorine Production to the Bank. http://oceana.org/fileadmin/oceana/uploads/mercury/FINAL_Cleaning_Up.pdf (accessed January 24, 2010).

McLees, L. 1995. Pollution Prevention: People Are the Key. Georgia Tech Research News. http://gtresearchnews.gatech.edu/newsrelease/PREVENT.html (accessed January 24, 2010).

Nikolova, N., and J. Jaworska. 2004. Approaches to Measure Chemical Similarity—A Review. *QSAR and Combinatorial Science,* 22(9–10): 1006–1026.

Regulation (EC) No 1907/2006 Of the European Parliament and of the Council of 18 December 2006 concerning the Registration, Evaluation, Authorisation and Restriction of Chemicals (REACH), Annex VII, p. 167. Brussels, BE: Council of the European Union.

U.S. Environmental Protection Agency. 2009a. Analog Identification Methodology. http://aim.epa.gov./ (accessed January 24, 2010).

U.S. Environmental Protection Agency. 2009b. The Analog Identification Methodology, a Fact Sheet. http://www.epa.gov/oppt/sf/pubs/aim_factsheet.pdf (accessed January 24, 2010).

U.S. Environmental Protection Agency. 2009c. *Nanomaterial Case Studies: Nanoscale Titanium Dioxide in Water Treatment and in Topical Sunscreen.* Washington, DC: EPA/600/R-09/057.

17 Costs and Benefits

INTRODUCTION

Implementing a program to reduce the usage of toxic chemicals will entail costs to

- Set up and administer the program (by federal and administrative agencies)
- Analyze, report on, and pay fees for usage of the chemicals (by companies)

Implementing a program to reduce the usage of toxic chemicals will also have associated economic and other benefits, consisting of

- Health benefits and reduced costs associated with reduced toxic chemical usage and, hence, exposure via direct releases and through use of products themselves.
- Environmental impacts. The ultimate goal of a toxic chemical reduction program is to have a positive impact on human health and the environment.
- Direct reduced business operating costs associated with the purchase, handling, disposal, and cleanup of toxic chemicals and reduced need, and therefore cost, to install control technologies to control their emissions.
- Human capacity building—development or growth of industries that support projects related to toxic chemical reduction.
- Technology transfer and productivity enhancement—new technologies and manufacturing techniques through the exploration of process improvements.
- Positive company publicity from being environmentally conscientious and associated increased revenue.
- Reduced regulatory burden (reduced company and regulatory agency engineering labor hours as well as other associated savings) by reducing the usage of toxic chemicals.
- Other benefits, such as the establishment of spin-off or demonstration projects.

In this chapter, we evaluate these costs and benefits of setting up a market-based toxic chemical usage reduction program.

COSTS OF THE PROGRAM

The European Commission (2006) did an analysis of the costs for setting up the Registration, Evaluation, and Authorization of Chemicals (REACH) program. The costs of running the REACH program were estimated at 0.4 billion euros over 11 yrs (cost of establishing and running the program). The costs of registration, including

TABLE 17.1
Estimated Costs for REACH Registration and Testing

Cost Items	Costs (Million Euros)
Registration	€500
Testing	€1,250
Safety data sheets	€250
Authorization procedures	€100
Reduced costs for new substances below 1 ton and so on	(€100)
Total testing and registration	€2,000
Agency fees (paid by chemicals sector)	€300
Total (including agency fees)	€2,300

Source: European Commission, Environment Fact Sheet: REACH—A New Chemicals Policy for the EU, 2006.http://ec.europa.eu/environment/pubs/pdf/factsheets/reach.pdf.

the necessary testing, were estimated at 2.3 billion euros over the 11 yrs that it would take for companies to register all the substances initially covered by REACH (Table 17.1).

The cost to downstream users at the introduction of the REACH program is assessed to be in the range of 2.8 to 5.2 billion euros. These costs will occur in the form of higher chemical prices resulting from the passing along testing and registration costs, as well as additional substitution costs for downstream users of chemicals finding potentially higher-cost or less-effective replacements for those substances removed from the market.

The REACH program is a command-and-control approach to reducing toxic chemicals. Companies are required to apply for permission to use each of the regulated chemicals in manufacturing processes and need to demonstrate that there is no reasonable alternative for using that chemical. The program sets up an agency to evaluate the requests and authorize usage. This is much more expensive than the market-based approach that we are recommending.

The costs for implementing the original Toxics Release Inventory (TRI) program were considerable. Costs included the cost of setting up programs to collect and analyze information on chemical usage and for estimating releases. Companies had to develop a system for collecting composition information on each product they used and keep track of the ever-changing formulations of these products.

The EPA has estimated that the annual cost of the TRI program is $650 million per year (American Chemistry Council 2006). The Environmental Protection Agency (EPA) has traditionally underestimated the costs of compliance, so it is likely that the total cost for the program exceeds $1 billion per year. The EPA has recognized that the burden was excessive and has been trying to reduce the work required to comply with the program without reducing the value of the program (EPA 2006). We have been unable to find a good estimate of the original cost of setting up the TRI program, but it is likely to have been billions of dollars as the cost of maintaining the program is likely to be a fraction of the cost of setting it up.

A program that is based on usage reporting and control would build on rather than replace the existing TRI program. Since the TRI program already exists and requires that companies collect data on composition and calculate usage of TRI chemicals prior to calculating releases, the data collection burden would be no more than the existing TRI program. The only additional costs would be for the addition of a line item on the TRI Form Rs to report usage and for the EPA and the states to compile this additional information in the TRI database. The additional costs would be a fraction of the existing program costs.

TOXIC CHEMICAL USE FEES

The most significant cost of the program would be the fees associated with the usage of TRI chemicals. If the fees were collected as indicated by the proposed fee schedule in Chapter 16 of this book, the initial fees would be on the order of $2.6 billion, as detailed in Table 17.2 (chemicals listed in the order of the highest to lowest total fees that would be incurred). This is based on the fee structure in Chapter 16, and the TRI releases reported in 2007. The actual fees would vary from this based on a number of factors, for example, actual usage in any given year, in addition to other factors, as detailed here.

The data for chromium and chromium compounds include releases of trivalent and metallic chromium, while the toxicity and fee structure is based on hexavalent chromium. In estimating toxicity of chromium, the EPA assumed that a sixth of the releases were hexavalent chromium. Since we have recommended that hexavalent chromium be reported (and regulated) separately, we based the fees on a sixth of reported releases of chromium and chromium compounds. We also recommend that a fee structure be based on fumes and dust forms of aluminum, cobalt, manganese, and nickel since these forms of these metals are associated with their toxic effects. As a result, these compounds are not expected to be used in the quantities reported for the total released amounts of these metals. Polychlorinated biphenyls (PCBs) are no longer manufactured or used. The reported releases are likely due to replacement of these compounds in transformers and other electrical equipment. Fees could be higher than reported since the release reporting does not include total usage.

HEALTH BENEFITS OF A SUCCESSFUL TOXIC CHEMICAL USAGE REDUCTION PROGRAM

The European Commission conducted an extended impact assessment of the REACH program. Effects on gross domestic product (GDP) are expected to be limited. REACH is predicted to yield business benefits such as innovation improvements, competitiveness, improved workers' safety, and health cost savings. This was estimated to be approximately 0.05–0.09 percent of the annual sales of the chemical industry. In addition, it was estimated that if REACH were to reduce chemical-related diseases by 10 percent, the health benefits would be 50 billion euros over 30 yrs (European Commission 2006).

TABLE 17.2
Estimated Chemical Usage Fees Based on 2007 TRI Reporting

Chemical	Lb	Fee ($/lb)	Fees ($)
Chromium, hexavalent	10,067,082	$100.00	$1,006,708,157
Cobalt and cobalt compounds	6,970,293	$100.00	$697,029,310
Arsenic and arsenic compounds	97,581,160	$6.38	$622,475,379
Lead and lead compounds	495,875,564	$0.52	$258,333,521
Polychlorinated biphenyls	2,090,371	$45.91	$95,966,874
Mercury and mercury compounds	6,935,622	$5.09	$35,312,873
Acrylamide	6,161,247	$4.43	$27,312,373
Acrylonitrile	7,059,836	$1.24	$8,729,516
1,3-Butadiene	1,788,084	$4.24	$7,580,082
Acrolein	1,696,876	$3.61	$6,132,876
Manganese and manganese compounds	245,353,348	$0.02	$5,628,686
Cadmium and cadmium compounds	3,907,391	$1.02	$3,992,130
Nickel and nickel compounds	37,902,077	$0.05	$1,894,651
Aluminum	39,901,864	$0.05	$1,819,888
Formaldehyde	21,933,684	$0.05	$1,111,555
Tetrachloroethylene	2,237,864	$0.25	$554,959
1-Chloro-1,1,2,2-tetrafluoroethane	8,626	$58.45	$504,201
2-Chloro-1,1,1,2-tetrafluoroethane	551,358	$0.87	$477,195
Ethylene oxide	305,961	$1.29	$394,973
Hexachlorobenzene	43,018	$9.05	$389,499
Benzene	8,465,367	$0.04	$297,564
Fomesafen	69,115	$4.19	$289,559
Vinyl chloride	372,635	$0.67	$250,527
Hydrazine	16,759	$14.38	$241,057
Beryllium and beryllium compounds	867,078	$0.24	$205,489
Chlorine	5,643,223	$0.03	$171,819
Acetaldehyde	11,309,426	$0.01	$155,781
2-Nitropropane	28,571	$3.57	$102,054
Propylene oxide	338,598	$0.25	$83,180
Carbon tetrachloride	308,357	$0.24	$74,636
Titanium tetrachloride	123,546	$0.58	$71,667
Chloroform	706,555	$0.10	$68,251
N-Methylolacrylamide	9,276	$6.45	$59,817
Trichloroethylene	4,485,202	$0.01	$57,198
Aniline	920,606	$0.06	$53,576
Dichloromethane	5,903,242	$0.01	$53,174
N-Nitrosodiethylamine	500	$100.00	$50,000
1,4-Dichloro-2-butene	5,727	$8.14	$46,601
2,4-Diaminotoluene	18,220	$2.35	$42,827
Heptachlor	1,133	$37.23	$42,188
Naphthalene	2,850,878	$0.01	$35,129
1,2-Dichloroethane	449,853	$0.08	$34,149
4,4′-Methylenedianiline	67,423	$0.48	$32,401

TABLE 17.2 (continued)
Estimated Chemical Usage Fees Based on 2007 TRI Reporting

Chemical	Lb	Fee ($/lb)	Fees ($)
Dioxin and dioxin-like compounds	319	$100.00	$31,913
Chlorine dioxide	545,291	$0.04	$23,095
Camphechlor	1,212	$13.15	$15,946
Diisocyanates	1,472,453	$0.01	$15,059
Aldrin	1,128	$11.93	$13,457
Hydroquinone	430,989	$0.03	$12,677
Nitrobenzene	601,120	$0.02	$9,229
Chloromethyl methyl ether	3,600	$2.19	$7,891
Hexachloro-1,3-butadiene	934	$7.13	$6,658
1,4-Dioxane	185,132	$0.04	$6,645
Allyl alcohol	526,216	$0.01	$6,172
2-Methylpyridine	39,138	$0.15	$5,816
1,2-Dichloropropane	115,710	$0.04	$4,052
1,2-Dibromoethane	4,236	$0.92	$3,879
1,2-Dibromo-3-chloropropane (Dbcp)	255	$15.15	$3,864
Pentachloroethane	27,513	$0.14	$3,826
Quinoline	15,825	$0.20	$3,180
N-Nitrosopiperidine	500	$6.08	$3,042
gamma-Lindane	1,555	$1.75	$2,722
1,2,3-Trichloropropane	1,474	$1.74	$2,566
Crotonaldehyde	4,008	$0.63	$2,519
2,4-Dinitrotoluene	13,594	$0.18	$2,472
4,4′-Methylenebis(2-chloroaniline)	6,233	$0.38	$2,399
1,4-Dichlorobenzene	79,266	$0.02	$1,892
Di-N-propylnitrosamine	751	$2.37	$1,782
Allyl chloride	35,188	$0.05	$1,611
Hexachlorophene (Hcp)	690	$2.13	$1,469
Aluminum phosphide	15,468	$0.07	$1,152
Benzidine	16	$65.44	$1,029
1,1,2-Trichloroethane	22,367	$0.05	$1,009
Epichlorohydrin	155,813	$0.01	$809
Chlorophenols	49,196	$0.02	$809
Ethyl acrylate	78,135	$0.01	$797
N-Nitrosodi-N-butylamine	500	$1.55	$773
Toluene-2,4-diisocyanate	18,955	$0.04	$760
O-Toluidine	16,348	$0.04	$688
Benzyl chloride	13,323	$0.04	$587
N-Nitroso-N-methylurea	10	$56.15	$561
2-Nitrophenol	33,232	$0.01	$482
1,3-Dichloropropene (mixed isomers)	5,695	$0.07	$422
1,1,1,2-Tetrachloroethane	2,249	$0.15	$344
Phosgene	15,290	$0.02	$287

continued

TABLE 17.2 (continued)
Estimated Chemical Usage Fees Based on 2007 TRI Reporting

Chemical	Lb	Fee ($/lb)	Fees ($)
3-Chloro-2-methyl-1-propene	6,536	$0.04	$239
Bis(2-chloroethyl) ether	347	$0.63	$218
1,2,4-Trichlorobenzene	26,339	$0.01	$215
2,4-D 2-Ethylhexyl ester	1,158	$0.18	$214
Quintozene	3,115	$0.07	$212
4-Dimethylaminoazobenzene	256	$0.82	$209
1,1,2,2-Tetrachloroethane	1,861	$0.11	$205
2-Acetylaminofluorene	750	$0.23	$172
Safrole	1,000	$0.17	$170
Benzoic trichloride	646	$0.26	$170
3,3′-Dichlorobenzidine dihydrochloride	1,565	$0.10	$150
Dinitrotoluene (mixed isomers)	4,103	$0.04	$150
Dichlorvos	1,715	$0.08	$138
Hexachloroethane	1,751	$0.07	$121
Boron trifluoride	13,391	$0.01	$118
Methyl isocyanate	1,259	$0.08	$106
1,1-Dichloroethane	7,253	$0.01	$104
O-Phenylenediamine	12,849	$0.01	$99
1,1-Dimethyl hydrazine	15	$5.02	$77
Chlorotetrafluoroethane	15	$3.60	$55
Dichlorobromomethane	296	$0.18	$53
4-Aminobiphenyl	11	$4.75	$52
N-Ethyl-N-nitrosourea	10	$4.54	$45
Trifluralin	7,295	$0.01	$44
Hexachlorocyclopentadiene	5,990	$0.01	$41
Thiourea	1,333	$0.03	$36
Pentachlorophenol	2,740	$0.01	$34
2,4-D Butoxyethyl ester	327	$0.08	$25
Bis(2-chloro-1-methylethyl) ether	788	$0.03	$25
Chloropicrin	3,081	$0.01	$23
trans-1,3-Dichloropropene	389	$0.05	$21
Pentachlorobenzene	1,464	$0.01	$20
Ethylene thiourea	1,945	$0.01	$19
P-Chloroaniline	761	$0.02	$16
Cyanazine	189	$0.08	$14
Acephate	1,736	$0.01	$13
2,4,6-Trichlorophenol	513	$0.02	$12
(1,1′-Biphenyl)-4,4′-diamine, 3,3′-dimethyl-	10	$1.10	$11
Bis(tributyltin) oxide	2,001	$0.01	$11
O-Anisidine	638	$0.01	$8
Diallate	255	$0.03	$8
Alachlor	373	$0.02	$7

TABLE 17.2 (continued)
Estimated Chemical Usage Fees Based on 2007 TRI Reporting

Chemical	Lb	Fee ($/lb)	Fees ($)
Chlorobenzilate	32	$0.19	$6
Cyfluthrin	34	$0.13	$5
trans-1,4-Dichloro-2-butene	1	$8.59	$4
1,2-Diphenylhydrazine	10	$0.12	$1
2,4-D Sodium salt	9	$0.13	$1

Source: U.S. Environmental Protection Agency, Toxics Release Inventory Burden Reduction Final Rule. *Federal Register*, 71(246), 76932, 2006.

Although potential cost savings cannot be quantified without a more detailed study, evidence in the United States points to the fact that the health and social benefits of enforcing tough new clean air regulations during the past decade were five to seven times greater in economic terms than were the costs of complying with the rules. One study ("Study Finds Net Gain from Pollution Rules; OMB Overturns Past Findings on Benefit" 2003, A1) noted the following:

> The value of reductions in hospitalization and emergency room visits, premature deaths and lost workdays resulting from improved air quality were estimated between $120 billion and $193 billion from October 1992 to September 2002. By comparison, industry, states and municipalities spent an estimated $23 billion to $26 billion to retrofit plants and facilities and make other changes to comply with new clean-air standards, which are designed to sharply reduce sulfur dioxide, fine particle emissions and other health-threatening pollutants.

In 2004, the direct cost of cancer treatment was estimated to be $72.1 billion. Direct medical expenditures are only one component of the total economic burden of cancer. In addition to the social impact, the indirect costs include losses in time and economic productivity resulting from cancer-related illness and death. The total economic cost of cancer in 2004 is estimated to have been $190 billion.

ENVIRONMENTAL IMPACTS

The ultimate goal of a toxic chemical reduction program is to have a positive impact on human health and the environment. As demonstrated throughout this book, various toxic chemical reduction programs in the United States and elsewhere have achieved these goals and had positive impacts on the environment. In addition, other environmental impacts can be realized, as detailed here.

Other environmental benefits from toxic chemical reduction program implementation could include changes in the level of energy and water consumption at the facility. Although some process improvements may result in increased water or energy usage at a facility, many of the case studies reviewed (including technical assistance programs in New Jersey and Washington) resulted in dramatic reductions in water and energy usage because potential efficiency opportunities were identified while the manufacturing processes were reviewed.

In addition, there may be opportunities to reduce greenhouse gas emissions.

Examples of links between toxic substance reduction with environmental quality improvement include the dichlorodiphenyltrichloroethane (DDT) ban on pelican/large-bird populations as well as the reduction of airborne lead with the implementation of lead-free gasoline in countries. DDT is the best known of a number of chlorine-containing pesticides used in the 1940s and 1950s. Among many other adverse effects, it became known to cause eggshell thinning in large-bird populations, including pelicans. After the ban of DDT in the early 1970s in the United States, brown pelican populations started to recover almost immediately in the southeast as residues and eggshell thinning declined and productivity increased (Blus 2007).

DIRECT BENEFITS TO BUSINESSES

Companies that have reduced or eliminated their use of toxic chemicals have noticed reduced operating costs associated with their purchase, handling, disposal, and cleanup. The most frequent reason given by attendees of the 1997 EPA Region III TRI workshop to undertake waste reduction activities was cost reduction (98 percent of respondents). Although results varied at each facility, most of the case studies showed substantial reductions in chemical use and chemical emissions, with a payback period of 2 yrs or less. Some other examples are also noted here.

The Haartz Corporation, located in Acton, Massachusetts, makes coated fabrics used in automobiles and estimated a savings of $200,000 annually by reducing its methyl ethyl ketone releases.

Recommendations provided by the state of Washington Technical Resources for Engineering Efficiency (TREE) program typically involve a capital investment that provides a payback within a few years. For example, the program has been in existence since 1998, and as of 2007, the program had identified potential savings representing over $1 million per year for Washington businesses. The estimated savings of past analyses have shown up to a sevenfold return of investment within the first year.

In the Lean and the Environment Program, as of September 2006, the cabinet manufacturing facility was expected to save $1,090,947 annually in cost, time, material, and environmental savings from actions implemented during the pilot project (May to August 2006) and was expected to save an additional $465,618 annually from actions pending implementation.

Also, as noted in the 2005 Indiana Pollution Prevention Annual Report (Office of Pollution Prevention and Technical Assistance 2005), Ryobi Die Casting Incorporated, a company that makes transmission castings for automobile manufacturers, voluntarily implemented various pollution prevention (P2) projects, including reduction of aluminum waste (dross) production. This in turn reduced the natural gas use of the company by 27 percent, or 34,440 million British thermal units (BTUs) per year, and reduced the dross production by 759,600 lb annually. Annual energy and material cost savings of this project amounted to nearly $900,000. The company received a loan to begin this project.

According to the Wisconsin Department of Natural Resources (DNR) Environmental Cooperation Pilot Program: 2007 Progress Report (October 31, 2007), from 2000 to 2005, the 3M Company (Menomonie site) reduced TRI releases per pound of good output by 54 percent, volatile organic compounds (VOCs) per pound of good output by 26 percent, waste per pound of good output by 18 percent, and reduced hazardous waste levels per pound by good output by 10 percent. They also implemented 14 Pollution Prevention Pays projects, resulting in savings of $2.67 million, and prevented 2,603 tons of pollution. Separately, they also prevented 24 tons of pollution (includes recycling, reuse, reformulation, and replacement), resulting in a total savings of $26,516 and energy savings of 19.32 million BTUs, among other beneficial outcomes.

Reichhold Chemical Incorporated manufactures polyester resin in Oxnard, California. The company installed equipment to condense, recover, and reduce leakages of methylene chloride. This reduced annual usage and emissions of this toxic chemical by over 49,000 lb from 1991 levels. Reichhold reported savings on raw materials and anticipates saving on long-term waste disposal.

In a survey of participating facilities conducted by the Massachusetts Toxic Use Reduction Program, 67 percent of the respondents saw some cost savings from the toxics use planning and reduction program in the state.

HUMAN CAPACITY BUILDING

One additional market consequence of implementing a toxic chemical usage reduction program could be the development or growth of industries that support projects related to toxic chemical reduction. These industries include toxic chemical usage reduction consulting services and nonprofit technical assistance (as described in Chapter 16) as well as equipment providers that would have the potential to provide their services to businesses in need, thereby further fueling economic growth. The development or growth of these support industries could further encourage participation in the toxic chemical usage reduction program because these companies would have a market incentive to promote the use of their products and services.

As part of the state of Michigan governor's economic plan, the Cool Cities Program helps revitalize Michigan cities by retaining and attracting the jobs and people critical to emerging economies while reducing pollution (Michigan Department of Environmental Quality 2005). Communities across the state create vibrant, attractive places for people to live, work, and play. P2 opportunities for Cool Cities include (but are not limited to) recycling, "green" building development, and landscaping for water quality. Additional P2 tools that Cool Cities can make use of are the Retired Engineer Technical Assistance Program (RETAP), RETAP Student Internship Program, and the Small Business P2 Loan Program. As of 2005, the first year participants in the program reported, the designation helped create 400 new jobs and retain 500 existing jobs and enabled 19 projects to have priority access to more than $100 million in existing grants, loans, and other resources.

In addition, the chloralkali industry is the largest user (and has the largest inventory of) mercury. This industry has moved away from mercury cells and into a membrane process. Making this type of change in Brazil would provide membrane and related equipment sales opportunities.

TECHNOLOGY TRANSFER AND PRODUCTIVITY ENHANCEMENT

The implementation of toxic chemical usage reduction programs will inevitably lead to new technologies and manufacturing techniques through the exploration of process improvements that could include

- New technologies required to implement material substitutions (new cleaning equipment, coating application methods, drying techniques, etc.).
- Process improvements for reducing the need for toxic chemicals (reducing the soiling of parts to reduce the need for cleaning).
- Beneficial substance substitution (replacement of toxic chemicals with ones that are less toxic and that do not generate additional pollution in transport, increased energy use, or the generation of more toxic waste products). An example includes the installation of aqueous parts washing equipment to replace vapor degreasers and eliminate the use of chlorinated solvents to clean parts before metal finishing (plating or painting).

Furthermore, process improvements examined during the review of manufacturing methods may lead to increased efficiency and productivity. One method by which efficiencies could be gained is the implementation of lean manufacturing techniques, by which all aspects of the manufacturing process are examined to improve efficiencies. Productivity enhancements from toxic chemical use reduction projects could include

- Reduced labor from the handling of toxic materials
- Improved health of employees due to reduced exposure to toxic chemicals (more productive on the job, fewer sick days, fewer trips to the doctor, etc.)
- Reduced need for personal protective equipment

There is a current debate about the potentials and limitations of using regulation to promote innovation and technology transfer and enhance productivity for the environment. A number of analysts have proposed models for how environmental regulation can promote innovation and be more effective in the process. The most famous hypothesis argues that the right kind of regulation can lead to competitive advantages for firms taking early and decisive action to improve "resource productivity." Others have criticized this limited view of "innovation-friendly regulation" as motivating only incremental innovations in pollution control, arguing instead that regulation should drive "radical innovation." Others argue that a key to moving forward to spur real innovation requires the development of policies for "social learning." These policies move beyond specific strategies (such as market mechanisms and P2) and should create or realign relationships to collectively solve environmental problems.

Taken together, these hypotheses provide insights into strategies to support innovation, technology transfer, and enhancement of productivity by focusing on performance outcomes (rather than standards). These insights mandate that firms conduct some form of self-evaluation and planning processes, that they employ market incentives (such as economic cost-benefit analyses of technology options), and promote "preventive" and proactive approaches to solving environmental problems. At a minimum, this literature argues that regulators can support firm innovations, technology transfer, and productivity enhancement through technical assistance and pooling of learning among firms.

POSITIVE PUBLICITY AND ASSOCIATED INCREASED REVENUE

Companies that have voluntarily reduced toxic chemicals used by their facilities have benefited economically by being viewed as environmentally conscientious. In a survey of participating facilities conducted by the Massachusetts Toxic Use Reduction Program, 39 percent of the respondents saw some benefit to the improved environmental image they received as a result of participation in the program. Although the majority of respondents did not see any benefit, increasing public awareness of environmental issues may result in the increased value of products that are deemed to be more environmentally responsible.

The 2005 Indiana Pollution Prevention Annual Report (Office of Pollution Prevention and Technical Assistance 2005) reported that a dry cleaning company in Evansville, Indiana, converted from a hazardous dry cleaning solvent, perchloroethylene, to a nonhazardous, environmentally friendly solvent. To do so, the company had to invest $225,000 in new equipment, and the new solvent costs the company an extra $9 per gallon. However, the company has absorbed the extra costs without raising prices for the customer, with the expectation that customers will appreciate the product benefits and do more business as a result. The company was a Pollution Prevention/Source Reduction award winner.

REDUCED REGULATORY BURDEN

There can be significant costs associated with regulatory compliance, both for individual companies and for regulatory agencies; however, the costs are typically more than recovered in other ways. One of us was recently working with Quantico Marine Corps Base on a treatment plant that needed to reduce total nitrogen discharges to the Chesapeake Bay. An organic substrate was to be added to the wastewater so that nitrates could be biologically converted to gaseous nitrogen. Typically, clients find that methanol is the most cost-effective chemical for this purpose. The client chose to use acetic acid, despite it costing twice the cost of methanol, because acetic acid is less toxic (and not regulated under the TRI program). However, the reduced regulatory cost more than made up for the additional cost for chemical for the base. As a sidelight on the TRI program, if the client were a municipality, rather than a Marine Corps Base, it would not have needed to report methanol releases because municipal treatment plants are exempt from TRI reporting.

Reducing the regulatory burden has been an effective approach for encouraging companies to reduce pollution. According to the Wisconsin DNR Environmental Cooperation Pilot Program: 2007 Progress Report (October 31, 2007), between 1996 and 2006, the Northern Engraving Corporation (NEC) reduced VOC use by

63 percent, hazardous air pollutants (HAPs) by 94 percent, water use by 74 percent, hazardous waste by 69 percent, and solid waste by 78 percent. In 2002, the company and the Environmental Cooperative Compliance Program entered into an agreement that included an innovative regulatory approach. This resulted in a decrease from roughly 2.5 construction permits being issued to the NEC annually prior to 2002 to an average of 1 permit per year after 2002. This resulted in a savings of over 1,000 hrs to state engineers and 3,000 h annually to the company in reduced paperwork during the first year of the agreement (as well as approximately 4,000 pages of paper and 2,500 million cubic feet of natural gas at two different locations during the first year of the agreement).

In addition, in 2004, due to the commitments of 3M to environmental improvement, the DNR and EPA approved a flexible permit (under Title V Part III) for 3M (Menomonie site). From 2005 to October 2007, 12 new construction projects were submitted and approved without going through traditional permitting processes, 18 mo of startup time were saved, and 315 hrs of 3M administrative time and over 1,500 h of DNR time were saved.

OTHER BENEFITS

Other benefits of a toxic chemical usage reduction program could be the establishment of spin-off or demonstration projects. There are numerous examples of these types of projects that are either in place or currently being set up in the United States (e.g., Oregon, Kentucky, Arizona, and New Hampshire). One of the projects that are highlighted here, however, has a different aspect to it than the others.

The California Assembly Bill (AB) 2588 Air Toxics "Hot Spots" program (California Environmental Resources Board 2010) is a spin-off of the air toxics program in California. The Air Toxics "Hot Spots" Information and Assessment Act (AB 2588, 1987, Connelly) was enacted in September 1987. This law and implementing regulations have done more to reduce air toxics air emissions than any other law or regulation. What makes the program unique is that the Air Toxics "Hot Spots" Act establishes a formal air toxic emission inventory risk quantification program for districts to manage. The goal of the Air Toxics "Hot Spots" Act is to collect emission data indicative of routine predictable releases of toxic substances to the air, to identify facilities having localized impacts, to evaluate health risks from exposure to the emissions, and to notify nearby residents of significant risks. Senate Bill (SB) 1731, which amends the "Hot Spots" program, added a requirement that facilities determined to have a significant risk must conduct an airborne toxic risk reduction audit and develop a plan to implement airborne toxic risk reduction measures. Due to SB 1731, another goal of the program is to reduce risk below the determined level of significance.

Information gathered from this program has complemented the California Air Resources Board (ARB) existing toxic air contaminant program by locating sources of substances that were not under evaluation and by providing exposure data needed to develop regulations for control of toxic pollutants. In addition, under the program, the reporting and the requirements to do the risk assessments are mandatory; the subsequent reductions are voluntary. The program has been a motivating factor for facility owners to voluntarily reduce the toxic emissions of the facilities and has demonstrated that the mandatory risk assessment/notification component has been a powerful tool in getting industry to produce results. It is important to note that skilled and sophisticated risk assessors or modelers are required under this type of program to conduct the risk analyses required.

The "Hot Spots" program is an example of how a law can be good for the environment and business. Some of the surveyed companies reported that the preparation of their air toxic "Hot Spots" emission inventory alerted them to the actual cost of waste in their processes and motivated them to look for ways of streamlining their operations. In addition, the "Hot Spots" emission inventory, combined with concern for worker safety and the possibility of future regulation, led companies to reduce emissions by substituting fewer toxic materials in their processes or by establishing more efficient operations. These emission reductions lower the health risk to workers and the public while helping to improve company performance and relations with the community.

Inherent in this program are two strong factors that motivated industry (as with the Massachusetts Toxics Use Reduction Act [TURA] program): (1) The plans allowed industry to analyze production processes that use toxic chemicals to see that cost-effective improvements can be made, and (2) there was public disclosure of the amounts of toxic chemicals used and the assessment of the risk of those chemicals on the residents and workers surrounding an industrial facility. Individually or combined, both these factors have resulted in improvements in toxic reduction technology, in productivity enhancement in industrial processes, and in encouraging the development of new technologies.

An ARB survey found 21 companies that voluntarily reduced air toxic emissions by almost 2 million pounds. This sample of the California air toxic sources suggests a larger, statewide trend and points to a significant benefit of the "Hot Spots" regulation. The state ARB has not compiled the emissions reductions from the AB 2588 program but rather left it up to the various local air pollution agencies to report progress. While it is difficult to document the precise emissions reductions of air toxics due to the Air Toxics "Hot Spots" program in California, the counties that reported actual emission reductions indicated overall air toxics reductions of 60 to 90 percent. Reductions of individual toxic air emissions varied up to 100 percent.

The ARB and the Metal Finishing Association of Southern California received a South Coast Air Quality Management District 1992 Clean Air Award for their work to advance P2 and the control of hexavalent chromium emissions. The two organizations cooperated in a demonstration project showing that simple process changes, combined with control equipment used successfully in other industries, consistently reduced hexavalent chromium emissions well below the required emission limit. The findings from this project will reduce exposure of the public to hexavalent chromium, thus preventing up to 2,600 potential cancer cases over the next 70 yrs.

BIBLIOGRAPHY

American Chemistry Council. 2006. *ACC Applauds EPA TRI Burden Reduction Measure.* Arlington, VA.

Becker, M., and K. Geiser. 1997. Evaluating Progress: A Report on the Findings of the Massachusetts Toxics Use Reduction Program Evaluation. Lowell, MA: Toxics Use Reduction Institute, University of Massachusetts at Lowell.

Blus, Lawrence J. 2007. Contaminants and Wildlife: The Rachel Carson Legacy Lives On. http://www.pwrc.usgs.gov/whatsnew/events/carson/Pres_Web/blus_pelicantalk051107.pdf (accessed January 20, 2010).

California Environmental Resources Board. 2010. AB 2588 Air Toxics "Hot Spots" Program. http://www.arb.ca.gov/ab2588/ab2588.htm (accessed January 20, 2010).

Cancer Trends Progress Report. 2007. Update, National Cancer Institute, U.S. National Institute of Health. http://progressreport.cancer.gov/index.asp (accessed January 20, 2010).

European Commission. 2006. Environment Fact Sheet: REACH—A New Chemicals Policy for the EU. http://ec.europa.eu/environment/pubs/pdf/factsheets/reach.pdf (accessed January 20, 2010).

General Laws of Massachusetts. 2007. Chapter 211, Massachusetts Toxics Use Reduction Act. http://www.mass.gov/legis/laws/mgl/21i-1.htm (accessed May 23, 2010).

Michigan Department of Environmental Quality. 2005. Protecting Michigan's Environment, Ensuring Michigan's Future, Preventing Pollution. http://www.michigan.gov/documents/deq/deq-ess-p2-2005p2rept_250542_7.pdf (accessed January 20, 2010).

Office of Pollution Prevention and Technical Assistance. 2005. 2005 Indiana Pollution Prevention Annual Report. http://www.in.gov/legislative/igareports/agency/reports/IDEM20.pdf (accessed January 20, 2010).

Study Finds Net Gains from Pollution Rules; OMB Overturns Past Findings on Benefits. 2003. *Washington Post,* September 27, A1.

Survey Evaluation of the Massachusetts Toxics Use Reduction Program. 1997 (February). Abt Associates Inc.

U.S. Environmental Protection Agency. 2006. Toxics Release Inventory Burden Reduction Final Rule. *Federal Register*, 71(246), 76932.

Washington TREE Program Brochure http://www.ecy.wa.gov/tree/TREE-08_PRINT-rev909.pdf Olympia, WA: Washington State Department of Ecology.

Wisconsin Department of Natural Resources. 2007. The Environmental Cooperation Pilot Program: 2007 Progress Report. http://dnr.wi.gov/org/caer/cea/ecpp/reports/2007annualreport.pdf (accessed October 31, 2007).

Appendix A: Chemicals List with CAS Numbers

Chemicals List with CAS Numbers

Chemical Name	CAS
(1,1′-Biphenyl)-4,4′-diamine, 3,3′-dimethyl-	119-93-7
1-(3-Chloroallyl)-3,5,7-triaza-1-azoniaadamantane chloride	4080-31-3
1,1,1,2-Tetrachloro-2-fluoroethane (Hcfc-121a)	354-11-0
1,1,1,2-Tetrachloroethane	630-20-6
1,1,1-Trichloroethane	71-55-6
1,1,1-Trifluoro-2,2-dichloroethane	306-83-2
1,1,2,2-Tetrachloro-1-fluoroethane	354-14-3
1,1,2,2-Tetrachloroethane	79-34-5
1,1,2-Trichloroethane	79-00-5
1,1′-Bi(ethylene oxide)	1464-53-5
1,1-Dichloro-1-fluoroethane	1717-00-6
1,1-Dichloroethane	75-34-3
1,1-Dichloroethylene	75-35-4
1,1-Dimethyl hydrazine	57-14-7
1,1′-Methylenebis(4-isocyanatobenzene)	101-68-8
1,2,3-Trichloropropane	96-18-4
1,2,4-Trichlorobenzene	120-82-1
1,2,4-Trimethylbenzene	95-63-6
1,2-Butylene oxide	106-88-7
1,2-Dibromo-3-chloropropane (Dbcp)	96-12-8
1,2-Dibromoethane	106-93-4
1,2-Dichloro-1,1,2-trifluoroethane	354-23-4
1,2-Dichloro-1,1-difluoroethane	1649-08-7
1,2-Dichlorobenzene	95-50-1
1,2-Dichloroethane	107-06-2
1,2-Dichloroethylene	540-59-0
1,2-Dichloropropane	78-87-5
1,2-Diphenylhydrazine	122-66-7
1,3-Butadiene	106-99-0
1,3-Dichloro-1,1,2,2,3-pentafluoropropane	507-55-1
1,3-Dichlorobenzene	541-73-1
1,3-Dichloropropene (mixed isomers)	542-75-6
1,4-Dichloro-2-butene	764-41-0
1,4-Dichlorobenzene	106-46-7
1,4-Dioxane	123-91-1

continued

Chemicals List with CAS Numbers (continued)

Chemical Name	CAS
1-Bromo-1-(bromomethyl)-1,3-propanedicarbonitrile	35691-65-7
1-Chloro-1,1,2,2-tetrafluoroethane	354-25-6
1-Chloro-1,1-difluoroethane	75-68-3
2,2′,6,6′-Tetrabromo-4,4′-isopropylidenediphenol	79-94-7
2,3-Dichloropropene	78-88-6
2,4,5-Trichlorophenol	95-95-4
2,4,6-Trichlorophenol	88-06-2
2,4,6-Trinitrophenol	88-89-1
2,4-D	94-75-7
2,4-D 2-Ethylhexyl ester	1928-43-4
2,4-D Butoxyethyl ester	1929-73-3
2,4-D Butyl ester	94-80-4
2,4-D Sodium salt	2702-72-9
2,4-Db	94-82-6
2,4-Diaminoanisole	615-05-4
2,4-Diaminotoluene	95-80-7
2,4-Dichlorophenol	120-83-2
2,4-Dimethylphenol	105-67-9
2,4-Dinitrophenol	51-28-5
2,4-Dinitrotoluene	121-14-2
2,6-Dinitrotoluene	606-20-2
2,6-Xylidine	87-62-7
2-Acetylaminofluorene	53-96-3
2-Aminonaphthalene	91-59-8
2-Chlor-1,3-butadiene	126-99-8
2-Chloro-1,1,1,2-tetrafluoroethane	2837-89-0
2-Chloro-1,1,1-trifluoroethane	75-88-7
2-Chloroacetophenone	532-27-4
2-Mercaptobenzothiazole	149-30-4
2-Methyllactonitrile	75-86-5
2-Methylpyridine	109-06-8
2-Nitrophenol	88-75-5
2-Nitropropane	79-46-9
2-Phenylphenol	90-43-7
3,3-Dichloro-1,1,1,2,2-pentafluoropropane	422-56-0
3,3′-Dichlorobenzidine	91-94-1
3,3′-Dichlorobenzidine dihydrochloride	612-83-9
3,3′-Dichlorobenzidine sulfate	64969-34-2
3,3′-Dimethoxybenzidine	119-90-4
3,3′-Dimethoxybenzidine dihydrochloride	20325-40-0
3-Chloro-2-methyl-1-propene	563-47-3
3-Chloropropionitrile	542-76-7
3-Iodo-2-propynyl butylcarbamate	55406-53-6

Chemicals List with CAS Numbers (continued)

Chemical Name	CAS
4,4′-Diaminodiphenyl ether	101-80-4
4,4′-Isopropylidenediphenol	80-05-7
4,4′-Methylenebis(2-chloroaniline)	101-14-4
4,4′-Methylenebis(N,N-dimethyl)benzenamine	101-61-1
4,4′-Methylenedianiline	101-77-9
4,6-Dinitro-O-cresol	534-52-1
4-Aminoazobenzene	60-09-3
4-Aminobiphenyl	92-67-1
4-Dimethylaminoazobenzene	60-11-7
4-Nitrophenol	100-02-7
5-Nitro-O-anisidine	99-59-2
5-Nitro-O-toluidine	99-55-8
Abamectin	71751-41-2
Acephate	30560-19-1
Acetaldehyde	75-07-0
Acetamide	60-35-5
Acetone	67-64-1
Acetonitrile	75-05-8
Acetophenone	98-86-2
Acifluorfen, sodium salt	62476-59-9
Acrolein	107-02-8
Acrylamide	79-06-1
Acrylic acid	79-10-7
Acrylonitrile	107-13-1
Alachlor	15972-60-8
Aldicarb	116-06-3
Aldrin	309-00-2
Allyl alcohol	107-18-6
Allyl amine	107-11-9
Allyl chloride	107-05-1
alpha-Naphthylamine	134-32-7
Aluminum	7429-90-5
Aluminum oxide (fibrous forms)	1344-28-1
Aluminum phosphide	20859-73-8
Ametryn	834-12-8
Amitraz	33089-61-1
Amitrole	61-82-5
Ammonia	7664-41-7
Ammonium nitrate (solution)	6484-52-2
Ammonium sulfate (solution)	7783-20-2
Aniline	62-53-3
Anthracene	120-12-7

continued

Chemicals List with CAS Numbers (continued)

Chemical Name	CAS
Antimony and antimony compounds	7440-36-0
Arsenic and arsenic compounds	7440-38-2
Asbestos (friable)	1332-21-4
Atrazine	1912-24-9
Auramine	492-80-8
Barium and barium compounds	7440-39-3
Bendiocarb	22781-23-3
Benfluralin	1861-40-1
Benomyl	17804-35-2
Benzal chloride	98-87-3
Benzene	71-43-2
Benzidine	92-87-5
Benzo(Ghi)perylene	191-24-2
Benzoic trichloride	98-07-7
Benzoyl chloride	98-88-4
Benzoyl peroxide	94-36-0
Benzyl chloride	100-44-7
Beryllium and beryllium compounds	7440-41-7
Bifenthrin	82657-04-3
Biphenyl	92-52-4
Bis(2-Chloro-1-methylethyl) ether	108-60-1
Bis(2-Chloroethoxy)methane	111-91-1
Bis(2-Chloroethyl) ether	111-44-4
Bis(2-Ethylhexyl) adipate	103-23-1
Bis(2-Ethylhexyl)phthalate	117-81-7
Bis(Chloromethyl) ether	542-88-1
Bis(Tributyltin) oxide	56-35-9
Boron trichloride	10294-34-5
Boron trifluoride	7637-07-2
Bromacil	314-40-9
Bromine	7726-95-6
Bromochlorodifluoromethane	353-59-3
Bromotrifluoromethane	75-63-8
Bromoxynil	1689-84-5
Bromoxynil octanoate	1689-99-2
Brucine	357-57-3
Butyl acrylate	141-32-2
Butyl benzyl phthalate	85-68-7
Butyraldehyde	123-72-8
C.I. acid red 114	6459-94-5
C.I. basic green 4	569-64-2
C.I. basic red 1	989-38-8
C.I. direct blue 218	28407-37-6
C.I. disperse yellow 3	2832-40-8

Chemicals List with CAS Numbers (continued)

Chemical Name	CAS
C.I. food red 15	81-88-9
C.I. solvent orange 7	3118-97-6
C.I. solvent yellow 14	842-07-9
C.I. solvent yellow 3	97-56-3
Cadmium and cadmium compounds	7440-43-9
Calcium cyanamide	156-62-7
Camphechlor	8001-35-2
Captan	133-06-2
Carbaryl	63-25-2
Carbofuran	1563-66-2
Carbon disulfide	75-15-0
Carbon tetrachloride	56-23-5
Carbonyl sulfide	463-58-1
Carboxin	5234-68-4
Catechol	120-80-9
Chloramben	133-90-4
Chlordane	57-74-9
Chlorendic acid	115-28-6
Chlorimuron ethyl	90982-32-4
Chlorine	7782-50-5
Chlorine dioxide	10049-04-4
Chloroacetic acid	79-11-8
Chlorobenzene	108-90-7
Chlorobenzilate	510-15-6
Chlorodifluoromethane	75-45-6
Chloroethane	75-00-3
Chloroform	67-66-3
Chloromethane	74-87-3
Chloromethyl methyl ether	107-30-2
Chlorophenols	25167-80-0
Chloropicrin	76-06-2
Chlorotetrafluoroethane	63938-10-3
Chlorothalonil	1897-45-6
Chlorotrifluoromethane	75-72-9
Chlorpyrifos methyl	5598-13-0
Chlorsulfuron	64902-72-3
Chromium and chromium compounds	7440-47-3
Cobalt and cobalt compounds	7440-48-4
Copper and copper compounds	7440-50-8
Creosotes	8001-58-9
Cresol (mixed isomers)	1319-77-3
Crotonaldehyde	4170-30-3
Cumene	98-82-8

continued

Chemicals List with CAS Numbers (continued)

Chemical Name	CAS
Cumene hydroperoxide	80-15-9
Cupferron	135-20-6
Cyanazine	21725-46-2
Cyanide compounds	1-07-3
Cycloate	1134-23-2
Cyclohexane	110-82-7
Cyclohexanol	108-93-0
Cyfluthrin	68359-37-5
Cyhalothrin	68085-85-8
Dazomet	533-74-4
Dazomet, sodium salt	53404-60-7
Decabromodiphenyl oxide	1163-19-5
Desmedipham	13684-56-5
Diallate	2303-16-4
Diaminotoluene (mixed isomers)	25376-45-8
Diazinon	333-41-5
Dibenzofuran	132-64-9
Dibromotetrafluoroethane [Halon 2402]	124-73-2
Dibutyl phthalate	84-74-2
Dicamba	1918-00-9
Dichloran	99-30-9
Dichlorobenzene (mixed isomers)	25321-22-6
Dichlorobromomethane	75-27-4
Dichlorodifluoromethane	75-71-8
Dichlorofluoromethane	75-43-4
Dichloromethane	75-09-2
Dichlorotetrafluoroethane (Cfc-114)	76-14-2
Dichlorotrifluoroethane	34077-87-7
Dichlorpentafluoro-propane	127564-92-5
Dichlorvos	62-73-7
Dicofol	115-32-2
Dicyclopentadiene	77-73-6
Diethanolamine	111-42-2
Diethyl phthalate	84-66-2
Diethyl sulfate	64-67-5
Diflubenzuron	35367-38-5
Diglycidyl resorcinol ether (Dgre)	101-90-6
Dihydrosafrole	94-58-6
Diisocyanates	EDF067
Dimethipin	55290-64-7
Dimethoate	60-51-5
Dimethyl chlorothiophosphate	2524-03-0
Dimethyl phthalate	131-11-3
Dimethyl sulfate	77-78-1

Chemicals List with CAS Numbers (continued)

Chemical Name	CAS
Dimethylamine	124-40-3
Dimethylamine dicamba	2300-66-5
Dimethylcarbamoyl chloride	79-44-7
Dinitrobutyl phenol	88-85-7
Dinitrotoluene (mixed isomers)	25321-14-6
Dinocap	39300-45-3
Di-N-propylnitrosamine	621-64-7
Dioxin and dioxin-like compounds	EDF018
Diphenylamine	122-39-4
Dipotassium endothall	2164-07-0
Dipropyl isocinchomeronate	136-45-8
Direct black 38	1937-37-7
Disodium cyanodithioimidocarbonate	138-93-2
Dithiobiuret	541-53-7
Diuron	330-54-1
Dodine	120-36-5
Dodine	2439-10-3
D-trans-Allethrin	28057-48-9
Epichlorohydrin	106-89-8
Ethoprop	13194-48-4
Ethyl acrylate	140-88-5
Ethyl chloroformate	541-41-3
Ethyl dipropylthiocarbamate	759-94-4
Ethylbenzene	100-41-4
Ethylene	74-85-1
Ethylene glycol	107-21-1
Ethylene glycol monoethyl ether	110-80-5
Ethylene glycol monomethyl ether	109-86-4
Ethylene oxide	75-21-8
Ethylene thiourea	96-45-7
Ethylenebisdithiocarbamic acid, salts, and esters	111-54-6
Ethyleneimine	151-56-4
Famphur	52-85-7
Fenarimol	60168-88-9
Fenbutatin oxide	13356-08-6
Fenoxycarb	72490-01-8
Fenpropathrin	39515-41-8
Fenthion	55-38-9
Fenvalerate	51630-58-1
Fluazifop-butyl	69806-50-4
Fluometuron	2164-17-2
Fluorine	7782-41-4
Fluoroacetic acid, sodium salt	62-74-8

continued

Chemicals List with CAS Numbers (continued)

Chemical Name	CAS
Fluorouracil	51-21-8
Fluvalinate	69409-94-5
Folpet	133-07-3
Fomesafen	72178-02-0
Formaldehyde	50-00-0
Formic acid	64-18-6
Freon 113	76-13-1
Gamma-lindane	58-89-9
Glycol ethers	EDF109
Heptachlor	76-44-8
Hexachloro-1,3-butadiene	87-68-3
Hexachlorobenzene	118-74-1
Hexachlorocyclopentadiene	77-47-4
Hexachloroethane	67-72-1
Hexachlorophene (Hcp)	70-30-4
Hexamethylphosphoramide	680-31-9
Hexazinone	51235-04-2
Hydramethylnon	67485-29-4
Hydrazine	302-01-2
Hydrazine sulfate	10034-93-2
Hydrochloric acid	7647-01-0
Hydrofluoric acid	7664-39-3
Hydrogen cyanide	74-90-8
Hydroquinone	123-31-9
Iron pentacarbonyl	13463-40-6
Isobutyraldehyde	78-84-2
Isodrin	465-73-6
Isofenphos	25311-71-1
Isopropyl alcohol	67-63-0
Isosafrole	120-58-1
Lactofen	77501-63-4
Lead and lead compounds	7439-92-1
Linuron	330-55-2
Lithium carbonate	554-13-2
Malathion	121-75-5
Maleic anhydride	108-31-6
Malononitrile	109-77-3
Maneb	12427-38-2
Manganese and manganese compounds	7439-96-5
M-Cresol	108-39-4
M-Dinitrobenzene	99-65-0
Mecoprop	93-65-2
Mercury and mercury compounds	7439-97-6
Merphos	150-50-5

Chemicals List with CAS Numbers (continued)

Chemical Name	CAS
Methacrylonitrile	126-98-7
Metham sodium	137-42-8
Methanamine, N-methyl-N-nitroso	62-75-9
Methanol	67-56-1
Methoxone	94-74-6
Methoxychlor	72-43-5
Methyl acrylate	96-33-3
Methyl bromide	74-83-9
Methyl chlorocarbonate	79-22-1
Methyl ethyl ketone	78-93-3
Methyl hydrazine	60-34-4
Methyl iodide	74-88-4
Methyl isobutyl ketone	108-10-1
Methyl isocyanate	624-83-9
Methyl isothiocyanate	556-61-6
Methyl methacrylate	80-62-6
Methyl parathion	298-00-0
Methyl tert-butyl ether	1634-04-4
Methylene bromide	74-95-3
Metribuzin	21087-64-9
Michler's ketone	90-94-8
Mixture	MIXTURE
Molinate	2212-67-1
Molybdenum trioxide	1313-27-5
Monochloropentafluoroethane	76-15-3
M-Phenylenediamine	108-45-2
M-Xylene	108-38-3
Myclobutanil	88671-89-0
N,N-Dimethylaniline	121-69-7
N,N-Dimethylformamide	68-12-2
Nabam	142-59-6
Naled	300-76-5
Naphthalene	91-20-3
N-Butyl alcohol	71-36-3
N-Dioctyl phthalate	117-84-0
N-Ethyl-N-nitrosourea	759-73-9
N-Hexane	110-54-3
Nickel and nickel compounds	7440-02-0
Nicotine and salts	54-11-5
Nitrapyrin	1929-82-4
Nitrate compounds	EDF038
Nitric acid	7697-37-2
Nitrilotriacetic acid	139-13-9

continued

Chemicals List with CAS Numbers (continued)

Chemical Name	CAS
Nitrobenzene	98-95-3
Nitrofen	1836-75-5
Nitroglycerin	55-63-0
N-Methyl-2-pyrrolidone	872-50-4
N-Methylolacrylamide	924-42-5
N-Nitrosodiethylamine	55-18-5
N-Nitrosodi-N-butylamine	924-16-3
N-Nitrosodiphenylamine	86-30-6
N-Nitrosomethylvinylamine	4549-40-0
N-Nitroso-N-methylurea	684-93-5
N-Nitrosopiperidine	100-75-4
Norflurazon	27314-13-2
O-Anisidine	90-04-0
O-Cresol	95-48-7
O-Dinitrobenzene	528-29-0
O-Phenylenediamine	95-54-5
O-Phenylphenate, sodium	132-27-4
Oryzalin	19044-88-3
Osmium oxide Oso4 (T-4)	20816-12-0
O-Toluidine	95-53-4
O-Toluidine hydrochloride	636-21-5
Oxydemeton methyl	301-12-2
Oxydiazon	19666-30-9
Oxyfluorfen	42874-03-3
O-Xylene	95-47-6
Ozone	10028-15-6
P-Anisidine	104-94-9
Paraldehyde	123-63-7
Paraquat	1910-42-5
Parathion	56-38-2
P-Chloroaniline	106-47-8
P-Cresidine	120-71-8
P-Cresol	106-44-5
P-Dinitrobenzene	100-25-4
Pebulate	1114-71-2
Pendimethalin	40487-42-1
Pentachlorobenzene	608-93-5
Pentachloroethane	76-01-7
Pentachlorophenol	87-86-5
Peracetic acid	79-21-0
Perchloromethyl mercaptan	594-42-3
Permethrin	52645-53-1
Phenanthrene	85-01-8
Phenol	108-95-2

Chemicals List with CAS Numbers (continued)

Chemical Name	CAS
Phenothrin	26002-80-2
Phenytoin	57-41-0
Phosgene	75-44-5
Phosphine	7803-51-2
Phosphorus (yellow or white)	7723-14-0
Phospohoric acid	7664-38-2
Phthalic anhydride	85-44-9
Picloram	1918-02-1
Piperonyl butoxide	51-03-6
Pirimiphos methyl	29232-93-7
P-Nitroaniline	100-01-6
P-Nitrosodiphenylamine	156-10-5
Polychlorinated alkanes (C10–C13)	EDF045
Polychlorinated biphenyls	1336-36-3
Polycyclic aromatic compounds	65996-93-2
Potassium bromate	7758-01-2
Potassium dimethyldithiocarbamate	128-03-0
Potassium N-methyldithiocarbamate	137-41-7
P-Phenylenediamine	106-50-3
Profenofos	41198-08-7
Prometryn	7287-19-6
Pronamide	23950-58-5
Propachlor	1918-16-7
Propane sultone	1120-71-4
Propanil	709-98-8
Propargite	2312-35-8
Propargyl alcohol	107-19-7
Propetamphos	31218-83-4
Propiconazole	60207-90-1
Propionaldehyde	123-38-6
Propoxur	114-26-1
Propylene	115-07-1
Propylene oxide	75-56-9
Propyleneimine	75-55-8
P-Xylene	106-42-3
Pyridine	110-86-1
Quinoline	91-22-5
Quinone	106-51-4
Quintozene	82-68-8
Quizalofop-ethyl	76578-14-8
Resmethrin	10453-86-8
S,S,S-Tributyltrithiophosphate	78-48-8
Saccharin	81-07-2

continued

Chemicals List with CAS Numbers (continued)

Chemical Name	CAS
Safrole	94-59-7
sec-Butyl alcohol	78-92-2
Selenium and selenium compounds	7782-49-2
Sethoxydim	74051-80-2
Silver and silver compounds	7440-22-4
Simazine	122-34-9
Sodium azide	26628-22-8
Sodium dicamba	1982-69-0
Sodium dimethyldithiocarbamate	128-04-1
Sodium hydroxide (solution)	1310-73-2
Sodium nitrite	7632-00-0
Strychnine	57-24-9
Styrene	100-42-5
Styrene oxide	96-09-3
Sulfuric acid	7664-93-9
Sulfuryl fluoride	2699-79-8
Sulprofos	35400-43-2
Tebuthiuron	34014-18-1
Temephos	3383-96-8
Terbacil	5902-51-2
Terephthalic acid	100-21-0
tert-Butyl alcohol	75-65-0
Tetrachloroethylene	127-18-4
Tetrachlorvinphos	961-11-5
Tetracycline hydrochloride	64-75-5
Tetramethrin	7696-12-0
Thallium and thallium compounds	7440-28-0
Thiabendazole	148-79-8
Thioacetamide	62-55-5
Thiobencarb	28249-77-6
Thiodicarb	59669-26-0
Thiophanate-methyl	23564-05-8
Thiosemicarbazide	79-19-6
Thiourea	62-56-6
Thiram	137-26-8
Thorium dioxide	1314-20-1
Titanium tetrachloride	7550-45-0
Toluene	108-88-3
Toluene diisocyanate (mixed isomers)	26471-62-5
Toluene-2,4-diisocyanate	584-84-9
Toluene-2,6-diisocyanate	91-08-7
Trade secret chemical	TRD SECRT
trans-1,3-Dichloropropene	10061-02-6
trans-1,4-Dichloro-2-butene	110-57-6

Chemicals List with CAS Numbers (continued)

Chemical Name	CAS
Triadimefon	43121-43-3
Triallate	2303-17-5
Tribenuron methyl	101200-48-0
Tribromomethane	75-25-2
Tributyltin methacrylate	2155-70-6
Trichlorfon	52-68-6
Trichloroacetyl chloride	76-02-8
Trichloroethylene	79-01-6
Trichlorofluoromethane	75-69-4
Triclopyr triethylammonium salt	57213-69-1
Triethylamine	121-44-8
Trifluralin	1582-09-8
Triforine	26644-46-2
Triphenyltin chloride	639-58-7
Triphenyltin hydroxide	76-87-9
Tris(2,3-dibromopropyl) phosphate	126-72-7
Trypan blue	72-57-1
Urethane	51-79-6
Vanadium and vanadium compounds	7440-62-2
Vinclozolin	50471-44-8
Vinyl acetate	108-05-4
Vinyl bromide	593-60-2
Vinyl chloride	75-01-4
Warfarin and salts	81-81-2
Xylene (mixed isomers)	1330-20-7
Zinc and zinc compounds	7440-66-6
Zineb	12122-67-7

Source: U.S. Environmental Protection Agency. TRI Explorer (version 4.5). http://www.epa.gov/triexplorer/ (accessed May 23, 2010).

Appendix B: CAS Numbers with Chemical Names

CAS Numbers with Chemical Names

CAS	Chemical Name
1-07-3	Cyanide compounds
50-00-0	Formaldehyde
51-03-6	Piperonyl butoxide
51-21-8	Fluorouracil
51-28-5	2,4-Dinitrophenol
51-79-6	Urethane
52-68-6	Trichlorfon
52-85-7	Famphur
53-96-3	2-Acetylaminofluorene
54-11-5	Nicotine and salts
55-18-5	N-Nitrosodiethylamine
55-38-9	Fenthion
55-63-0	Nitroglycerin
56-23-5	Carbon tetrachloride
56-35-9	Bis(tributyltin) oxide
56-38-2	Parathion
57-14-7	1,1-Dimethyl hydrazine
57-24-9	Strychnine
57-41-0	Phenytoin
57-74-9	Chlordane
58-89-9	gamma-Lindane
60-09-3	4-Aminoazobenzene
60-11-7	4-Dimethylaminoazobenzene
60-34-4	Methyl hydrazine
60-35-5	Acetamide
60-51-5	Dimethoate
61-82-5	Amitrole
62-53-3	Aniline
62-55-5	Thioacetamide
62-56-6	Thiourea
62-73-7	Dichlorvos
62-74-8	Fluoroacetic acid, sodium salt
62-75-9	Methanamine, N-methyl-N-nitroso
63-25-2	Carbaryl
64-18-6	Formic acid
64-67-5	Diethyl sulfate

continued

CAS Numbers with Chemical Names (continued)

CAS	Chemical Name
64-75-5	Tetracycline hydrochloride
67-56-1	Methanol
67-63-0	Isopropyl alcohol
67-64-1	Acetone
67-66-3	Chloroform
67-72-1	Hexachloroethane
68-12-2	N,N-Dimethylformamide
70-30-4	Hexachlorophene (Hcp)
71-36-3	N-Butyl alcohol
71-43-2	Benzene
71-55-6	1,1,1-Trichloroethane
72-43-5	Methoxychlor
72-57-1	Trypan blue
74-83-9	Methyl bromide
74-85-1	Ethylene
74-87-3	Chloromethane
74-88-4	Methyl iodide
74-90-8	Hydrogen cyanide
74-95-3	Methylene bromide
75-00-3	Chloroethane
75-01-4	Vinyl chloride
75-05-8	Acetonitrile
75-07-0	Acetaldehyde
75-09-2	Dichloromethane
75-15-0	Carbon disulfide
75-21-8	Ethylene oxide
75-25-2	Tribromomethane
75-27-4	Dichlorobromomethane
75-34-3	1,1-Dichloroethane
75-35-4	1,1-Dichloroethylene
75-43-4	Dichlorofluoromethane
75-44-5	Phosgene
75-45-6	Chlorodifluoromethane
75-55-8	Propyleneimine
75-56-9	Propylene oxide
75-63-8	Bromotrifluoromethane
75-65-0	tert-Butyl alcohol
75-68-3	1-Chloro-1,1-difluoroethane
75-69-4	Trichlorofluoromethane
75-71-8	Dichlorodifluoromethane
75-72-9	Chlorotrifluoromethane
75-86-5	2-Methyllactonitrile
75-88-7	2-Chloro-1,1,1-trifluoroethane
76-01-7	Pentachloroethane

CAS Numbers with Chemical Names (continued)

CAS	Chemical Name
76-02-8	Trichloroacetyl chloride
76-06-2	Chloropicrin
76-13-1	Freon 113
76-14-2	Dichlorotetrafluoroethane (Cfc-114)
76-15-3	Monochloropentafluoroethane
76-44-8	Heptachlor
76-87-9	Triphenyltin hydroxide
77-47-4	Hexachlorocyclopentadiene
77-73-6	Dicyclopentadiene
77-78-1	Dimethyl sulfate
78-48-8	S,S,S-Tributyltrithiophosphate
78-84-2	Isobutyraldehyde
78-87-5	1,2-Dichloropropane
78-88-6	2,3-Dichloropropene
78-92-2	sec-Butyl alcohol
78-93-3	Methyl ethyl ketone
79-00-5	1,1,2-Trichloroethane
79-01-6	Trichloroethylene
79-06-1	Acrylamide
79-10-7	Acrylic acid
79-11-8	Chloroacetic acid
79-19-6	Thiosemicarbazide
79-21-0	Peracetic acid
79-22-1	Methyl chlorocarbonate
79-34-5	1,1,2,2-Tetrachloroethane
79-44-7	Dimethylcarbamoyl chloride
79-46-9	2-Nitropropane
79-94-7	2,2′,6,6′-Tetrabromo-4,4′-isopropylidenediphenol
80-05-7	4,4′-Isopropylidenediphenol
80-15-9	Cumene hydroperoxide
80-62-6	Methyl methacrylate
81-07-2	Saccharin
81-81-2	Warfarin and salts
81-88-9	C.I. food red 15
82-68-8	Quintozene
84-66-2	Diethyl phthalate
84-74-2	Dibutyl phthalate
85-01-8	Phenanthrene
85-44-9	Phthalic anhydride
85-68-7	Butyl benzyl phthalate
86-30-6	N-Nitrosodiphenylamine
87-62-7	2,6-Xylidine
87-68-3	Hexachloro-1,3-butadiene

continued

CAS Numbers with Chemical Names (continued)

CAS	Chemical Name
87-86-5	Pentachlorophenol
88-06-2	2,4,6-Trichlorophenol
88-75-5	2-Nitrophenol
88-85-7	Dinitrobutyl phenol
88-89-1	2,4,6-Trinitrophenol
90-04-0	O-Anisidine
90-43-7	2-Phenylphenol
90-94-8	Michler's ketone
91-08-7	Toluene-2,6-diisocyanate
91-20-3	Naphthalene
91-22-5	Quinoline
91-59-8	2-Aminonaphthalene
91-94-1	3,3′-Dichlorobenzidine
92-52-4	Biphenyl
92-67-1	4-Aminobiphenyl
92-87-5	Benzidine
93-65-2	Mecoprop
94-36-0	Benzoyl peroxide
94-58-6	Dihydrosafrole
94-59-7	Safrole
94-74-6	Methoxone
94-75-7	2,4-D
94-80-4	2,4-D butyl ester
94-82-6	2,4-Db
95-47-6	O-Xylene
95-48-7	O-Cresol
95-50-1	1,2-Dichlorobenzene
95-53-4	O-Toluidine
95-54-5	O-Phenylenediamine
95-63-6	1,2,4-Trimethylbenzene
95-80-7	2,4-Diaminotoluene
95-95-4	2,4,5-Trichlorophenol
96-09-3	Styrene oxide
96-12-8	1,2-Dibromo-3-chloropropane (Dbcp)
96-18-4	1,2,3-Trichloropropane
96-33-3	Methyl acrylate
96-45-7	Ethylene thiourea
97-56-3	C.I. solvent yellow 3
98-07-7	Benzoic trichloride
98-82-8	Cumene
98-86-2	Acetophenone
98-87-3	Benzal chloride
98-88-4	Benzoyl chloride
98-95-3	Nitrobenzene

CAS Numbers with Chemical Names (continued)

CAS	Chemical Name
99-30-9	Dichloran
99-55-8	5-Nitro-O-toluidine
99-59-2	5-Nitro-O-anisidine
99-65-0	M-Dinitrobenzene
100-01-6	P-Nitroaniline
100-02-7	4-Nitrophenol
100-21-0	Terephthalic acid
100-25-4	P-Dinitrobenzene
100-41-4	Ethylbenzene
100-42-5	Styrene
100-44-7	Benzyl chloride
100-75-4	N-Nitrosopiperidine
101-14-4	4,4′-Methylenebis(2-chloroaniline)
101-61-1	4,4′-Methylenebis(N,N-dimethyl)benzenamine
101-68-8	1,1′-Methylenebis(4-isocyanatobenzene)
101-77-9	4,4′-Methylenedianiline
101-80-4	4,4′-Diaminodiphenyl ether
101-90-6	Diglycidyl resorcinol ether (Dgre)
103-23-1	Bis(2-ethylhexyl) adipate
104-94-9	P-Anisidine
105-67-9	2,4-Dimethylphenol
106-42-3	P-Xylene
106-44-5	P-Cresol
106-46-7	1,4-Dichlorobenzene
106-47-8	P-Chloroaniline
106-50-3	P-Phenylenediamine
106-51-4	Quinone
106-88-7	1,2-Butylene oxide
106-89-8	Epichlorohydrin
106-93-4	1,2-Dibromoethane
106-99-0	1,3-Butadiene
107-02-8	Acrolein
107-05-1	Allyl chloride
107-06-2	1,2-Dichloroethane
107-11-9	Allyl amine
107-13-1	Acrylonitrile
107-18-6	Allyl alcohol
107-19-7	Propargyl alcohol
107-21-1	Ethylene glycol
107-30-2	Chloromethyl methyl ether
108-05-4	Vinyl acetate
108-10-1	Methyl isobutyl ketone
108-31-6	Maleic anhydride

continued

CAS Numbers with Chemical Names (continued)

CAS	Chemical Name
108-38-3	M-Xylene
108-39-4	M-Cresol
108-45-2	M-Phenylenediamine
108-60-1	Bis(2-chloro-1-methylethyl) ether
108-88-3	Toluene
108-90-7	Chlorobenzene
108-93-0	Cyclohexanol
108-95-2	Phenol
109-06-8	2-Methylpyridine
109-77-3	Malononitrile
109-86-4	Ethylene glycol monomethyl ether
110-54-3	N-Hexane
110-57-6	trans-1,4-Dichloro-2-butene
110-80-5	Ethylene glycol monoethyl ether
110-82-7	Cyclohexane
110-86-1	Pyridine
111-42-2	Diethanolamine
111-44-4	Bis(2-chloroethyl) ether
111-54-6	Ethylenebisdithiocarbamic acid, salts, and esters
111-91-1	Bis(2-chloroethoxy)methane
114-26-1	Propoxur
115-07-1	Propylene
115-28-6	Chlorendic acid
115-32-2	Dicofol
116-06-3	Aldicarb
117-81-7	Bis(2-ethylhexyl)phthalate
117-84-0	N-Dioctyl phthalate
118-74-1	Hexachlorobenzene
119-90-4	3,3′-Dimethoxybenzidine
119-93-7	(1,1′-Biphenyl)-4,4′-diamine, 3,3′-dimethyl-
120-12-7	Anthracene
120-36-5	Dodine
120-58-1	Isosafrole
120-71-8	P-Cresidine
120-80-9	Catechol
120-82-1	1,2,4-Trichlorobenzene
120-83-2	2,4-Dichlorophenol
121-14-2	2,4-Dinitrotoluene
121-44-8	Triethylamine
121-69-7	N,N-Dimethylaniline
121-75-5	Malathion
122-34-9	Simazine
122-39-4	Diphenylamine
122-66-7	1,2-Diphenylhydrazine

CAS Numbers with Chemical Names (continued)

CAS	Chemical Name
123-31-9	Hydroquinone
123-38-6	Propionaldehyde
123-63-7	Paraldehyde
123-72-8	Butyraldehyde
123-91-1	1,4-Dioxane
124-40-3	Dimethylamine
124-73-2	Dibromotetrafluoroethane (Halon 2402)
126-72-7	Tris(2,3-dibromopropyl) phosphate
126-98-7	Methacrylonitrile
126-99-8	2-Chlor-1,3-butadiene
127-18-4	Tetrachloroethylene
128-03-0	Potassium dimethyldithiocarbamate
128-04-1	Sodium dimethyldithiocarbamate
131-11-3	Dimethyl phthalate
132-27-4	O-Phenylphenate, sodium
132-64-9	Dibenzofuran
133-06-2	Captan
133-07-3	Folpet
133-90-4	Chloramben
134-32-7	alpha-Naphthylamine
135-20-6	Cupferron
136-45-8	Dipropyl isocinchomeronate
137-26-8	Thiram
137-41-7	Potassium N-methyldithiocarbamate
137-42-8	Metham sodium
138-93-2	Disodium cyanodithioimidocarbonate
139-13-9	Nitrilotriacetic acid
140-88-5	Ethyl acrylate
141-32-2	Butyl acrylate
142-59-6	Nabam
148-79-8	Thiabendazole
149-30-4	2-Mercaptobenzothiazole
150-50-5	Merphos
151-56-4	Ethyleneimine
156-10-5	P-Nitrosodiphenylamine
156-62-7	Calcium cyanamide
191-24-2	Benzo(Ghi)perylene
298-00-0	Methyl parathion
300-76-5	Naled
301-12-2	Oxydemeton methyl
302-01-2	Hydrazine
306-83-2	1,1,1-Trifluoro-2,2-dichloroethane
309-00-2	Aldrin

continued

CAS Numbers with Chemical Names (continued)

CAS	Chemical Name
314-40-9	Bromacil
330-54-1	Diuron
330-55-2	Linuron
333-41-5	Diazinon
353-59-3	Bromochlorodifluoromethane
354-11-0	1,1,1,2-Tetrachloro-2-fluoroethane (Hcfc-121a)
354-14-3	1,1,2,2-Tetrachloro-1-fluoroethane
354-23-4	1,2-Dichloro-1,1,2-trifluoroethane
354-25-6	1-Chloro-1,1,2,2-tetrafluoroethane
357-57-3	Brucine
422-56-0	3,3-Dichloro-1,1,1,2,2-pentafluoropropane
463-58-1	Carbonyl sulfide
465-73-6	Isodrin
492-80-8	Auramine
507-55-1	1,3-Dichloro-1,1,2,2,3-pentafluoropropane
510-15-6	Chlorobenzilate
528-29-0	O-Dinitrobenzene
532-27-4	2-Chloroacetophenone
533-74-4	Dazomet
534-52-1	4,6-Dinitro-O-cresol
540-59-0	1,2-Dichloroethylene
541-41-3	Ethyl chloroformate
541-53-7	Dithiobiuret
541-73-1	1,3-Dichlorobenzene
542-75-6	1,3-Dichloropropene (mixed isomers)
542-76-7	3-Chloropropionitrile
542-88-1	Bis(chloromethyl) ether
554-13-2	Lithium carbonate
556-61-6	Methyl isothiocyanate
563-47-3	3-Chloro-2-methyl-1-propene
569-64-2	C.I. basic green 4
584-84-9	Toluene-2,4-diisocyanate
593-60-2	Vinyl bromide
594-42-3	Perchloromethyl mercaptan
606-20-2	2,6-Dinitrotoluene
608-93-5	Pentachlorobenzene
612-83-9	3,3′-Dichlorobenzidine dihydrochloride
615-05-4	2,4-Diaminoanisole
621-64-7	Di-N-propylnitrosamine
624-83-9	Methyl isocyanate
630-20-6	1,1,1,2-Tetrachloroethane
636-21-5	O-Toluidine hydrochloride
639-58-7	Triphenyltin chloride
680-31-9	Hexamethylphosphoramide

CAS Numbers with Chemical Names (continued)

CAS	Chemical Name
684-93-5	N-Nitroso-N-methylurea
709-98-8	Propanil
759-73-9	N-Ethyl-N-nitrosourea
759-94-4	Ethyl dipropylthiocarbamate
764-41-0	1,4-Dichloro-2-butene
834-12-8	Ametryn
842-07-9	C.I. solvent yellow 14
872-50-4	N-Methyl-2-pyrrolidone
924-16-3	N-Nitrosodi-N-butylamine
924-42-5	N-Methylolacrylamide
961-11-5	Tetrachlorvinphos
989-38-8	C.I. basic red 1
1114-71-2	Pebulate
1120-71-4	Propane sultone
1134-23-2	Cycloate
1163-19-5	Decabromodiphenyl oxide
1310-73-2	Sodium hydroxide (solution)
1313-27-5	Molybdenum trioxide
1314-20-1	Thorium dioxide
1319-77-3	Cresol (mixed isomers)
1330-20-7	Xylene (mixed isomers)
1332-21-4	Asbestos (friable)
1336-36-3	Polychlorinated biphenyls
1344-28-1	Aluminum oxide (fibrous forms)
1464-53-5	1,1′-Bi(ethylene oxide)
1563-66-2	Carbofuran
1582-09-8	Trifluralin
1634-04-4	Methyl tert-butyl ether
1649-08-7	1,2-Dichloro-1,1-difluoroethane
1689-84-5	Bromoxynil
1689-99-2	Bromoxynil octanoate
1717-00-6	1,1-Dichloro-1-fluoroethane
1836-75-5	Nitrofen
1861-40-1	Benfluralin
1897-45-6	Chlorothalonil
1910-42-5	Paraquat
1912-24-9	Atrazine
1918-00-9	Dicamba
1918-02-1	Picloram
1918-16-7	Propachlor
1928-43-4	2,4-D 2-Ethylhexyl ester
1929-73-3	2,4-D Butoxyethyl ester
1929-82-4	Nitrapyrin

continued

CAS Numbers with Chemical Names (continued)

CAS	Chemical Name
1937-37-7	Direct black 38
1982-69-0	Sodium dicamba
2155-70-6	Tributyltin methacrylate
2164-07-0	Dipotassium endothall
2164-17-2	Fluometuron
2212-67-1	Molinate
2300-66-5	Dimethylamine dicamba
2303-16-4	Diallate
2303-17-5	Triallate
2312-35-8	Propargite
2439-10-3	Dodine
2524-03-0	Dimethyl chlorothiophosphate
2699-79-8	Sulfuryl fluoride
2702-72-9	2,4-D sodium salt
2832-40-8	C.I. disperse yellow 3
2837-89-0	2-Chloro-1,1,1,2-tetrafluoroethane
3118-97-6	C.I. solvent orange 7
3383-96-8	Temephos
4080-31-3	1-(3-Chloroallyl)-3,5,7-triaza-1-azoniaadamantane chloride
4170-30-3	Crotonaldehyde
4549-40-0	N-Nitrosomethylvinylamine
5234-68-4	Carboxin
5598-13-0	Chlorpyrifos methyl
5902-51-2	Terbacil
6459-94-5	C.I. acid red 114
6484-52-2	Ammonium nitrate (solution)
7287-19-6	Prometryn
7429-90-5	Aluminum
7439-92-1	Lead and lead compounds
7439-96-5	Manganese and manganese compounds
7439-97-6	Mercury and mercury compounds
7440-02-0	Nickel and nickel compounds
7440-22-4	Silver and silver compounds
7440-28-0	Thallium and thallium compounds
7440-36-0	Antimony and antimony compounds
7440-38-2	Arsenic and arsenic compounds
7440-39-3	Barium and barium compounds
7440-41-7	Beryllium and beryllium compounds
7440-43-9	Cadmium and cadmium compounds
7440-47-3	Chromium and chromium compounds
7440-48-4	Cobalt and cobalt compounds
7440-50-8	Copper and copper compounds
7440-62-2	Vanadium and vanadium compounds

CAS Numbers with Chemical Names (continued)

CAS	Chemical Name
7440-66-6	Zinc and zinc compounds
7550-45-0	Titanium tetrachloride
7632-00-0	Sodium nitrite
7637-07-2	Boron trifluoride
7647-01-0	Hydrochloric acid
7664-38-2	Phospohoric acid
7664-39-3	Hydrofluoric acid
7664-41-7	Ammonia
7664-93-9	Sulfuric acid
7696-12-0	Tetramethrin
7697-37-2	Nitric acid
7723-14-0	Phosphorus (yellow or white)
7726-95-6	Bromine
7758-01-2	Potassium bromate
7782-41-4	Fluorine
7782-49-2	Selenium and selenium compounds
7782-50-5	Chlorine
7783-20-2	Ammonium sulfate (solution)
7803-51-2	Phosphine
8001-35-2	Camphechlor
8001-58-9	Creosotes
10028-15-6	Ozone
10034-93-2	Hydrazine sulfate
10049-04-4	Chlorine dioxide
10061-02-6	Trans-1,3-dichloropropene
10294-34-5	Boron trichloride
10453-86-8	Resmethrin
12122-67-7	Zineb
12427-38-2	Maneb
13194-48-4	Ethoprop
13356-08-6	Fenbutatin oxide
13463-40-6	Iron pentacarbonyl
13684-56-5	Desmedipham
15972-60-8	Alachlor
17804-35-2	Benomyl
19044-88-3	Oryzalin
19666-30-9	Oxydiazon
20325-40-0	3,3′-Dimethoxybenzidine dihydrochloride
20816-12-0	Osmium oxide Oso4 (T-4)
20859-73-8	Aluminum phosphide
21087-64-9	Metribuzin
21725-46-2	Cyanazine
22781-23-3	Bendiocarb

continued

CAS Numbers with Chemical Names (continued)

CAS	Chemical Name
23564-05-8	Thiophanate-methyl
23950-58-5	Pronamide
25167-80-0	Chlorophenols
25311-71-1	Isofenphos
25321-14-6	Dinitrotoluene (mixed isomers)
25321-22-6	Dichlorobenzene (mixed isomers)
25376-45-8	Diaminotoluene (mixed isomers)
26002-80-2	Phenothrin
26471-62-5	Toluene diisocyanate (mixed isomers)
26628-22-8	Sodium azide
26644-46-2	Triforine
27314-13-2	Norflurazon
28057-48-9	D-trans-Allethrin
28249-77-6	Thiobencarb
28407-37-6	C.I. direct blue 218
29232-93-7	Pirimiphos methyl
30560-19-1	Acephate
31218-83-4	Propetamphos
33089-61-1	Amitraz
34014-18-1	Tebuthiuron
34077-87-7	Dichlorotrifluoroethane
35367-38-5	Diflubenzuron
35400-43-2	Sulprofos
35691-65-7	1-Bromo-1-(bromomethyl)-1,3-propanedicarbonitrile
39300-45-3	Dinocap
39515-41-8	Fenpropathrin
40487-42-1	Pendimethalin
41198-08-7	Profenofos
42874-03-3	Oxyfluorfen
43121-43-3	Triadimefon
50471-44-8	Vinclozolin
51235-04-2	Hexazinone
51630-58-1	Fenvalerate
52645-53-1	Permethrin
53404-60-7	Dazomet, sodium salt
55290-64-7	Dimethipin
55406-53-6	3-Iodo-2-propynyl butylcarbamate
57213-69-1	Triclopyr triethylammonium salt
59669-26-0	Thiodicarb
60168-88-9	Fenarimol
60207-90-1	Propiconazole
62476-59-9	Acifluorfen, sodium salt
63938-10-3	Chlorotetrafluoroethane
64902-72-3	Chlorsulfuron

CAS Numbers with Chemical Names (continued)

CAS	Chemical Name
64969-34-2	3,3′-Dichlorobenzidine sulfate
65996-93-2	Polycyclic aromatic compounds
67485-29-4	Hydramethylnon
68085-85-8	Cyhalothrin
68359-37-5	Cyfluthrin
69409-94-5	Fluvalinate
69806-50-4	Fluazifop-butyl
71751-41-2	Abamectin
72178-02-0	Fomesafen
72490-01-8	Fenoxycarb
74051-80-2	Sethoxydim
76578-14-8	Quizalofop-ethyl
77501-63-4	Lactofen
82657-04-3	Bifenthrin
88671-89-0	Myclobutanil
90982-32-4	Chlorimuron ethyl
101200-48-0	Tribenuron methyl
127564-92-5	Dichlorpentafluoro-propane
EDF018	Dioxin and dioxin-like compounds
EDF038	Nitrate compounds
EDF045	Polychlorinated alkanes (C10–C13)
EDF067	Diisocyanates
EDF109	Glycol ethers
MIXTURE	Mixture
TRD SECRT	Trade secret chemical

Index

Q

R

S

T